Stable Analysis Patterns for Software and Systems

T0143464

Stable Analysis Patterns for Software and Systems

M. E. Fayad

CRC Press is an imprint of the
Taylor & Francis Group, an **informa** business
AN AUERBACH BOOK

CRC Press
Taylor & Francis Group
6000 Broken Sound Parkway NW, Suite 300
Boca Raton, FL 33487-2742

First issued in hardback 2019

First issued in paperback 2022

© 2017 by Taylor & Francis Group, LLC
CRC Press is an imprint of Taylor & Francis Group, an Informa business

No claim to original U.S. Government works

ISBN 13: 978-1-03-247680-3 (pbk)
ISBN 13: 978-1-4987-0274-4 (hbk)

DOI: 10.1201/9781315198774

This book contains information obtained from authentic and highly regarded sources. Reasonable efforts have been made to publish reliable data and information, but the author and publisher cannot assume responsibility for the validity of all materials or the consequences of their use. The authors and publishers have attempted to trace the copyright holders of all material reproduced in this publication and apologize to copyright holders if permission to publish in this form has not been obtained. If any copyright material has not been acknowledged please write and let us know so we may rectify in any future reprint.

Except as permitted under U.S. Copyright Law, no part of this book may be reprinted, reproduced, transmitted, or utilized in any form by any electronic, mechanical, or other means, now known or hereafter invented, including photocopying, microfilming, and recording, or in any information storage or retrieval system, without written permission from the publishers.

For permission to photocopy or use material electronically from this work, please access www.copyright.com (http://www.copyright.com/) or contact the Copyright Clearance Center, Inc. (CCC), 222 Rosewood Drive, Danvers, MA 01923, 978-750-8400. CCC is a not-for-profit organization that provides licenses and registration for a variety of users. For organizations that have been granted a photocopy license by the CCC, a separate system of payment has been arranged.

Trademark Notice: Product or corporate names may be trademarks or registered trademarks, and are used only for identification and explanation without intent to infringe.

Publisher's Note
The publisher has gone to great lengths to ensure the quality of this reprint but points out that some imperfections in the original copies may be apparent.

Visit the Taylor & Francis Web site at
http://www.taylorandfrancis.com

and the CRC Press Web site at
http://www.crcpress.com

To all Software Professionals and Developers,
especially the ones who apply my work.

Contents

PART II SAPs: Detailed Documentation Templates

Preface

Software analysis patterns play an important role in reducing the overall cost, and in abridging and compressing the time of software project lifecycles. However, building reusable and stable analysis patterns (SAPs) is still considered a major and sensitive challenge. This book proposes the novel concept of SAPs based on software stability as a modern approach for building stable and highly reusable and widely applicable analysis patterns. This book also intends to provide a true and definite understanding of the problem space and focus users' requirements analysis accurately. It shows that this new formation approach of discovering/creating SAPs accords with Alexander's current understanding of architectural patterns. This agreement is not accidental, but truly fundamental. The SAPs are a kind of knowledge patterns that underline human problem-solving methods, and appeal to the pattern community to look for patterns from building a foundation of architectures on demand from a broader system perspective.

This book presents a new and pragmatic approach for understanding the problem domain and in utilizing SAPs for any field of knowledge and modeling the right and stable software systems, components, and frameworks. Along with the core value of reusing the presented patterns, this book also helps readers attain the basic knowledge that is needed to analyze and extract analysis patterns for their own domains of interests. Moreover, readers will also learn and master ways to document their own patterns in an effective, easy, and comprehensible manner.

The work reported in this book brings truly significant contributions to the computing field for several important reasons. It is the first and the only *complete reference manual* on the topic of SAPs. It is also the first book on handling the true understanding of the problem space, and it will teach a reader the methods and processes to analyze user's requirements accurately, and ways to build a myriad of cost-effective and highly maintainable systems by using SAPs.

THE SCOPE OF THIS BOOK

Software analysis patterns play a major and decisive role in reducing the overall cost and in condensing the time duration of software project lifecycles. However, building reusable and SAPs is still a major challenge. SAPs are new and fresh tools for building stable and reusable analysis patterns based on the concept of software stability.

Although SAPs are known to lend the much needed solidity and stability to a software system, nagging doubts still persist in today's domains of contemporary design and analysis patterns and their applications in software programming industry. For example, one of the pertinent questions raised by pattern developers is the perceived concerns on the stability factors of a software system and its ability to adapt to the issue of reuse over time. Similarly, another question that might be posed elates to a SAP's capability to capture and model the core knowledge of a given problem. Furthermore, it is also pertinent to seek answers to the question of achieving necessary levels of abstraction that usually makes software systems robust and reusable over a period. Often, pattern designers link the aspect of smooth transition from the analysis phase to design phase as a key factor to build effective software systems. In this regard, analysis patterns will need to provide necessary details that reveal the overall structure of an analysis pattern. Additionally, another key question that arises is the practicality of methods and processes that are used to describe and narrate analysis patterns from the initial phase of conception to designing aspects to creation.

One of the major pitfalls or weaknesses in developing meaningful and convenient analysis and design patterns is the perceived factor of immaturity; thus, most of the patterns developed are yet to fulfill the expectations for their use in facilitating the development of software systems. As a result, it becomes a major concern to investigate, why software patterns have not yet developed the level of maturity that is so much needed for establishing the stability of a given software system.

Even with the best of software architecture knowledge, a piece of software system is bound to face some problems in its lifetime. This book attempts to highlight and deliberate a number of common problems found in today's traditional software patterns, like the Gang of Four, Siemens Group, and others.

While Christopher Alexander's earlier work on patterns (e.g., "Timeless Way of Building" and "A Pattern Language") influenced the development of software (analysis, design, and architectural) patterns, his latest work on fundamental properties of patterns (presented in the four book series called "The Nature of Order") seems to be less influential to software patterns, in general, and analysis patterns in particular. Thus, the most obvious solution would be to focus more on developing effective patterns that contributes positively to the future development of software systems. This book is expected to overcome all the pitfalls of traditional or existing analysis patterns.

The main purpose of this book proposal is therefore to fill in the important gap in the current analysis pattern book, by adding new sets of analysis knowledge, and by providing a very well-defined paradigm of discovering/creating analysis patterns, based on software stability. This book also proposes the novel concept of SAPs based on software stability as a modern approach for building stable and highly reusable and widely applicable analysis patterns. This book overcomes the factor of immaturity in existing software patterns, in general, and analysis patterns, in particular, presented in many different existing software patterns pitfalls.

Throughout this book, we have furnished a number of answers to questions raised before, and practical approaches to follow clear-cut processes that arise from these answers. The Software Stability Concepts acted as the major backbone to all these questions. By applying concepts of stability model to the assumptions of analysis patterns, we suggest the concept of SAPs. The main idea behind using SAPs is to analyze the overall problem under question, in terms of its EBTs and the BOs, mainly with the goal of increased stability and broader reuse. By examining ensuing problem in terms of their EBTs and the BOs, the resulting pattern will form the core knowledge of the problem. The ultimate goal of this new concept is achieving ample stability. Accordingly, anyone could easily comprehend, understand, and build reusable models that tackle solving similar problems occurring under different contexts and domains. The real essence of SAPs is twofold: A clear paradigm and a precise visual representation. For the paradigm approach, we have presented a set of guidelines, heuristics, and quality factors that will ease the process of creating SAPs, along with a well-defined and reusable documentation. On the other hand, for visual representation, we have offered visual gadgets or symbols that convey what the SAPs are and how to apply them.

In addition to an introduction of twin and narrative essences of SAPs, this book also attempts to analyze reasons and factors that might affect designing, sustained life cycle, durability, generality, traceability, and quality of composition. In order to overcome all these perceived lacunae, this book provides a standard way of creating SAPs by using the principle of software stability that has been covered in depth in our earlier book titled *Software Patterns, Knowledge Maps and Domain Analysis*. The scope of this book extends beyond just designing an SAP. Being the only complete reference manual on SAPs, this book documents and highlights several aspects, techniques, and processes that are linked to the problem of understanding software systems that exist today. In addition, this book also presents a diversity of domain-less SAPs that one can easily comprehend and reuse to model similar problems in any given context.

Furthermore, tips from this book show the reader how to link the analysis pattern to the design phase; the main design issues necessary for a smooth transition between analysis and design phases are some of the useful contributions provided by this book. This book also offers a brand new template for improving the communication of analysis patterns among all developers; incidentally, this template aims to capture the static and dynamic behavior of the pattern, while maintaining the simplicity of reading and understanding the pattern. The main goal of this book is to ensure reusability of the pattern and its stability. Eventually, the overall scope of this book extends beyond the usual paraphernalia that is attached to pattern designing and making without taking into account the factors of traceability, robustness, stability, enhanceability, and reusability.

WHOM THIS BOOK IS FOR

We always write for the mass and public—This book could be of great help (frankly speaking is necessary). This book is specially written for a large community of computing and modeling academics, students, software technologists and methodologists, software patterns' communities, component developers, and software reuse communities and software professionals (analysts, designers, architects, programmers, testers, maintainers, developer), who are involved in the management, research, and development of methodologies and software patterns. Industry agents, who work on any technology project and want to improve the project's reliability and cost-effectiveness, will also benefit hugely by reading this book.

This book is also specially designed for a big community of computer and software professionals, who are involved in the management, research, and development of software systems, components, and enterprise and application frameworks. Software researchers, system architects, software designers, and software engineers (both systems level and application level) will also greatly benefit from this book. We anticipate that the new concepts presented in this book will definitely impact the development of new software systems, components, enterprise frameworks, and application frameworks for the next one or two decades.

Potential buyers for this book will be from established software firms and those people who are currently analyzing, designing, and developing different types of software for different domains. In addition, the UML and communities, who are active in designing software architecture, components, software engineering, enterprise frameworks, application frameworks, and the analysis pattern and design pattern, may also evince a keen interest in this book.

In a bookshop and library, the librarian would index and store this book in the Computer Science/Software Engineering/Object-Orientation/Component Software/Enterprise Frameworks/Application Frameworks category. This book is very suitable for anyone dealing with knowledge in their domain, software engineering academic and software development, sciences, engineering, business, and other communities.

All large software firms, people who are studying, researching, analyzing, designing, and developing any systems, and educators, academicians, and students in many other fields are also the potential customers for this book. In addition, a number of communities related to UML, requirements engineering software architecture, components, software engineering, enterprise frameworks, application frameworks, knowledge modeling and management, ontology, domain analysis, software reuse, architectural patterns, and the analysis pattern and design pattern would benefit by reading this book as well.

Further, people and experts, who attend workshops on object-oriented technology, software engineering, requirements engineering, robotics systems, and domain-specific conferences, will also be potentially interested in reading this book. These conferences and workshops may include OOPSLA, ECOOP, TOOLS, UML, ICSE, ICSR, PLOP, COTS, ICRA, Robotics, IROS, ICAR, ASRO, CORR, AIM, RO-MAN, ARAS, ISR, and many others. In a bookstore or a library shelf, this book would be, indexed and shelved in the Computer Science/Software Engineering categories.

We believe that concepts presented in this book will act as a catalyst to urge academicians, software students, software technologists, methodologists, pattern designers, software reuse proponents, analysts, designers, testers, and developers, to pursue rigorous research in the areas of SAPs. Industry agents, developers, and enthusiasts, who want to improve project life cycles, cost-effectiveness, and project reliability, will also benefit after reading this book. We also perceive that concepts presented in this book may positively affect the development of new software systems and application framework systems for the next decade or two. This book is also expected to be a knowledge source for creating future graduate course on software engineering, pattern modeling, domain analysis, knowledge modeling, advanced software architecture, and designing.

HOW TO USE THIS BOOK

The main goal of writing this book is to allow readers to understand and master the basics of SAPs from their theoretical underpinning to practical applications of principles to create robust and stable patterns. To enable easy understanding and comprehension of the subject matter, this book is divided into numerous sections, each of which deals with separate aspects of SAPs.

BOOK SUPPLEMENT

The author provides a book supplement of 400+ pages containing solutions to the exercises and projects in each chapter. In addition to problem statements for team projects, exams, quizzes, and a modeling tips and heuristics are also given. The supplement also provides over 20–30 numbers of special SAPs. Static, dynamic, and behavior models for each of the patterns are also given. Similarly, the author also provides a number of scenarios to illustrate how to do them in the supplement section. The book supplement also has several private links for power point presentation of all the sections of the book.

Acknowledgments

This book would not have been completed without the help of many great people; *I thank them all.* Special thanks to my friend and one of my best student Dr. Haitham Hamza, Cairo University, Egypt for his excellent thesis work on Stable Analysis Patterns. Special thanks to my friend Srikanth G. K. Hegde. I also would like to thank all of my student assistants, Vishnu Sai Reddy Gangireddy, Mansi Joshi, Siddharth Jindal, Hema Veeraragavathatham, and Pavan Pavuluri for their work on the figures and diagrams and long discussions on some of the topics of this book.

This was a great and fun project because of your tremendous help and extensive patience. Special thanks to my San Jose State University students for forming teams and work on some of the exercises and projects in this book. Special thanks to my wife Raefa, my lovely daughters Rodina and Rawan, and my son Ahmad for their great patience and understanding. Special thanks to Srikanth's wife, Kumuda Srikanth, for help with reviewing some of the chapters. Special thanks to all my friends all over the world for their encouragement and long discussions about the topics and the issues in this book. Thanks to all my students and coauthors of many articles related to this topic with me, in particular Ahmed Mahdy (Texas A&M University at Corpus Christi, Corpus Christi, TX), Shasha Wu (Spring Arbor University, Michigan), and Shivanshu Singh, to my friends Davide Brugali and Ahmed Yousif for their encouragement during this project, to the *Communications of the ACM* staff—my friends Diana Crawford, the executive editor, Thomas E. Lambert, the manging editor, and Andrew Rosenbloom, the senior editor.

I would like to acknowledge and thank all of those who have had a part in the production of this book. First, and foremost, we owe our families a huge debt of gratitude for being so patient while we put their world in a whirl by injecting this writing activity into their already busy lives. We also thank the various reviewers and editors who have helped in so many ways to get the book together. We thank our associates who offered their advice and wisdom in defining the content of the book. We also owe special thanks to those who have worked on the various projects covered in the case studies and examples.

Finally we would like to acknowledge and thank the work of some of the people and who helped us in this effort. John Wyzalek, acquisition editor, Stephanie Place, editorial assistant, Glenon Butler, project editor, and Michelle Rivera-Spann the marketing manager at CRC Press, Taylor & Francis Group for their excellent and quality support and work done to produce this book and a special acknowledgment and thanks to Karthick Parthasarathy, assistant manager at Nova Techset, who did a tremendous job for proofreading and copyediting of all the chapters in detail and the elegant and focused ways of taking care of day-to-day handling of this book, and special thanks to all the people in marketing, design, and support staff at CRC Press, Taylor & Francis Group.

Author

Dr. M. E. Fayad is a full professor of computer engineering at San Jose State University from 2002 to present. He was a J. D. Edwards professor of computer science and engineering at the University of Nebraska, Lincoln, from 1999 to 2002, and an associate professor of the computer science and computer engineering faculty at the University of Nevada, from 1995 to 1999. He has over 15 years of industrial experience. Dr. Fayad is an IEEE distinguished speaker, an associate editor, editorial advisor, and a columnist for *The Communications of the ACM* where his column name is Thinking Objectively. He is also a columnist for *Al-Ahram Egyptians Newspaper*, and an editor-in-chief for IEEE Computer Society Press—Computer Science and Engineering Practice Press (1995–1997). He also served as General Chair of IEEE/Arab Computer Society International Conference on Computer Systems and Applications (AICCSA 2001), Beirut, Lebanon, June 26–29, 2001, and he is the Founder and President of Arab Computer Society (ACS) from April 2004 to April 2007.

Dr. Fayad is a known and well-recognized authority in the domain of theory and the applications of software engineering. Fayad's publications are in the very core, archival journals and conferences in the software engineering field. Dr. Fayad was a guest editor on 11 theme issues: CACM's OO Experiences, Oct. 1995, IEEE Computer's Managing OO Software Development Projects, Sept. 1996, CACM's Software Patterns, Oct. 1996, CACM's OO Application Frameworks, Oct. 1997, ACM Computing Surveys—OO Application Frameworks, March 2000, IEEE Software—Software Engineering in-the-small, Sept./Oct. 2000, and *International Journal on Software Practice and Experiences*, July 2001, *IEEE Transaction on Robotics and Automation*—Object-Oriented Methods for Distributed Control Architecture, October 2002, *Annals of Software Engineering Journal*—OO Web-Based Software Engineering, October 2002, *Journal of Systems and Software*, Elsevier, *Software Architectures and Mobility*, July 2010, and *Pattern Languages: Addressing the Challenges*, the *Journal of Software, Practice and Experience*, March–April 2012.

Dr. Fayad has published more than 300 high-quality articles, which include profound and well-cited reports (more than 50 in number) in reputed journals, and over 100 advanced articles in refereed conferences, more than 25 well-received and cited journal columns, 16 blogged columns, 11 well-cited theme issues in prestigious journals and flagship magazines, 24 different workshops in very respected conferences, over 125 tutorials, seminars, and short presentations in more than 30 different countries, such as Hong Kong (April 96), Canada (12 times), Bahrain (2 times), Saudi Arabia (3 times), Egypt (25 times), Lebanon (04 & 05), UAE (2 times), Qatar (2 times), Portugal (Oct. 96, July 99), Finland (2 times), UK (3 times), Holland (3 times), Germany (4 times), Mexico (Oct. 98), Argentina (3 times), Chile (00), Peru (02), and Spain (02), Brazil (04), a founder of 7 new online journals, NASA Red Team Review of QRAS and NSF-USA Research Delegations' Workshops to Argentina and Chili and four authoritative books, of which three of them are translated into different languages such as Chinese and over 5 books currently in progress. Dr. Fayad is also filling for eight new, valuable, and innovative patents and has developed over 800 stable software patterns. Dr. Fayad earned an MS and a PhD in computer science from the University of Minnesota at Minneapolis. His research topic was OO Software Engineering: Problems and Perspectives. He is the lead author of several classic Wiley books: *Transition to OO Software Development*, August 1998, *Building Application Frameworks*, September, 1999, *Implementing Application Frameworks*, September, 1999, *Domain-Specific Application Frameworks*, October, 1999, and several classic books by CRC Press, Taylor & Francis Group: *Software Patterns, Knowledge Maps, and Domain*

Analysis, December 2014 and several new books in Progress with Taylor & Francis Group—*Stable Analysis Pattern for Software and Systems* (May 2017), *Stable Design Pattern for Software and Systems* (July 2017), *Unified Business Rules Standard* (UBRS), *Software Architecture On Demand, Unified Software Engineering Reuse (USER) and Unified Software engineering (USE) in Progress.*

Part I

Introduction

Stable analysis patterns (SAPs) are conceptual models that model the core knowledge of the problem. Therefore, it is expected that the pattern that model the specific problem should be easily and successfully used to model the same problem, regardless of the context in which the problem appears. SAPs are generated using the stability model [1–4]. In this subsection, we will provide its structure, mantra, and the rationale-driven language used to discover and visualize elemental pieces of knowledge (patterns), how to organize them, and how to relate them to formulate an accurate solution in contexts, that shares the same core knowledge (rationale or goals, and capabilities).

Building SAPs using software stability [1–4] for a specific problem involves myriad skills, knowledge and steps beyond the identification of the tangible artifacts that are bound to a specific context of applicability. It also requires a systematic capture and full understanding of the domain, where our solution would be laid down and expanded. That includes, describing the problem not from its tangible side, but focusing more on its conceptual side, describing underlying affairs with respect to the problem, and the elements required to fulfill them. Part I comprises 5 chapters and 20 sidebars.

Chapter 1 titled "Stable Analysis Patterns Overview" introduces the key concepts, and technologies of the development of SAPs, such as stability model and knowledge maps [5], where the enduring business themes (EBTs) or goals are the SAPs, and its business objects (BOs) are their own capabilities. It also discusses related work and different analysis patterns' development approaches, the challenges that analysis patterns face with examples, the overview of SAPs, the development processes of SAPs, and examples of SAPs.

Chapter 2 titled "Applying Analysis Patterns through Analogy: Problems and Solutions" focuses on two of the qualities: *traceability* and *generality*. *Generality* means that the pattern that analyzes a specific problem can be successfully reused to analyze the same problem whenever it appears, even within different applications or across different domains. It discusses analysis patterns as templates: (1) The problems with using analysis patterns through analogy and (2) SAPs with examples are discussed. It also shows how to apply SAPs in different application.

Chapter 3 titled "Pattern Language for Building Stable Analysis Patterns" provides an overview of the proposed pattern language, shows how a list of patterns build SAPs, and displays the relations between the different patterns. It also provides a detailed description of each pattern that is part of the proposed pattern language.

Chapter 4 titled "Model-based Software Reuse Using Stable Analysis Patterns" treats SAPs as model-based architectures and proposes nine essential properties (or metrics) to measure pattern reusability that are used to compare the three different analysis patterns' development approaches.

Chapter 5 titled "Stable Patterns Documentation: Templates, UML Forms, Rules, and Heuristics" addresses key questions and issues about each stable pattern. The template is suitable for documenting stable atomic patterns and stable architectural patterns where stable atomic patterns are SAPs (also called Enduring Business Themes [EBTs], goals, rationales, aims, purposes, and objectives) and stable design patterns (also called Business Objects [BOs] or capabilities used to achieve each of the goals). This chapter provides three different templates with rules, UML forms, and heuristics to document any pattern: (1) detailed, (2) mid-size, and (3) short templates.

Each of the chapters concludes with a summary and an open research issue. This chapter also provides a number of review questions, exercises, and projects.

Chapter 1 has four sidebars (1–4):

Sidebar 1.1: titled "The Roots of Patterns—Historical Perspectives" explores the patterns from ancient civilizations [6,7], such as Egyptians, Chinese, Indian, etc. and concludes with ALEXANDRINE Patterns as Inherited Architectural Solutions [8].

Sidebar 1.2: titled "Common Stable Design Patterns" lists the most common design patterns that will be utilized in developing SAPs as part of their capabilities to achieve them [9].

Sidebar 1.3: titled "Analysis Patterns' References" presents a brief overview of the existing journal and book literature on analysis patterns.

Sidebar 1.4: titled "Martin Fowler's Analysis Patterns" discusses briefly Martin Fowler's analysis patterns [10] and highlights their pitfalls with an example.

Chapter 5 has 8 sidebars (5–12):

Sidebar 5.1: titled "Context/Scenario Template."
Sidebar 5.2: titled "NonFunctional Requirements."
Sidebar 5.3: titled "Challenges Template."
Sidebar 5.4: titled "Fayad's Stability Model Templates."
Sidebar 5.5: titled "Fayad's Class Responsibility and Collaborations (CRC) Card."
Sidebar 5.6: titled "Fayad's Use Case Template."
Sidebar 5.7: titled "Model Essential Properties."
Sidebar 5.8: titled "Model Adequacies."

A reference to sidebar on Functional Requirements in the Knowledge Maps Book [5].

REFERENCES

1. M.E. Fayad, and A. Altman, Introduction to software stability, *Communications of the ACM*, 44(9), 2001, 95–98.
2. M.E. Fayad, Accomplishing software stability, *Communications of the ACM*, 45(1), 2002a, 111–115.
3. M.E. Fayad, How to deal with software stability, *Communications of the ACM*, 45(4), 2002b, 109–1112.
4. M.E. Fayad, and S. Wu, Merging multiple conventional models into one stable model, *Communications of the ACM*, 45(9), 2002c.

5. M.E. Fayad, H.A. Sanchez, S.G.K. Hegde, A. Basia, and A. Vakil, *Software Patterns, Knowledge Maps, and Domain Analysis*, Boca Raton, FL: Auerbach Publications, 2014.
6. C. Somers, and R. Engelbach, *Ancient Egyptian Construction and Architecture*, New York: Dover Publications, 1990.
7. C. Eric, and H. O'Connor, *David Kevin Amenhotep III: Perspectives on His Reign*, Ann Arbor, Michigan: University of Michigan Press, 2001, 273.
8. C. Alexander et al., *A Pattern Language*, New York: Oxford University Press, 1977.
9. M.E. Fayad, *Stable Design Patterns for Software and Systems: Ultimate Solutions*, Auerbach Publications, 2017.
10. M. Fowler, *Analysis Patterns—Reusable Object Models*, Reading, MA: Addison-Wesley Publishing, 1997.

1 Stable Analysis Patterns Overview

You look at where you're going and where you are and it never makes sense, but then you look back at where you've been and a pattern seems to emerge.

Robert M. Pirsig

Zen and the Art of Motorcycle Maintenance: An Inquiry Into Values

Domain models are conceptual models that capture the main concepts within the domain and their embedded relationships. These models play an important role in software development to facilitate understanding and communicating the problem to be solved among developers and stakeholders. However, developing conceptual models requires both domain knowledge and modeling skills, which can be challenging for novice as well as experienced practitioners. Analysis patterns are conceptual models that can be used to model and share domain knowledge, and hence, they can aid in developing domain models. However, most existing analysis patterns do not separate the core concepts of the domain from business-specific concepts, limiting the scope of the pattern usability. In this chapter, the main challenges in analysis patterns reuse are identified and elucidated. To overcome some of these challenges, the concept of *Stable Analysis Pattern* is proposed as a new approach for developing patterns that separate domain concepts from business-specific concepts. The proposed approach is illustrated through several examples.

1.1 INTRODUCTION

Developing software systems involves several activities, starting from domain and requirement analysis to system implementation and deployment [1–7]. Numerous requirements, techniques, and object-oriented design methods that are widely used in industry, start invariably with the development of a domain model that captures the core concepts in the domain [8]. Developing domain models, however, requires domain knowledge and modeling skills. But domain experts and analysts may have limited access to skill and knowledge, faced with under-tied "time-to-market" constraints and restricted by condensed budgets.

It has been long realized that software systems do share many requirements. For example, in a health-care system and an e-commerce infrastructure, there is a common need for techniques to attain "Information Privacy" in order to protect user information. At a conceptual level, the concept of "Privacy" does not change in both systems, although, realizing privacy at the implementation level is likely to be different in both systems. Nonetheless, conceptual modeling can greatly benefit from capturing commonly recurring problems at a reusable, yet meaningful abstraction level.

Software community has long realized the advantage of reuse [1,2], which is evident by the emergence of several reuse communities over the last few decades. Among these communities is the patterns community that advocates reuse by extracting and documenting solutions of recurring problems throughout the different phases of software development, for example, design (design patterns) and analysis (analysis patterns). In this book, we will focus on the important topic of analysis patterns. Analysis patterns are known to shape the knowledge of the problem domain. They support developers in understanding the underlying problem, rather than showing how to design a solution. (Please refer to Sidebar 1.2 on Common Stable Design Patterns for creating stable analysis patterns.)

Despite the fact that analysis patterns have been around for almost a decade now, their reuse is still limited when compared to design patterns and other reuse techniques. Here, we will argue that this limited reuse is due to several shortcomings in existing analysis patterns structures including the lack of stability, the lack of proper abstraction levels, and inadequate documentation. In this chapter, we will first identify the main challenges that adversely impact analysis patterns' reuse. Then, we will propose the concept of *Stable Analysis Pattern* as a new approach for separating problem concepts from business-specific concepts, in order to overcome some of the main shortcomings in existing analysis patterns. We will also demonstrate the proposed approach through several examples.

The remainder of this chapter is organized as follows: related work is discussed in Section 1.2; challenges that face analysis patterns are discussed in Section 1.3. The concept of stable analysis patterns is introduced in Section 1.4. In Section 1.5, an approach for developing stable analysis patterns is described. Examples of stable analysis patterns are discussed in Section 1.6; conclusion in Section 1.7. Please refer to Sidebar 1.3 (Analysis Patterns References), which is useful for more knowledge about analysis patterns.

1.2 RELATED WORK

The increasing interest in analysis patterns over the last few years have led to several analysis patterns that span a wide spectrum of domains, such as healthcare, finance, business enterprise, and system security [9–13]. Several types of analysis patterns have been proposed over the last decade (Refer to Sidebar 1.4: Quick Survey of Analysis Patterns). We will briefly review these techniques by classifying them based on their *development approaches. In addition,* we will also distinguish between three different development approaches: *Direct Approach, Specialization Approach,* and *Analogy Approach.*

In the *Direct Approach,* analysts identify analysis patterns they encounter in similar and related projects. Identified patterns are not further generalized or abstracted; instead, they are presented the way they were found. All developers, who follow this approach, believe that it is unsafe to further abstract patterns generated within certain projects in order to make them reusable in other contexts, as there is no guarantee that these abstracted patterns will successfully apply in other contexts. Examples of pattern in this category are the analysis patterns given in Reference [13].

In the *Specialization approach,* analysts usually document patterns they come across; however, in this approach, identified patterns are further abstracted, so that they can be applied to similar and related applications through the process of specialization (i.e., by instantiating the abstracted pattern).

In the *Analogy approach*, analysts document patterns, by following previously stated two approaches; however, here patterns are abstracted to create templates that are applied to other applications through the process of analogy. Incidentally, this approach is widely used in the domain of analysis patterns. In fact, in Reference [14], a pattern is defined as follows: *A pattern is a template of interacting objects, one that may be used again and again by analogy.* However, applying analysis patterns through analogy usually suffers from several significant limitations [15].

1.3 CHALLENGES THAT ANALYSIS PATTERNS FACE

In this section, we will identify and discuss the main challenges that analysis patterns face in general. We will also summarize the main challenges of analysis patterns as follows:

1. *Pattern instability:* Here, we will define stability, informally, as the degree by which the system can accommodate future changes and evolving requirements while preserving most of its original design. Software systems, by nature, evolve or change over time as new

requirements emerge; yet, it is always desired that such changes do not result in higher cost or force the system to be redesigned from scratch. If analysis patterns used in the system are unstable themselves, then this instability will ripple or cascade through other development phases, eventually leading to the development of unstable systems. Current analysis patterns approaches do not consider the stability as a goal when designing and developing analysis patterns.

2. *Pattern redundancy:* The increasing growth of the pattern community has also increased the likelihood of finding several patterns that address similar problems. Analyzing any problem with specific applications in mind might result in analysis models that are bound to specific applications and that are hard to apply, when similar problems appear in other applications. Consequently, several patterns that address similar problems, which are positioned within different applications will be derived and developed. For instance, different patterns have been proposed to model accounts.
3. *Analysis patterns documentations:* Describing pattern is perhaps as crucial as the pattern itself. A fine thin line should be drawn between the pattern's details and their depths. Regarding a pattern's details, it is understood that the quantum of information presented to describe or highlight the pattern in hand, while a pattern's depth is closely linked to the technical complexity of the solution the pattern usually presents. Too great a ratio of details to depth, or too small a ratio, might leave the user lose the intended track eventually making the pattern very hard to understand and reuse [1,2]. Efforts on exploring how analysis patterns are documented are rather rare, while design pattern templates are the most commonly used ways to document analysis patterns.
4. *Identification of the problem boundary:* One of the major challenges that developers face while developing analysis patterns are to precisely identify the boundary of the problems that these patterns try to analyze and evaluate. Without clear identification of problem boundary, it is more likely that the pattern will embody many other problems that are invariably out of the scope of the main problem. For example, the word "account" alone becomes a vague concept, if it is not allied with a word that is related to a certain context. Besides all of the traditionally well-known business and banking accounts, we also have e-mail accounts, online shopping accounts, etc. The account pattern models has two significantly different problems: the "account" problem and the "entry" problem. These are closely related, although they are independent problems. It is quite possible that we come across numerous entries that do not have any accounts or they are accounts without any entries. Tables 1.1 and 1.2 show some examples of such situations.
5. *Pattern traceability:* This type of problem is very common in analysis patterns that belong to the analogy approach (Section 1.3). When patterns are used as thorough analogy approach, the original patterns are no longer extractable. This un-traceability factor may complicate the maintainability aspect of the developed software.

TABLE 1.1
Accounts without Entries

1. *Free online services account:* Many online sites provide free goods or services. For example, some companies provide learning software packages or instructional documents. In order to access these materials, you will need to create an account. This account is merely a passport provided to the users to access their services; in fact, there is nothing in this account that can be considered as your property.
2. *Access account to the copy machine:* Suppose, you have an account to access the copy machine in your school or work. This account is no more than a passport for you to use the copier. There are no entries in this case.

TABLE 1.2
Entries without Accounts

The following table contains information about class schedules, at the University of Nebraska-Lincoln. In this table, each piece of information forms an entry to the table. Here, we do not need any accounts where user entries could be maintained.

Call #	Course Title	Cr Hrs	Time	Day	Room
2850	Computer Architecture	003	0230–0320p	M W F	Freg 112
2855	Software Engineering	003	0930–1045p	T R	Freg 111

1.4 STABLE ANALYSIS PATTERNS

In order to confront some of the problems that were discussed in Section 1.3, we will propose the concept of *Stable analysis pattern* as a new approach for developing analysis patterns. Stable analysis patterns are broadly based on the generic concept of stability model [3–6] and Knowledge Maps [7,14]. Before we proceed to discuss the details of stable analysis patterns, we will initially review the main concepts of stability model.

Stability model is a layered approach for designing and developing robust software systems. In this approach, the classes of the system are classified into three layers: the enduring business themes (EBTs) layer [3,16], the business objects (BOs) layer [4], and the industrial objects (IOs) layer [4]. Based on its innate nature, each class in the system is classified into one of these three layers. EBTs are the classes that present the enduring and core knowledge of the underlying industry or business. Therefore, they are extremely stable and form the nucleus of the stability model. BOs are the classes that map the EBTs of the system into more concrete and precise objects. BOs are semi-conceptual and externally stable, but they are internally adaptable. IOs are the classes that map the BOs of the system into physical objects. The detailed properties of EBTs, BOs, and IOs are discussed in References [15,17].

Figure 1.1 illustrates the basic concepts of stable models and their uses. The EBTs and BOs in the left-hand side of the figure form and create a stable core. By *externally* adapting the BOs through hooks and by adding the necessary IOs, two new systems were constructed, that is, system

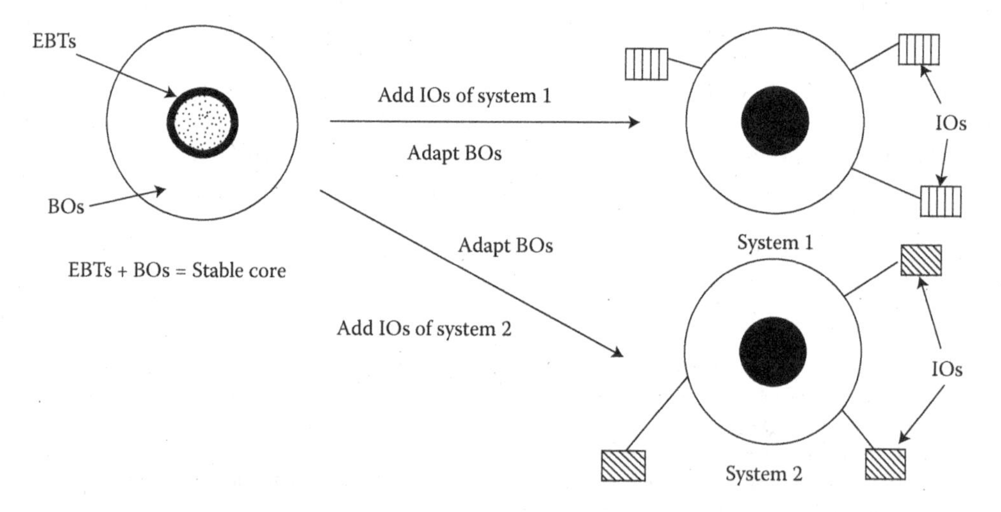

FIGURE 1.1 Software stability concepts. The stable core (in the left) has been adapted to construct two systems (in the right).

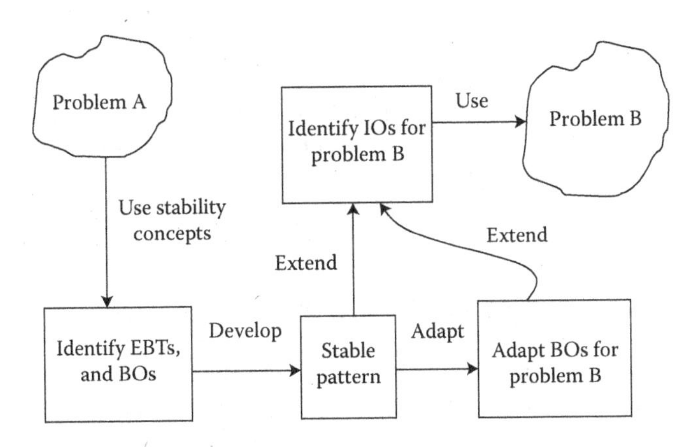

FIGURE 1.2 Stable analysis patterns approach.

1 and system 2, which are shown in the right-hand side of the figure. It is important to point out the assumption that these two systems are closely related (e.g., share the same domain) so that we can identify common EBTs and BOs.

The main goal of stable analysis patterns is to develop models that capture the core knowledge of the problem and present it in terms of the EBTs and the BOs of that problem. Consequently, the resultant pattern will inherit stability features and hence it can be reused to analyze the same problem whenever it appears. Figure 1.2 shows the generic approach of developing and reusing stable analysis patterns.

Since the EBTs and BOs are stable by their nature, it is guaranteed that they will compose a stable pattern, and hence, one can overcome the instability problem that was discussed in Section 1.3. Stable pattern can be used by directly extending it by attaching most appropriate IOs for the problem B (Figure 1.2). Stable patterns may also be used by first adapting the hooks configurations of its BOs and then attaching all appropriate IOs for problem B (Figure 1.2).

1.5 DEVELOPMENT OF STABLE ANALYSIS PATTERNS

In this section, we will describe a high-level iterative approach for developing stable analysis patterns. Figure 1.3 illustrates the main activities in the course of the development approach and essential relations between them.

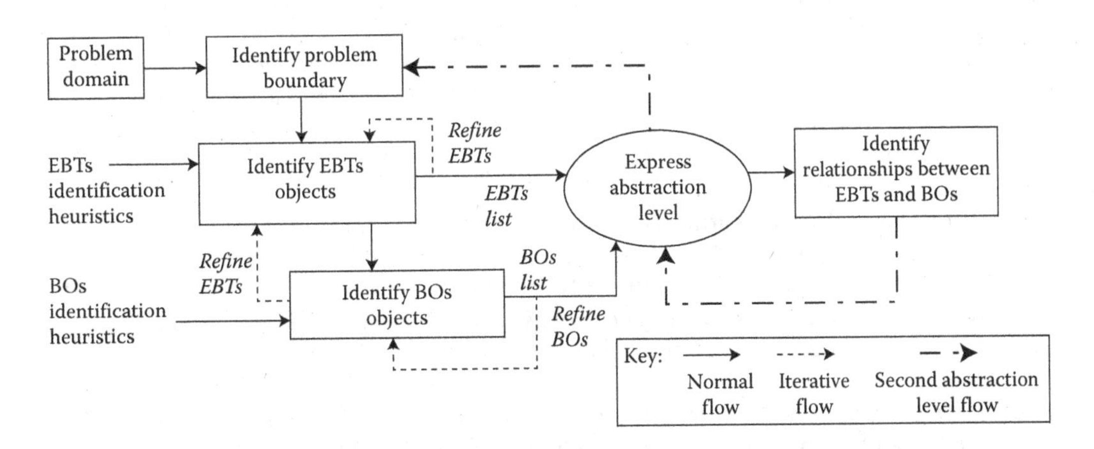

FIGURE 1.3 The high-level approach for developing stable analysis patterns.

In the following section, we will also explain different phases of the integration approach:

Identify problem boundary: The reusability of a pattern is related to the number of problems it addresses and tackles. If a pattern is used to model an overly broad portion of a system, the generality of resulting patterns is sacrificed—the maxim holds here: "the probability of occurrence of all the problems together is far lesser than the probability of the occurrence of each problem when considered individually."

Identifying problem boundaries is not a trivial activity for many important reasons. First, certain groups of problems often appear together, and as a result, they will be naturally considered as one, solitary problem. Here, the resultant model may or may not be correctly modified to sculpt these problems whenever they appear separately. Additionally, in practice, not all of the small problems that are separable are qualified to form practically capable and stand-alone problems. There is a subtle tradeoff between dividing the problem and the complexity of integrating smaller problems to model a larger problem.

This phase can be accomplished by checking whether the problem could be further divided into smaller and practical problems. Answering the following questions help us to achieve the above-mentioned task: "What is the exact problem that we want to solve?" "Can we split this problem further into a list of smaller problems?" "Are there any known and identified possible scenarios, where these smaller problems might appear?" If one is able to find practical scenarios for each of the smaller problems that are separated, then there is an urgent need to model each of them separately. If smaller problems have no practical use, they should be grouped together and considered as combined entities.

Identifying objects of EBT: We will need to identify core elements of the problem while we are developing a stable pattern. Certain issues are known to complicate the identification of EBTs in a given problem. For instance, domain experts may not always be able to identify accurate, precise, and relevant EBTs [14]. Thus, experience is absolutely essential, but not sufficient enough for extracting the right EBTs of the problem. The 3-step heuristic shown in Table 1.1 will help us in identifying the most appropriate EBTs of the problem:

As an example for applying the 3-step heuristic that is shown in Table 1.3, it is important to consider the identification of the EBTs in a modeling account:

Step 1 Create initial EBTs list. We will need to answer the question: "What is the 'Account' made for?" The initial EBTs list might include the following EBTs: Storage, Ownership, Tractability, and Recording.

Step 2 Filter EBTs list. In this step, we may need to dig out the most appropriate EBTs for the "account" problem. To do so, we will require examining each EBT given in the list and see whether or not it reflects the enduring concepts of the problem modeled. Since our main focus is centered on modeling the "account" problem alone, we will realize that most of the EBTs that we have defined are mainly related to accounts that have entries. For instance, "Storage," "Tractability," and "Recording" are all important concepts that are dependent upon the existence of entries within the account. Suppose the account has no entries, "Storage," "Tractability" and "Recording" is unneeded. Therefore, we may need to eliminate the EBTs like "Storage," "Tractability," and "Recording" from the list. Now, we have only one EBT in the arrangement that is, "Ownership," instead of four.

Step 3 Check the Main EBTs Properties. We will also answer the following questions: "Can we replace the 'Ownership' with other EBT?"—*No.* "Is 'Ownership' stable internally and externally?"—*Yes.* "Does 'Ownership' belong to a specific application or domain?"—*No.* "Can we have direct physical representation for 'Ownership'?"—*No.*

Identifying business objects: During the development of a stable analysis pattern, and after identifying the EBTs of the problem, we will want to identify the BOs of the problem. In some cases, it is not readily obvious whether the object is an EBT or BO. After the EBTs of the problem have

TABLE 1.3
A 3-Step Heuristic for Identifying EBTs

Step	Step Name	Details
1	*Create Initial EBTs List*	To create the initial list of the EBTs of the problem, answer the question: "What is the 'problem' designed for?" In other words: "What are the reasons for the existence of the 'problem?'" The output of this step is the list of the initial EBTs of the problem. These EBTs are still tentative and some of them are not as strongly related to the problem as they might appear.
2	*Filter the EBTs List*	Eliminate the redundant and irrelevant EBTs from the initial list. People unintentionally construct the initial EBTs list with a specific application in mind. The output of this step is a modified EBTs list, which should contain less EBTs than the initial list.
3	*Check the Main EBTs Properties*	Examine the EBTs obtained in previous steps against the main essential properties of the EBTs. The typical procedure is to answer the following questions for each EBT in the list. The desired answer is written in **bold** for each question: Can we replace this EBT with another one? **No**. Is this EBT stable internally and externally? In other words, does this EBT reflect the core aspects of the problem we are trying to model? **Yes**. Can we directly represent this EBT physically? **No**. It is important to note that the EBTs should not have direct physical representations (IO); otherwise they should be considered BOs instead. For example: "Agreement" is a concept and one can see it as an EBT. However, "Agreement" also has a direct physical representation (for instance "Contract"). Therefore, "Agreement" is not an EBT, it is a BO.

been identified, the conceptualization becomes comprehensible, since the BOs of the problem must be based on the defined EBTs.

In addition to the main BOs identified for the problem, it is also possible to obtain some hidden BOs that have no direct relationship with the defined EBTs. Instead, they are related to the main BOs and to the other hidden BOs in the problem. The 4-step, heuristic shown in Table 1.4 can help us in identifying the most appropriate BOs of the problem:

As an example of applying the 4-step heuristic shown in Table 1.4, suppose we want to identify the possible BOs for the given "account" problem. In order to identify these BOs, we will apply all the four steps that we have proposed earlier. The input of the first step is the EBTs list that contains one essential EBT, "Ownership."

Step 1 Identify the main BO of the problem: We will also answer the following questions: How can we approach the goal of "Ownership"? In the account problem given here, to achieve "Ownership," we need to have something to own. The "Account" itself is what makes the meaning of "Ownership."

What are the results of doing/using "Ownership"? By having "Ownership," we also have "Privacy," but this is not a BO. Rather, it is a redundant EBT, and hence we would exclude it. "Agreement" is a possible BO for the problem, but, when you own an account, you will have to agree with its policy. Who should do/use "Ownership"? For the "Owner" to be able to use the "Account," he or she should follow the responsibilities that are defined by the "Ownership" policy. Hence, the BOs main list is "Owner," "Account," and "Agreement."

Step 2 Filter the main BOs List. Now, the following BOs still remain: "Owner," "Account," and "Agreement." While modeling this unique problem, we have debated the accuracy of using the BO "Agreement" in our model. "Agreement" is a more general term that appears

TABLE 1.4
A 4-Step Heuristic for Identifying BOs

Step	Step Name	Details
1	*Identify the main BOs of the problem*	We will identify the main set of BOs that are directly related to each of the EBTs that we have in the problem. There could be one or more BOs corresponding to each EBT in the problem. However, some of the EBTs may have no corresponding BOs. The main set of BOs of the problem could be identified by answering one or more of the following questions for each EBT: How can we approach the goal that this EBT presents? (e.g., To achieve the goal of the EBT Organization, we can use, the BO Schedule. Another example: for the EBT Negotiation, we need the BOs AnyContext, and AnyMedia to perform the negotiation). What are the results of doing/using this EBT? (e.g., for the EBT Negotiation, the eventual result is to reach an Agreement, so this is one possible BO that maps this EBT). Who should do/use this EBT? (e.g., The BO Party does/ uses Negotiation. This Party can be a person, a company, or an organization. Therefore, Party is one possible BO that maps the EBT Negotiation).
2	*Filter the main BOs List*	Purify the main BOs identified in the previous step. The objective of this step is to eliminate the redundant and irrelevant BOs from the initial list. One way to achieve this goal is to debate the listed BOs with a group. (Please refer to Sidebar 1.2)
3	*Identify the hidden BOs of the problem*	Identify the hidden BOs of the problem. These BOs are named "hidden," because they have no direct relationships with any of the EBTs of the problem. Thus, we cannot extract them in the first two steps we have performed. For example, let us assume that we need to model a simple transportation system that offers transportation services for different types of materials (e.g., gas, water, etc.). One possible EBT is Transportation. One possible BO that maps this EBT is Transport. A possible IO that can physically represent this BO is Trucks. In this problem, one possible hidden BO is Materials. We do not have a direct EBT that the BO Materials can be mapped to; however, there is a clear relationship between the two BOs: Transport and Materials. Before thinking about the hidden BOs in the problem, just visualize a provisional scenario for each EBT and its corresponding BOs. Then, answer the question, "What is still missing in the problem?" Usually, the answer to this question is the list of the hidden BOs of the problem. Some problems do not have any hidden BOs, especially in the case of the small-scale problems.
4	*Check the characteristics of the BOs*	This step is to ensure that the identified BOs satisfy the main BOs characteristics. BOs are partially tangible, internally stable and should remain stable throughout the life of the problem, externally adaptable through hooks, and can be physically represented by IOs.

in many contexts and under different scenarios. For instance, in the "negotiation" problem, we will need the BO "Agreement." Therefore, possessing the BO "Agreement" as part of the "Account" pattern will offset the simplicity property of the pattern. "Agreement" is a stand-alone problem that occurs in many contexts, and hence it is more appropriate to model the "Agreement" problem wholly independent of the context of this problem. Consequently, we have excluded the BO "Agreement" from the main list. The filtered BOs list thus contains "Account" and "Owner."

Step 3 Identify the hidden BOs of the problem. After identifying and filtering the main BOs list, now answer this question, "Does the account problem need anything else so as to be complete in every sense?" We have an "Account," its "Owner," and the concept of "Ownership" that regulates the responsibilities and benefits of the "Owner." This is all that is needed to model any basic account.

Step 4 Check the characteristics of the BOs. BO should be partially tangible. "Owner" and "Account" are partially tangible. BOs are internally stable, and they should be so throughout the life span of the problem. "Owner" and "Account" are always stable. In other words, we cannot have any account without having these two objects in its model. Here, BOs are adaptable; thus, they might change externally.

Express abstraction level: Stable patterns can be generally classified into two main categories: simple and composite patterns. A simple analysis pattern is a pattern that consists of classes and no sub-patterns exist. On the other hand, composite stable analysis pattern consists of both classes and subpatterns. Simple patterns are said to have one level of abstraction (such as the conventional class diagram), while composite analysis patterns may have several levels of abstraction depending on the structure of its subpatterns. Figure 1.4 shows the negotiation stable analysis pattern, which is actually a composite pattern. Notice that the AnyMedia box in the displayed pattern is a pattern in itself. Two different applications of the negotiation stable analysis pattern are illustrated in Figures 1.7 and 1.8.

In stable patterns, we usually differentiate between two main participants in the pattern model: classes, and patterns. Classes are defined as in the case of any traditional object-oriented class diagram. On the other hand, patterns also present a second level of abstraction in the model, where each pattern is by itself another model that contains classes and in some cases, other patterns.

A class within a stable pattern could be one of the five following kinds: an EBT, a BO, an IO, a sub-pattern and EBT, or a sub-pattern and BO. Therefore, each class in the stable pattern should have one of the following tags: EBTs, BOs, IOs, Pattern-EBT, or Pattern-BO.

Note that the second abstraction level flow shown in Figure 1.4 will restart the modeling process again from the first step. The negotiation pattern has resulted from the first execution of the development process in Figure 1.4, the second abstraction level is the result of repeating the development process for all the patterns that are tagged with <P-Bo>, but for different components (classes) in

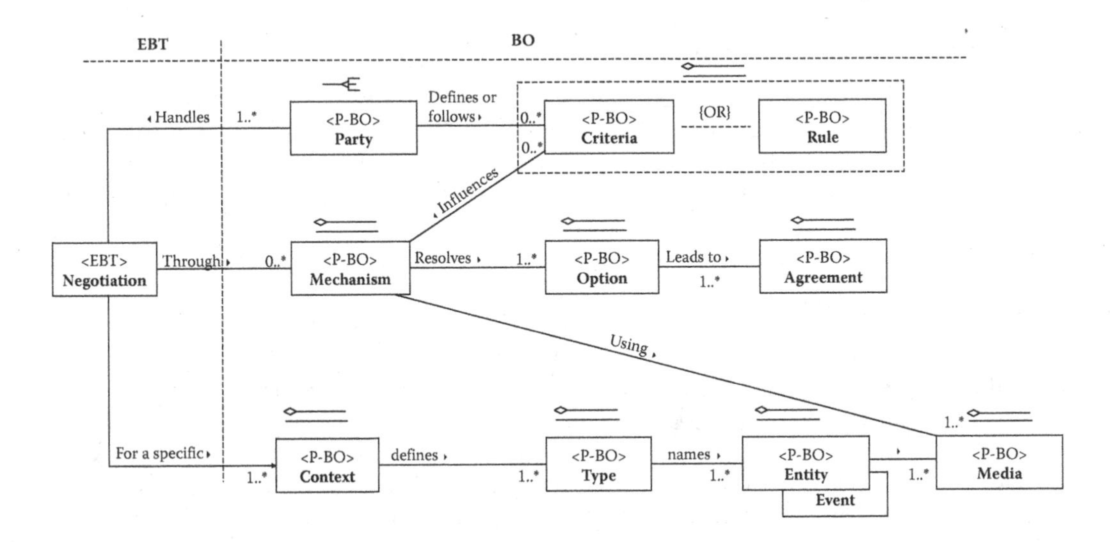

FIGURE 1.4 *Negotiation* pattern stable object model.

the negotiation pattern (in Figure 1.4), the result of repeating the development process over the AnyMedia, the AnyParty has resulted in the second abstraction level all of the patterns which are tagged with <P-Bo>.

1.6 EXAMPLE OF STABLE ANALYSIS PATTERNS

We would illustrate the concept of stable analysis patterns by presenting two stable analysis patterns: negotiation analysis patterns (Figure 1.4), and trust analysis patterns (Figure 1.7) [11].

1.6.1 The Negotiation Analysis Pattern

Negotiation may take place in different situations and under different context. For instance, buying or selling activity usually involves some sort of negotiation (e.g., buying or selling a home or a car). Similarly, in the domain of software systems, negotiation appears frequently in the process of development of different applications. For example, developing a piece of software for online auctions and e-shopping might involve the negotiation for the price and/or the negotiation for different product aspects and issues. More technically, negotiation is an essential part in the development of next generation web-based devices and appliances. Today, numerous devices that need to access the web usually diverge greatly in their capabilities, making them highly desirable for similar resources to be made available in several different representations (e.g., different languages). Negotiation algorithms also play a fundamental role in assisting servers to decide the mode of representation of a device or gadget should be provided. In this case, the browser (client agent) will indicate its preferences and options.

The important fact that negotiation concept embed a wide range of spectrum of heterogeneous applications, along with the reality that the negotiation concept itself does not change whenever it appears, make the development of a model that captures the core knowledge of the negotiation very dicey and truly challenging.

We may also apply the negotiation pattern in two different applications: negotiation of buying a car, and content negotiation using composite capability/preference profile (CC/PP) (for simplicity, these examples do not present the complete model for the problem. The details of these examples can be found in Reference [18]). Figure 1.5 shows the model of the negotiation used in buying a car, and Figure 1.6 shows another instance of the negotiation pattern [15].

1.6.2 The Trust Analysis Pattern

The emerging boom of extremely useful web-based applications, like e-commerce, e-business, and e-negotiation, makes the development of an accurate model, which captures that the core aspect of trust is of great importance and immense interest. The issue of trust has been studied intensively in the literature from different perspectives [16,19,20]. In addition, several studies have also focused on the deployment issues of trust in e-commerce and web-based applications. For instance, in Reference [21], a summary of the three trust-building measures for e-commerce applications is given.

Even though different applications impose different requirements for establishing, different applications may still share a vital part of trust requirements, although they differ in some aspects. We will develop an analysis pattern that captures the core knowledge of the trust concept that is independent of any specific application. This eventually leads to the trust analysis pattern, which is shown in Figure 1.7.

Here, we will use the patterns presented in Reference [21] as an example of trust pattern applicability. Three sample patterns to illustrate the idea of building a pattern language for online consumer trust have also been proposed in Reference [21], namely, "Contact Us," "Consumer Testimonials,"

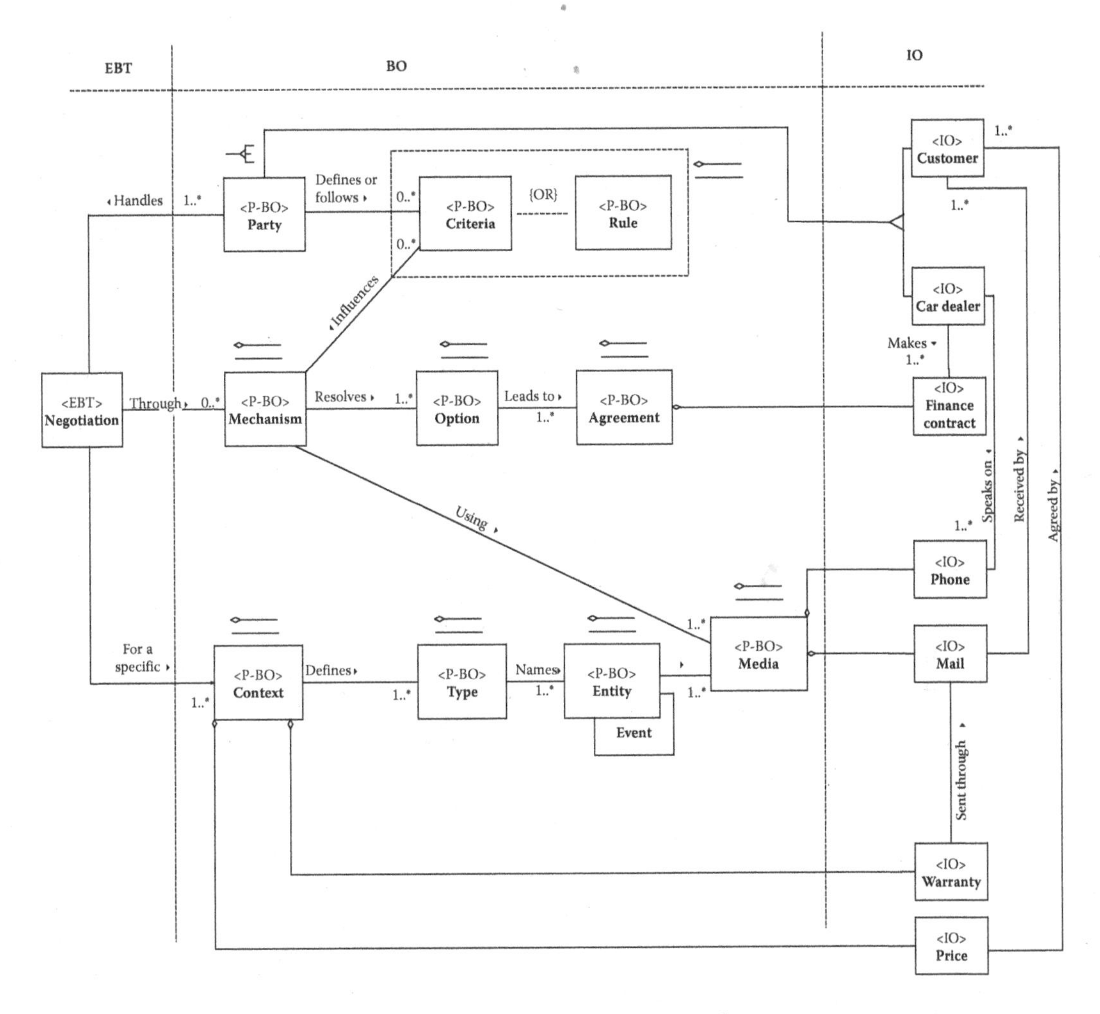

FIGURE 1.5 Examples of the negotiation pattern: buying a car example.

and "Return Policy." Each of these three patterns represents, in the given order, an example of one of the following trust-building recommendations presented in Reference [22] and summarized in Reference [21]: *information policy*, *reputation policy*, and *warranty policy*.

Based solely on the stability concepts (Section 1.3), it is easy to see the three main trust-building recommendations: *information policy*, *reputation policy*, and *warranty policy*, could be viewed as BOs, while the example pattern for each category "Contact Us," "Consumer Testimonials," and "Return Policy," respectively, can be regarded as IOs. This simple, yet effective mapping allows us to better understand the structure of the given patterns and to develop a relationship between them and the trust stable analysis pattern. It is also possible to plot the three categorizes and their corresponding patterns that are proposed in Reference [21]. Notice that the AnyLog <P-Bo> shown in the Figure 1.7 represents the AnyLog in the trust analysis patterns in the second level.

We can also use trust analysis pattern to gain more insight into how comprehensive the patterns in Figure 1.2 are. It really turns out that all the proposed patterns are related to the AnyLog pattern, which is part of the trust analysis pattern. For example, we can argue that a consumer would use the three proposed policies together in order to perform the rating process for trusting the corresponding e-commerce web site (which in this case could be just a matter of deciding, whether to shop in

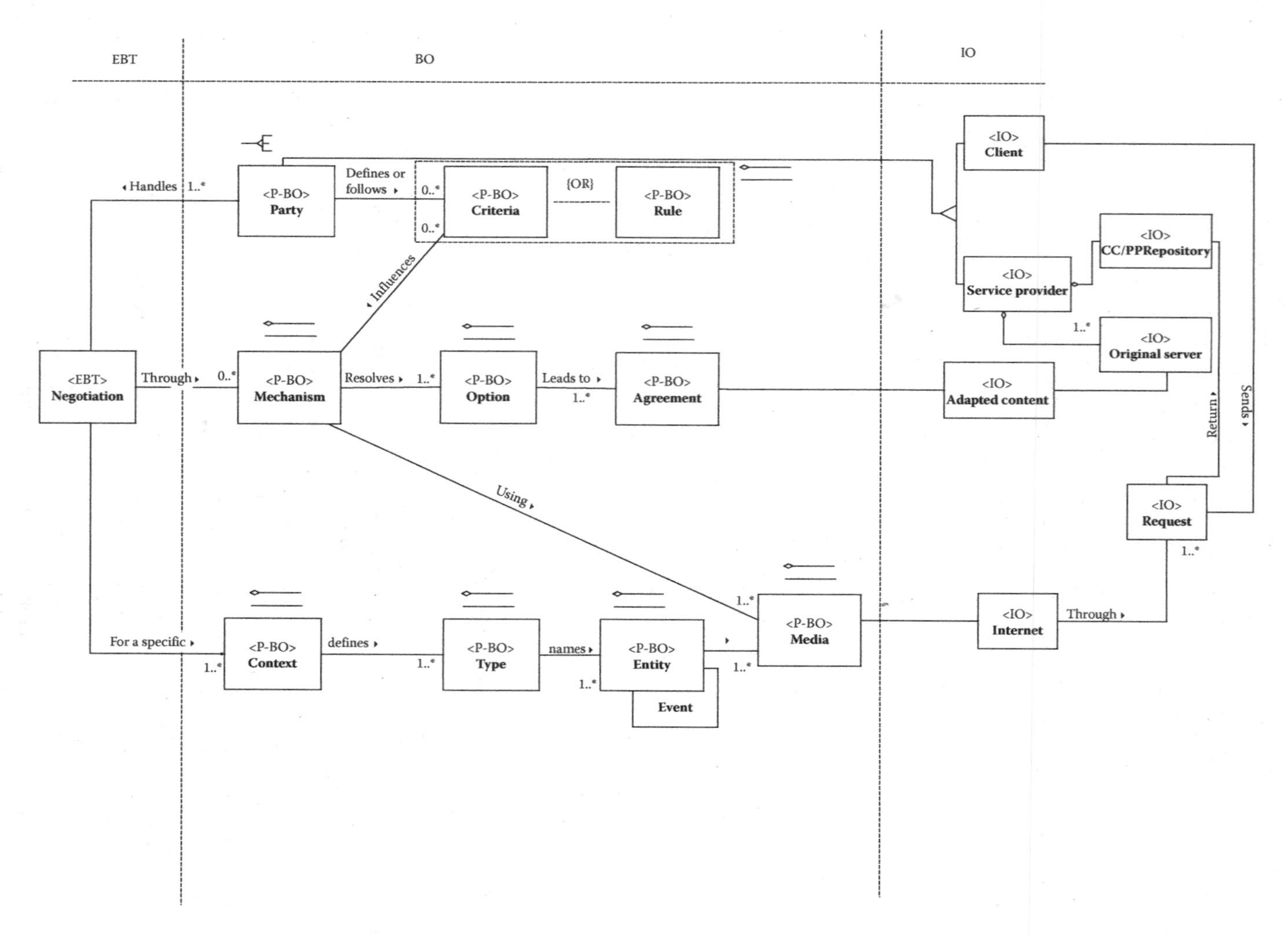

FIGURE 1.6 Examples of the negotiation pattern: negotiation example.

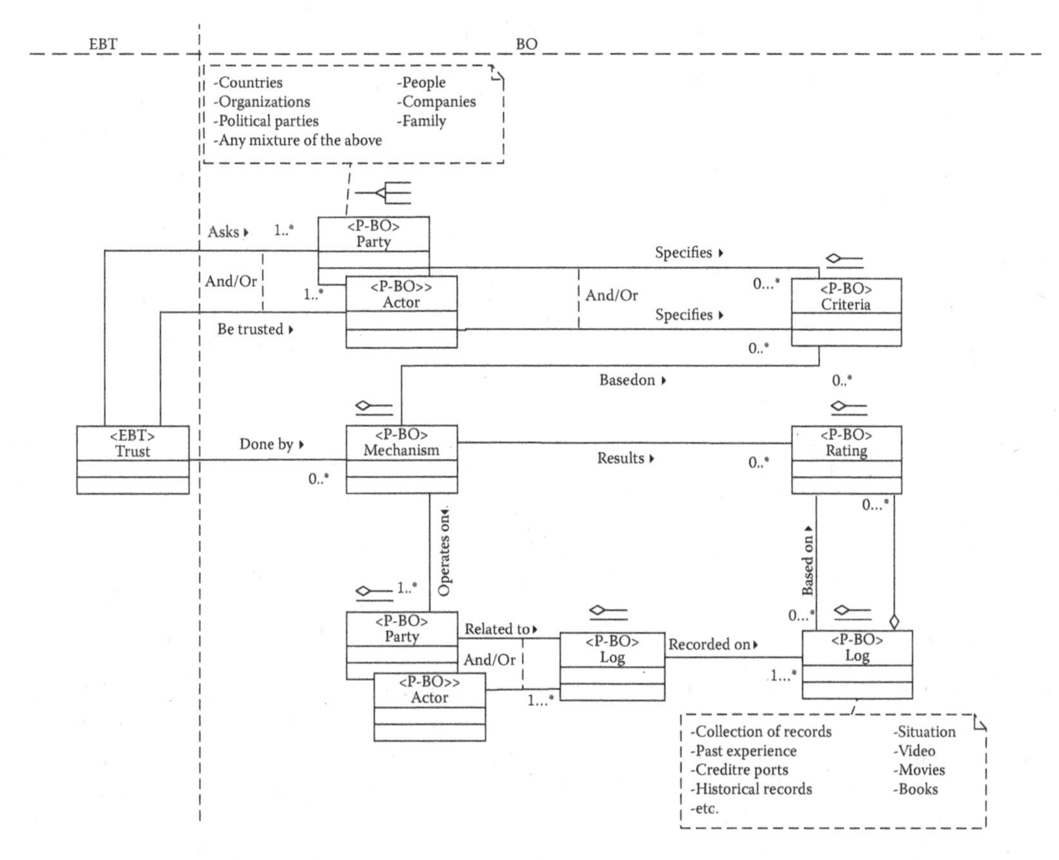

FIGURE 1.7 The Trust analysis pattern. (From R.T. Vaccare Braga et al., A confederation of patterns for business resource management, *Pattern Language of Programs Conference*, Monticello, IL, 1998.)

this site or not). In that sense, all the three proposed patterns can fit in our analysis patterns as an instance of AnyLog.

As shown in Figure 1.8, even though the three sample patterns proposed in Reference [21] are not claimed to be complete or sufficient, analysis pattern can be used to show that the three categories of the trust-measures (information policy, reputation policy, warranty policy) that were used as drivers to the three sample patterns (Contact Us, Customer Testimonials, Return Policy) are not perfect and does not fully cover the core knowledge of trust concept. In the meantime, we can argue that none of the three categorizes covers the rating issues, which is presented as AnyRating in our analysis pattern. This rating process is a fundamental step for any trust process, but the way that the rating process is conducted may greatly differ though.

Another important aspect in the analysis pattern is that it covers future applications that are likely to emerge in the near future, which is why we claim that this trust pattern is robust and stable. One can argue that in the case of e-commerce consumer trust, the rating process is subjective and will not be implemented as part of the system. The user, a human being, will take the decision of whether to trust an e-commerce application or not. However, recent applications, such as e-negotiation, may involve a trust agent that will be responsible for automatically understanding the AnyLog records and performing the rating process, and based on that it will take the trust decision. Such visible application is beyond the capabilities of the patterns categorizes given in Reference [21]; however, our analysis pattern can still capture the aspects of the trust in these systems, and hence, can be still applied in a wide spectrum of domains.

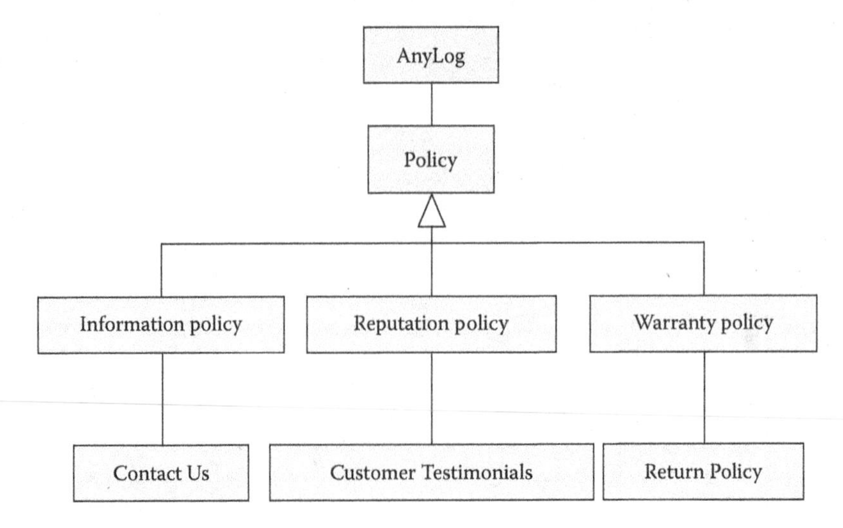

FIGURE 1.8 A diagram o represent patterns and categories given in Reference [20].

1.7 SUMMARY

In this chapter, we have successfully identified the main challenges that limit the use of analysis patterns. To overcome some of these problems, we have proposed the concept of stable analysis patterns. Stable analysis patterns separate the core concepts of the problem from the business-specific concepts, thereby leading to conceptual models that are more reusable compared to existing analysis patterns. We have illustrated the approach with a number of examples and showed how stable analysis patterns can help not only to model the problem, but also to develop a deeper understanding of the essence of the problem.

1.8 OPEN AND RESEARCH ISSUES

Generate a catalog of stable analysis patterns fully documented and implemented to be used as part of new stable programming environments, and show how we can easily generate sophisticated applications very quickly and perform dynamic analysis of each of the generated applications on top of the stable analysis patterns. This will ultimately lead to comparative studies and real-time data about the dynamic analysis, and allow the developers and users of the applications to give concrete results based on real running systems or applications.

OVERVIEW QUESTIONS

1. What is a domain model?
2. What are the properties of domain models?
3. T/F: Developing conceptual models require both domain knowledge and modeling skills.
4. T/F: Analysis patterns are conceptual models that can be used to model and share domain knowledge.
5. T/F: Stable analysis patterns (SAPs) separate the core concepts of the domain from business-specific concepts.
6. T/F: Traditional analysis patterns do not separate the core concepts of the domain from business-specific concepts.

7. T/F: A traditional analysis pattern has a very limited reuse.
8. T/F: SAPs model the knowledge of the problem domain.
9. T/F: SAPs aid the understanding of the problem, rather than showing how to design a solution.
10. Traditional analysis patterns have limited reuse. Why?
11. Identify and discuss the traditional analysis patterns development approaches.
12. Name three samples of traditional analysis patterns developed using direct approach.
13. Name three samples of traditional analysis patterns developed using specialization approach.
14. Name three samples of traditional analysis patterns developed by using analogy approach.
15. T/F: SAPs are based on stability model.
16. Stability model is a layered approach for developing software systems.
17. Stability models have three layers. Name them?
18. T/F: EBT layer is the nucleus layer of stability model.
19. EBTs are enduring concepts and extremely stable.
20. T/F: EBTs are classes and they could be considered as goals, aims, rationales, and objectives.
21. T/F: BOs is the middle layer of stability model.
22. T/F: BOs are internally stable and externally adaptable.
23. T/F: BOs are semi-tangible and mostly conceptual.
24. T/F: BOs are the capabilities of achieving the goals (EBTs).
25. T/F: BOs are classes that map the EBTs of the system into more concrete objects.
26. T/F: IOs are classes that map the BOs of the system into physical objects.
27. T/F: BOs are externally adaptable through hooks or extension points by adding the necessary IOs.
28. EBTs + BOs = Knowledge Map of any domain or any system or any pattern.
29. T/F: IOs can be connected directly to the EBTs
30. Identifying the problem boundaries is not trivial for many means. Discuss.
31. Describe the process of identifying the EBTs.
32. Describe the process of identifying the BOs.
33. What are the differences between EBTs and BOs?
34. T/F: The product of the first-level abstraction is design.
35. T/F: The product of the second-level abstraction is architecture.
36. What are the differences between SAPs and traditional analysis patterns?

EXERCISES

1. Take the negotiation pattern and add a second abstraction level for AnyAgreement.
2. Take the trust pattern and add a second abstraction level for AnyRating.
3. Apply the Negotiation pattern to purchasing an item online.
    - Draw a class diagram of using negotiation patterns to purchase an item online.
    - Generate a significant use case for this context.
    - Map the use case to a sequence diagram.

PROJECTS

1. List the properties of EBTs and each of the properties as Stable Analysis Patters.
2. For negotiation stable analysis pattern.
    a. Name a few context of negotiation.
    b. Use one of the context and draw a class diagram.
    c. Document a detailed and significant use case selected context in b.
    d. Create a sequence diagram of the created use case of c.

3. For trust stable analysis pattern.
   a. Name a few context of Trust.
   b. Use one of the context and draw a class diagram.
   c. Document a detailed and significant use case selected context in b.
   d. Create a sequence diagram of the created use case of c.

SIDEBARS

SIDEBAR 1.1 THE ROOTS OF PATTERNS: HISTORICAL PROSPECTIVE

The roots of patterns in general include ancient Egyptians, the ancient Chinese and Indian civilization, Muslim civilization and architectures, and western civilization, and the influence of mathematics and architecture from many other civilizations. This sidebar will conclude with ALEXANDRINE Patterns as Inherited Architectural Solutions.

Ancient Egyptians and Imhotep: The Architect

The original intention of patterns was introduced by Ancient Egyptians, as shown in Figure SB1.1.1. The ancient Egyptians built their pyramids, tombs, temples, and palaces out of stone, the most durable of all building materials. Although earthquakes, wars, and the forces of nature ravaged these memorable structures, the remnants of Egypt's monumental architectural achievements are clearly visible across the ancient land, a tribute to the greatness of this wonderful civilization. These building projects took a high degree of architectural and engineering skills and techniques, and the accumulation of a large labor force consisting of highly trained craftsmen and laborers [1,2].

Apart from the great pyramids, Egyptian buildings were decorated with beautiful paintings, carved stone images, hieroglyphs, and other three-dimensional statues. The superb art of yesteryears tells the undying story of the pharaohs, the gods, the common people, and the natural world of plants, birds, and animals. The eternal beauty and grandeur of these sites are beyond any comparison. How the ancient Egyptians were so successful in constructing these massive structures using primitive tools is still a big mystery.

Pattern Mystery: How to build massive structures using primitive tools?

Imhotep, the world's first well-known architect who built Egypt's first pyramids, is often recognized as the first of many skills and techniques: a doctor, a priest, a scribe, and a vizier, among others [1,2]. He is known to have worked for Djoser, the second great king of Egypt's third dynasty. A historical inscription calls him "the chancellor of the king of lower Egypt," the "first one under the king," the "administrator of the great mansion," and the "chief carpenter."

FIGURE SB1.1.1 Ancient Egyptians architectures.

Mathematics and Architecture in Several Civilizations

Historically, architecture was actually an integral part of mathematics, and in many periods of the past, the two disciplines were indistinguishable from each other [3]. In the ancient world, mathematicians were architects, whose great constructions—the pyramids, ziggurats, temples, stadia, and irrigation projects—we still gape and marvel at today. In the classical period of Greece and ancient Rome, architects were also accomplished mathematicians. When the Byzantine emperor Justinian wanted an architect to build the Hagia Sophia as a building that surpassed everything ever built before, he sought the help of two professors of mathematics (geometers), Isidoros and Anthemios, to perform the task [4]. This practice continued throughout the great Islamic civilizations. Well-known Islamic architects created a wealth of two-dimensional tiling patterns many centuries before western mathematicians started giving a comprehensive and complete classification [5].

Check the Architectural History

In the past, architecture was considered an amalgamation of several significant influences such as artistic, cultural, political, technological, and social. There was also a direct relationship between the build environment and the overall meaning of the construction. Themes like functions, purposes, objectives, goals, and other intangible functions were directly related to the concept of architecture. Being a prisoner of limitations, potentialities, and restrictions of history, architecture offered a wider perspective of understanding things. One such example of this perception is the wide belief that architecture is very formal with wider association with many morphological characteristics like form, technique, and materials.

Pattern Mystery: Patterns exist in all walks of life.

Check the Type of Architectures

Our world has witnessed a rapid advancement in the development of architecture, right from the Neolithic era to present-day modern architecture. Neolithic or "stone age" architecture includes many old structures that are considered the oldest in the history of humankind. Every well-known civilizations and cultures around the world has contributed immensely to the growth and progress of architecture. Each of the succeeding periods in the timeline of history had its own type of architectural concepts and principles that were based on the finer aspects of form, shape, concepts, vision, goal, materials, techniques, purposes, and objectives. As time progressed, man's thinking process underwent a radical evolution too! Currently the concept of architecture goes beyond the abovementioned principles to a wider scheme of things such as geometry, symmetry, usability, stability, life-span, tangibility, user comfort, styles, and a concept of integrating several things that are nonarchitecture in nature.

Pattern Mystery: Patterns can be categorized or classified in different category level and different classes.

- Assyrian architecture
- Babylonian architecture
- Etruscan architecture
- Minoan architecture
- Maya architecture
- Mycenaean architecture
- Persian architecture
- Sumerian architecture
- African architecture
- Chinese architecture
- Indian architecture

- Islamic architecture
- Japanese architecture
- Mesoamerican architecture
    - Incan architecture
- Classical architecture
    - Architecture of ancient Greece
    - Roman architecture
- Medieval architecture
    - Byzantine architecture
    - Sassanid architecture
    - Romanesque architecture
    - Gothic architecture
    - Tudor and Jacobean architecture
- Renaissance architecture
- Baroque architecture
- Regency architecture
- Neo-Classical architecture
- Neo-Gothic architecture
- Neo-Byzantine architecture
- Neo-Romanesque architecture
- Jacobethan architecture
- Tudorbethan architecture
- Beaux-Arts architecture
- Modern architecture
    - Expressionist architecture
    - Futurist architecture
    - Functionalism (architecture)
    - De Stijl
    - Bauhaus
    - Art Nouveau
    - Art Deco
    - International style
- Postmodern architecture
    - Googie architecture
    - Deconstructivist architecture
- Australian architectural styles
- Canadian architecture
- Indonesian architecture
- Origamic architecture (OA)
- Spanish architecture
- Temple architecture (Latter-day Saints)

Alexandrine Patterns as Inherited Architectural Solutions

Christopher Alexander and his associates first attempted to define patterns in solution space by collecting architectural and urban solutions into a *Pattern Language* [6]. These distill timeless archetypes such as the need for light from two sides of a room; a well-defined entrance; interaction of footpaths and roads; a hierarchy of privacy in different rooms of a house; etc. The value of Alexander's Pattern Language is that it is not about specific building types, but about building blocks that can be combined in an infinite number of ways. This implies a more mathematical, combining approach to design in general. Unfortunately, this book is not yet used for a required course in architecture schools [3].

Alexandrine patterns represent solutions repeated in time and space, and are thus akin to visual patterns transposed into other dimensions [7]. Fortunately, the structural solutions that architects depend upon remain part of engineering, which preserves its accumulated knowledge for reuse [3].

Pattern Mystery: Patterns capture the essence of successful solutions to recurring design problems in urban architecture.

References

1. S. Clarke, and R. Engelbach, *Ancient Egyptian Construction and Architecture*, New York, NY: Dover, 2014.
2. E.H. Cline, and D.K. O'Connor, *Amenhotep III: Perspectives on His Reign*, Ann Arbor, Michigan: University of Michigan Press, 2001, p. 273.
3. N.A. Salingaros, Architecture, patterns, and mathematics, originally published in the *NNJ* 1(2), 1999, has become widely read and referred to on the Internet, we have decided to republish an updated version of it, included new Internet links. It is also now available in print in the Nexus Network Journal vol. 1 (1999).
4. N.A. Salingaros, The laws of architecture from a physicist's perspective, *Physics Essays*, 8, 1995, 638–643.
5. R.J. Mainstone, *Hagia Sophia*, New York, NY: Thames and Hudson, 1988.
6. B. Grünbaum, and G.C. Shephard, *Tilings and Patterns*, New York, NY: Freeman, 1987.
7. C. Alexander et al., *A Pattern Language*, New York, NY: Oxford University Press, 1977.

SIDEBAR 1.2 COMMON STABLE DESIGN PATTERNS

1. *AnyActor:* The AnyActor design pattern models the concept of actor by using Software Stability Model [1–5] and knowledge maps [6]. Since AnyActor design pattern captures the core knowledge, it is easy to model AnyActor in any application, by just hooking in the dynamic components of the application. AnyActor can represent a human, smart hardware device, external software package, or creatures depending on the context, where the term AnyActor is being used. The AnyActor design pattern can be applied to numerous applications and within diverse disciplines such as Software systems, Entertainment Business, and other fields like renting, lending, engineering, and science applications, etc. This pattern can be applied to different applications where AnyActor performs diverse roles.
2. *AnyParty:* AnyParty is the most commonly used entity in the entire system. Generally, AnyParty refers to an external legal user of the system. However, AnyParty can also be an internal entity, when it is part of the system or the system itself. AnyParty is used in numerous applications and with in diverse disciplines such as Software systems, Entertainment Business, etc., each having its own context and rationale for using the term Party. AnyParty can represent a human, any organization, any country, or even any political party depending on the context, where the term AnyParty is being used. For example, in context of politics, AnyParty refers to a group of persons with common political opinions, beliefs, and purposes, whose goal is to gain political influence, seize initiative, and governmental control for directing future government policy. Examples of political parties include the Republican Party, the Democratic Party, etc. In other scenarios like peace treaty signing, AnyParty may represent a country. Example of such scenario includes Camp David Accord. However, a party may represent a social gathering of humans to celebrate some events or occasion like birthday, promotion, etc. Again, party might also mean a person or group of persons who are working as an organization having specific goals like United Nations, NGO's, etc. The AnyParty design pattern tries to capture the

core knowledge of AnyParty that is common to all these application scenarios to create stable, extendable, adaptable, and reusable pattern.

3. *AnyCriteria:* The Stable design pattern for AnyCriteria describes the term Criteria. AnyParty or AnyActor, who wants evaluate AnyEntity, uses this term. EBT [2,3,6] that describes the ultimate goal of "AnyCriteria" is Evaluation. AnyCriteria stable sesign pattern can be applied to any context where an AnyEntity needs to be evaluated under AnyCriteria for AnyReasons. Criteria serve as a principle or a standard by which something can be judged or evaluated.
4. *AnyConstraint:* Constraint stable design pattern can be applied to any domain, where limitation is applied or forced. Constraint limits an entity to produce a certain output that can be positive or negative depending on the situation. It restricts certain activity, by imposing noticeable limitations. Constraints are usually negative forces that can stop someone to take some actions. Constraints also impose severe restrictions by way of unforeseen circumstances and scenarios.
5. *AnyRule:* A rule is a set of regulations or laws that informs what needs to be done or allowed and eventually the best way of doing something. It is generally defined by a legislative body, an authority or a head and abides to the masses under them. A rule is defined, either to make an orderly system or to obtain uniformity for reaching the target.
6. *AnyMechanism:* AnyMechanism or AnyService design pattern deals with the concept of service, mechanism, assistance, and the interactions between various parties involved with services. The concept of assistance and using a mechanism or a service to fulfill a functionality is used in multiple contexts, though each of these contexts use different services and the parties dealing with services are unique to the domain or the context. The AnyMechanism or AnyService design pattern makes it easier to capture underlying concept of service and related interactions in different application domains in a generic way using the Stability Model. The pattern presents a generic model for services that can be extended for different domains and contexts, by capturing the core knowledge of services and assistance.
7. *AnyConsequence:* A consequence is a result of a course of action (or of a decision) taken by the decision maker. In an analysis, the consequences of a course of action are determined (predicted) by the use of models. The AnyConsequence pattern deals with the core of problem in depth to find a solution for the aftermath of consequence which is incidentally, the impact of consequence. The pattern designed after the term consequence is reusable, extendable, and robust under any context and usage scenario. Once this pattern is created, a developer need not work again and repeatedly to recreate and rebuild it from the scratch.
8. *AnyOutcome:* The Outcome Design Pattern represents the return result of some of the AnyMechanism or AnyService.
9. *AnyImpact:* The concept *impact* suggests that any change in a given situation, when an impact is applied. Impact can be positive or negative. The AnyImpact design pattern is used in many ways, for different situations and is applied to different scenarios. Impact can also be stated as the striking of one thing to other, a forceful action or collision. For example, social media helps the youths update with what is happening around the world, help them stay connected with friends and interact with their family members even if they are distance apart. This will strengthen relationship among them, even if they work and live in different locations and this is considered to be a positive impact.
10. *AnyReason:* The stable design pattern for Reason describes the term "Reason" that is used by AnyParty or AnyActor to get motivated. We give reasons to get motivated and hence EBT for "Reason" is said to be "Motivation".

11. *AnyCause:* A cause is the motive for some human action. The word cause is also used to mean explanation or answer to a why question. The key idea is that a cause helps to identify the reason for behaving in a particular way or for feeling a particular emotion. This will lead to a stable design pattern for Cause by using it as one of BO and Justification as its EBT. If implemented in such a fashion, this pattern can be used in all the areas where the concept of justification and cause are present.
12. *AnyType:* AnyType design pattern signifies the classification of entities in a problem domain. It could be the classification of entities into various *data types* in a class diagram or various data items in a database. The *AnyType* design pattern generalizes the concept of classifying entities into *types* in different domains and numerous contexts.
13. *AnyEntity:* The AnyEntity design pattern abstracts the existence of an entity based on certain properties inherent to the context in which entity is used and thereby achieves a general pattern that can be used across any application. This pattern is required to model the core knowledge of AnyEntity without tying the pattern to a specific application or domain; hence, the name AnyEntity is chosen. AnyEntity design pattern is one of the most common pattern and it is used by the majority of stable analysis and design patterns.
14. *AnyEvent:* Event is an occurrence; something that happens. The wide range of context of AnyEvent in many applications makes it necessary to have a stable pattern that can be reused and extended further depending on the application. An event exists in all walks of life. An event may refer to: (1) Events in gathering of people or social activities, such as ceremony, convention, festival, sport, and social event, etc. (2) Events in science, technology, mathematics, such as software events, synchronization, impact events, etc. (3) Events in art and entertainment. (4) Other events, such as competition, news, phenomenon, etc.

    AnyEvent design pattern is one of the most common patterns and it is used by the majority of stable analysis and design patterns.
15. *AnyMedia:* AnyMedia is a very general concept with wide range of application in many different contexts. The AnyMedia Stable Design Pattern aims at analyzing the general and important concept of AnyMedia. Since AnyMedia pattern is introduced based on the stable design pattern, it makes it easier to employ this pattern in many different applications. This is possible by just hooking [6] the unstable IOs to the stable business objects according to the application under study. Here we introduce the general AnyMedia design pattern based on stability model and introduce few scenarios where this pattern can be used.
16. *AnyLog:* The AnyLog design pattern models the core knowledge of any Log, as a written record. The Log finds extensive use in the computing industry. The pattern makes it easy to model different kinds of logs rather than thinking of the problem each time from scratch. This pattern can be utilized to model any kind of log in any application and it can be reused as part of a new model [7].

References

1. M.E. Fayad, *Stable Design Pattern for Software and Systems*, Boca Raton, FL: Auerbach Publications, 2016.
2. M.E. Fayad, and A. Altman, Introduction to software stability, *Communications of the ACM*, 44(9), 2001, 95–98.
3. M.E. Fayad, Accomplishing software stability, *Communications of the ACM*, 45(1), 2002, 111–115.
4. M.E. Fayad, How to deal with software stability, *Communications of ACM*, 45(4), 2002, 109–112.

5. M. E. Fayad, and S. Wu, Merging multiple conventional models into one stable model, *Communications of the ACM*, 45(9), 2002, 95–98.
6. M.E. Fayad, H.A. Sanchez, S.G.K. Hegde, A. Basia, and A. Vakil, *Software Patterns, Knowledge Maps, and Domain Analysis*, Boca Raton, FL: Auerbach Publications, 2014.
7. M.E. Fayad, J. Rajagopalan, and A. Arun, The AnyLog design pattern, The 2003 IEEE International Conference on Information Reuse and Integration, Las Vages, NV, October 2003.

SIDEBAR 1.3 ANALYSIS PATTERNS' REFERENCES

(Only Journals and Book Publications)

This survey briefly summarizes the present state of literature of analysis patterns. It presents a brief overview of the existing journal and book literature on analysis patterns. It seems that the development of design patterns is more advanced than that of analysis patterns, as there is more literature available on the subject of design patterns.

1. C. Alexander, S. Ishikawa, M. Silverstein, *A Pattern Language. Towns, Buildings, Construction*, New York, NY: Oxford University Press, 1977.
2. J. Arlow, and I. Neustadt, *Enterprise Patterns and MDA—Building Better Software with Archetype Patterns and UML*, Boston, MA: Pearson Education, 2004.
3. N. Bolloju, Improving the quality of business object models using collaboration patterns. *Communications of the ACM*, 47(7), 2004, 81–86.
4. W.J. Brown, R.C. Malveau, H.W.S. McCormick, and T.J. Mowbray, *AntiPatterns: Refactoring Software, Architectures, and Projects in Crisis*, New York, NY: John Wiley & Sons Ltd, 1998.
5. F. Buschmann, R. Meunier, H. Rohnert, P. Sommerlad, and M. Stal, *Pattern-Oriented Software Architecture, A System of Patterns*, Chichester, England: John Wiley & Sons Ltd, 1996.
6. P. Coad, M. Mayfield, and D. North, *Object Models—Strategies, Patterns & Applications*, Raleigh, NC: Prentice Hall Publishing, 1995.
7. P. Coad, Object-oriented patterns, *Communications of the ACM*, 35(9), 1992, 63–74.
8. V. Devedzic, and A. Harrer, Software patterns in ITS architectures. *International Journal of Artificial Intelligence in Education (IJAIED)*, 15(2), 2005, 63–94.
9. M.E. Fayad, and A. Altman, Introduction to software stability, *Communications of the ACM*, 44(9), 2001, 95–98.
10. M.E. Fayad, Accomplishing software stability, *Communications of the ACM*, 45(1), 2002, 95–98.
11. M.E. Fayad, How to deal with software stability, *Communications of ACM*, 45(4), 2002, 109–112.
12. M.E. Fayad, and S. Wu, Merging multiple conventional models into one stable model, *Communications of the ACM*, 45(9), 2002, 102–106.
13. M.E. Fayad, H.A. Sanchez, S.G.K. Hegde, A. Basia, and A. Vakil, *Software Patterns, Knowledge Maps, and Domain Analysis*, Boca Raton, FL: Auerbach Publications, 2014.
14. M.E. Fayad, and S. Jindal, Legality stable analysis pattern (pattern based model driven software development), *i-manager's Journal on Software Engineering (ISE)*, 9(4), 2015, 1–6.
15. M.E. Fayad, and C. Flood III, A pattern for stress and its resolution, *i-manager's Journal on Software Engineering (ISE)*, 9(5), 2015.
16. M. Fowler, *Analysis Patterns—Reusable Object Models*, Mountain View, CA: Addison-Wesley Publishing, 1997.
17. E. Gamma, R. Helm, R. Johnson, and J. Vlissides, *Design Patterns—Elements of Reusable Object-Oriented Software*, Mountain View, CA: Addison-Wesley Publishing, 1995.
18. H.S. Hamza, and M.E. Fayad, Applying analysis patterns through analogy: Problems and solutions, *Journal of Object Technology*, 3(4), 2004, 197–208.
19. D.C. Hay, *Data Model Patterns: Convention of Thoughts*, New York, NY: Dorset House Publishing, 1995.
20. S. Konrad, B.H.C. Cheng, and L.A. Campbell, Object analysis patterns for embedded systems, *IEEE Transactions on Software Engineering*, 2004, 970–992.
21. F.J. Lisboa, C. Iochpe, and K.A.V. Borges, Analysis patterns for GIS data schema reuse on urban management applications, *CLEI Electronic Journal*, 5(2), 2002.

22. A. Manjarrés, G. Sunyé, D. Pollet, Pickin, F.D. Mossé, Modeling roles: A practical series of analysis patterns, *Journal of Object Technology*, 1(4), 2002.
23. J. Nicola, M. Mayfield, and M. Abney, *Streamlined Object Modeling—Patterns, Rules and Implementation*, Upper Saddle River, NJ: Prentice Hall Publishing, 2002.
24. D.C. Schmidt, M.E. Fayad, and R.E. Johnson, Software patterns, *Communications of the ACM* 39(10), 1996, 37–39.
25. I. Seruca, and P. Loucopoulos, Towards a systematic approach to the capture of patterns within a business domain, *Journal of Systems and Software*, 67(1), 2002, 1–18.
26. L. Sesera, Hierarchical patterns: A way to organize (analysis) patterns. *Journal of Systemics, Cybernetics and Informatics*, 3(4), 2004, 37–40.

SIDEBAR 1.4 MARTIN FOWLER'S ANALYSIS PATTERNS

Martin Fowler defines pattern as "it is commonly said that a pattern, however it is written, has four essential parts: a statement of the context where the pattern is useful, the problem that the pattern addresses, the forces that play in forming a solution, and the solution that resolves those forces. ... it supports the definition of a pattern as 'a solution to a problem in a context,' a definition that [unfortunately] fixes the bounds of the pattern to a single problem-solution pair" [1] and "A pattern is an idea that has been useful in one practical context and will be probably useful in others" [1]. We tried to use the majority of Fowler's patterns and we found that each of his pattern is created for a particular context as he indicated and it is suitable for other context. It is obviously to say Martin Fowler' Analysis Patterns has nothing to do with analysis.

Most of Fowler's patterns can be used across domains or reusable object models [2], which is not true and his analysis patterns are mostly domain-specific patterns, are not reusable, and not well-documented as indicated by Reference [2].

The diagram in Figure SB1.4.1 shows one of Fowler's most well-known analysis patterns, the so-called party pattern.

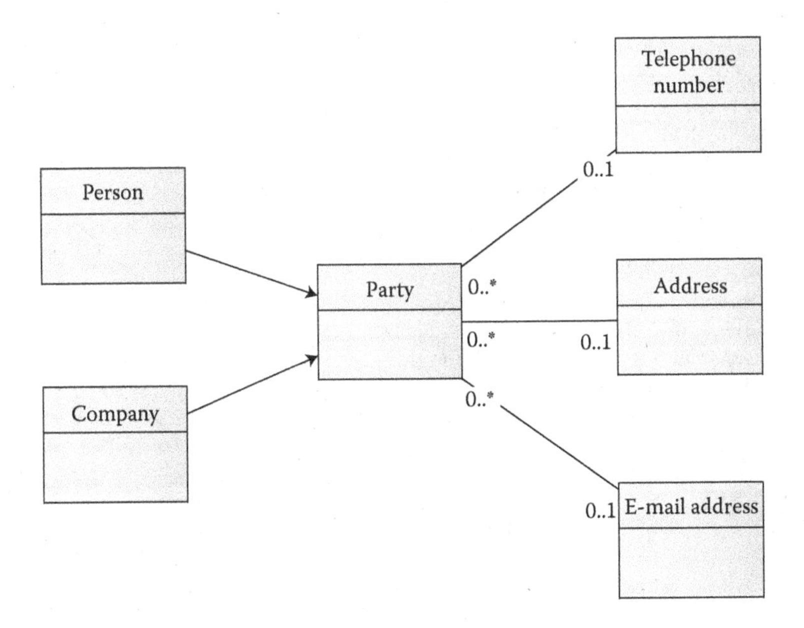

FIGURE SB1.4.1 "Party" pattern [2]. (Based on M. Fowler, Analysis Patterns—Reusable Object Models, Addison-Wesley Publishing, 1997, p. 18.)

References

1. M. Fowler, *Analysis Patterns—Reusable Object Models,* Reading, MA: Addison-Wesley Publishing, 1997.
2. N. Blaimer, A. Bortfeldt, and G. Pankratz, Patterns in Object-Oriented Analysis; Working Paper No. 451, 2010.

REFERENCES

1. A. Mahdy, M.E. Fayad, H. Hamza, and P. Tugnawat, Stable and reusable model-based architectures, ECOOP 2002, Workshop on Model-Based Software Reuse, Malaga, Spain, June 2002.
2. H. Hamza, and M.E. Fayad, Model-based software reuse using stable analysis patterns, ECOOP 2002, Workshop on Model-Based Software Reuse, Malaga, Spain, June 2002.
3. M.E. Fayad, and A. Altman, Introduction to software stability, *Communications of the ACM,* 44(9), 2001, 95–98.
4. M.E. Fayad, Accomplishing software stability, *Communications of the ACM,* 45(1), 2002a, 111–115.
5. M.E. Fayad, How to deal with software stability, *Communications of the ACM,* 45(4), 2002b, 109–1112.
6. H. Hamza, Building stable analysis patterns using software stability, 4th European GCSE Young Researchers Workshop, Germany, 2002.
7. M.E. Fayad, H.A. Sanchez, S.G.K. Hegde, A. Basia, and A. Vakil, *Software Patterns, Knowledge Maps, and Domain Analysis.* Boca Raton, FL: Auerbach Publications, 2014.
8. P.T. Devanbu, Software engineering for security: A roadmap, 22nd International Conference on Software Engineering, Limerick Ireland, June 4–11, 2000, pp. 227–239.
9. D. Hay, *Data Model Patterns-Conventions of Thoughts,* 1st ed., New York, NY: Dorset House, 1996.
10. E.B. Fernandez, X. Yuan, and S. Brey, Analysis patterns for the order and shipment of a product, Pattern Languages of Programs Conference, Monticello, IL, 2000.
11. M.E. Fayad, and H. Hamza, The trust analysis pattern, Pattern Languages of Programs Conference, Monticello, IL, 2004.
12. R.T. Vaccare Braga et al., A confederation of patterns for business resource management, *Pattern Language of Programs Conference,* Monticello, IL, 1998.
13. M. Fowler, *Analysis Patterns: Reusable Object Patterns,* 1st ed., Boston, MA: Addison-Wesley, 1997.
14. P. Coad et al., *Object Models: Strategies, Patterns, and Applications,* 1st ed., New Jersey, NJ: Prentice Hall, 1995.
15. H. Hamza, and M.E. Fayad, Applying analysis patterns through analogy: Problems and solutions, *Journal of Object Technology,* 3(4), 197–208, 2004.
16. T. Grandison, and M. Sloman, Specifying and analyzing trust for internet applications, 2nd IFIP Conference E-Commerce, E-Business, E-Government, Lisbon, 2002.
17. E. Fernandez, and Y. Liu, The account analysis pattern, European Pattern Languages of Programs Conference, Germany, 2002.
18. H. Hamza, A foundation for building stable analysis patterns, MS thesis, Department of Computer Science, University of Nebraska-Lincoln, 2002.
19. A. Kini, and J. Choobineh, Trust in electronic commerce: Definition and theoretical considerations, *Thirty-First Annual Hawaii International Conference on System Sciences,* Kohala Coast, Hawaii, USA, January 6–9, 1998. IEEE Computer Society 1998.
20. T. Grandison, and M. Sloman, A survey of trust in internet applications, IEEE Communications Surveys, Fourth Quarter 2000. http://www.comsoc.org/pubs/surveys
21. E.A. Kaluscha, and S. Grabner-Kräuter, Towards a pattern language for consumer trust in electronic commerce, European Pattern Languages of Programs Conference, Germany, 2003.
22. S. Grabner-Krauter, and E.E. Kaluscha, Engendering consumer trust in E-commerce: Conceptual clarification and empirical findings, in Petrovic, O. et al., eds. *Trust in the Network Economy,* Wien, New York, NY: Springer.

2 Applying Analysis Patterns through Analogy
Problems and Solutions

It is often in our Power to obtain an Analogy where we cannot have an Induction.

David Hartley

Traceability and generality are among the main qualities that determine the effectiveness of developed analysis patterns. However, satisfying them at the same time is a real challenge. Today, most of the analysis patterns are thought of as templates, where they can be instantiated, and hence, reused through an analogy between the original pattern and the application on hand. Developing analysis patterns as templates might maintain the appropriate level of generality; however, it scarifies patterns' traceability, once they are applied in a developed system. This chapter illustrates the main problems with developing analysis patterns as templates and reusing them through analogy. In addition, it also shows through examples how stable analysis patterns can satisfy both the generality and the traceability, and therefore, enhance the role of analysis patterns in software development.

2.1 INTRODUCTION

In the last decade, patterns have emerged as a promising technique for improving the quality and reducing the cost and time of software development [1,2]. A pattern can be defined as: *An idea that has been useful in one practical context and will probably be useful in others* [3].

The obscurity of doing accurate analysis along with the fact that analysis is a tedious and time-consuming process increases the interest in the development of effective and reusable analysis artifacts of special interest. Analysis patterns form a promising base for facilitating and improving the quality of performing analysis. For analysis patterns to fulfill their expectation as a major contributor in software development, patterns should maintain some essential quality factors.

This chapter focuses on two of these qualities: *traceability* and *generality*. *Generality* means that the pattern that analyzes a specific problem can be successfully reused to analyze the same problem whenever it appears, even within different applications or across different domains. This quality factor is essential due to the fact that the analysis of a specific problem should be the same if the problem remains similar. There is no benefit of having different analysis models for the same exact problem. Analysis patterns that fail to model the exact same problem in different applications may diminish the aim of developing patterns as reusable artifacts.

On the other hand, *traceability* means that a pattern that is used in the development of a specific system can be successfully traced back to the original analysis pattern that has been used. Untraceable patterns will disappear once the developer has instantiated it in their current system, thereby imposing further complications in the maintainability of the developed system.

Satisfying both generality and traceability is a factual challenge in the case of current analysis patterns. This challenge is mainly because most current techniques for developing analysis patterns are based on dealing with patterns as templates that form a general model for the problem; thus, it can be reused repeatedly through analogy [4–7]. As we would soon discuss in the following section, this approach can maintain the generality factor of patterns to some extent, while at the same time sacrificing their traceability.

This chapter illustrates important problems with developing analysis patterns as templates and reusing them through analogy. It also proposes the use of the concept of *Stable Analysis Patterns* [8–10], which are analysis patterns that are built based on the software stability concepts [11], and that provide a solution to accomplish both generality and traceability while developing analysis patterns.

The chapter is organized as follows: Section 2.2 describes the use of analysis patterns through analogy approach; Section 2.3 illustrates the problems associated with this approach; Section 2.4 provides an overview of stable analysis patterns; and Section 2.5 provides examples of using stable analysis patterns. The conclusions are presented in Section 2.6.

2.2 ANALYSIS PATTERNS AS TEMPLATES

Today, most of analysis patterns are thought of as templates [4–7]. In Reference [12], Peter Code has defined patterns in general as follows: *A pattern is a template of interacting objects, one that may be used again and again by analogy.* That is, the pattern that is extracted from a specific project can be put into an appropriate abstract level, such that it can be used to model the same problem in a wide range of applications and domains. The abstracted pattern is then considered to be a template, by which it could be used through an analogy. Developing patterns as templates, while providing an appropriate level of generality, will sacrifice their traceability when using these patterns through analogy.

As an example of this approach, Figure 2.1 shows the class diagram of the *Resource Rental* pattern taken from Reference [8], which forms the abstract template of the Resource Rental problem. The objective of the pattern is to provide a model that can be reused to model the problem of renting

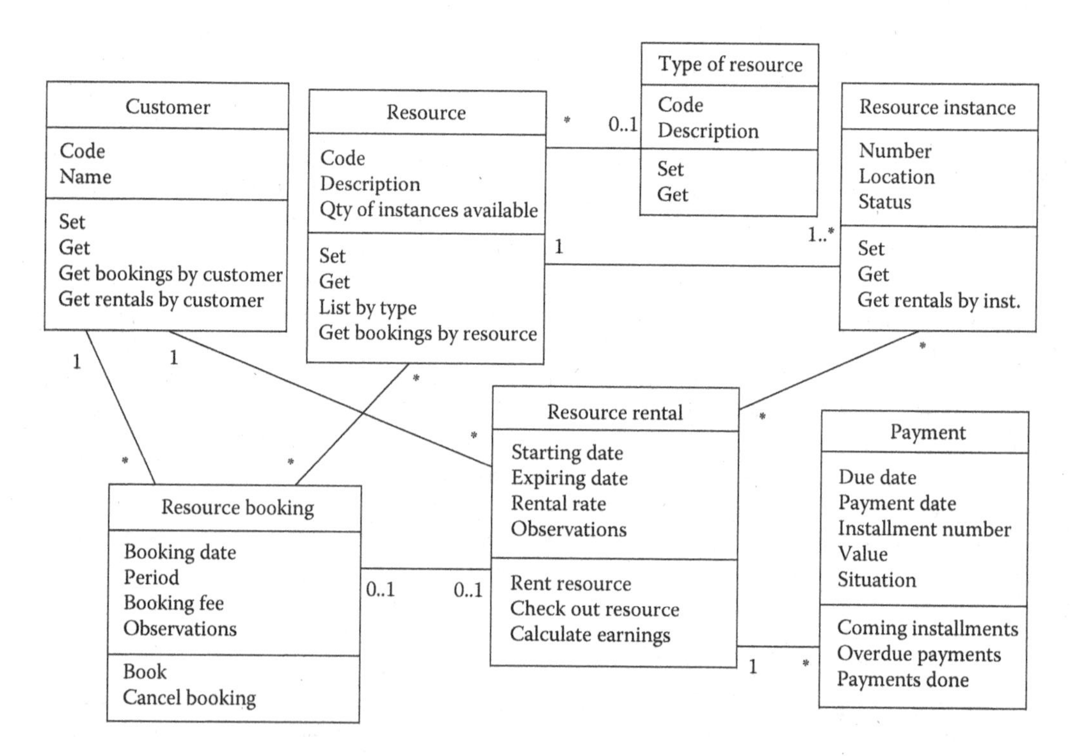

FIGURE 2.1 Resource rental pattern.

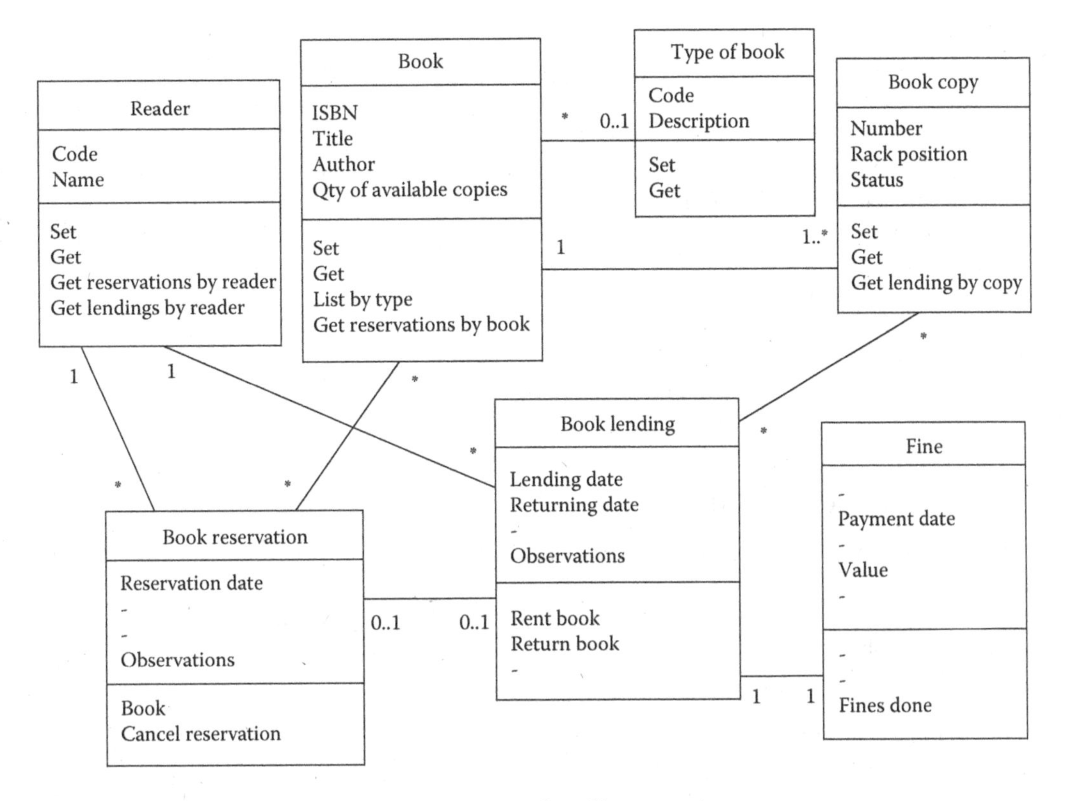

FIGURE 2.2 Instantiation of resource rental pattern for a library service.

any resource; therefore, the class diagram does not tie the renting to a specific recourse. Figure 2.2 shows an example of using the *Resource Rental* pattern in the application of the library service taken from Reference [8]. Simply through an analogy, one can apply the original abstract pattern into a specific application.

2.3 PROBLEMS WITH USING ANALYSIS PATTERNS THROUGH ANALOGY

Although using analysis pattern through analogy looks to be a straightforward technique for maintaining a good level of generality in the pattern, this technique raises some critical problems that should be further investigated and scrutinized. Among these critical problems, we are more interested in those that are related to the focus of this chapter, some of which are summarized as follows:

- *Generate untraceable systems:* Once analysis pattern templates have been instantiated in the developed system, the original patterns will no longer be extractable. For example, consider the instance of the Resource pattern shown in Figure 2.2 and imagine it as part of a complete library service system; it would be hard, if not impossible, to extract the original pattern after such instantiation.
- *Complicate system maintainability:* Software maintenance is considered to be one of the most costly phases in the system development life cycle. Therefore, complicating system maintainability is expected to further increase this cost. One can imagine a very simple situation, where we need to update the developed system documentation due to some modification in system requirements. Since the developed system is using several patterns in its development, identifying which patterns these changes apply to will be tedious and time consuming.

- *Trivialize classes' roles of the pattern:* To better discuss this issue, we will use an example from Reference [5], where a class diagram for designing a computer repair shop is used, by an analogy, to build the class diagram of a hospital registration project. Thus, instead of a shop that fixes broken computers, we have a hospital that fixes sick people. We can simply replace the class named computer in the first project by the new class named patient in the next project. Even though such an analogy seems doable, it is impractical. There is a great difference between the computer as a machine and the patient as a human. These two classes might look analogous, since they both have problems that need to be fixed; however, their behaviors within the system are completely different. The role of the computer class is completely different from that of the patient. Hence, such analogy in inaccurate. There would be even more occurrences of differences, if we try to generate the dynamic behavior of these two systems by using an analogy as suggested in Reference [8].

2.4 STABLE ANALYSIS PATTERNS

Stable analysis patterns introduced in References [8–10] are analysis patterns constructed based on software stability concepts [11]. Before we describe how stable analysis patterns can satisfy both the generality and the traceability quality factors, a brief overview of software stability concepts, and an example of stable analysis patterns are provided in this section.

2.4.1 Stability Model

Software stability stratifies the classes of the system into three layers: EBTs layer [11,13], BOs layer, and IOs layer [11]. Based on its nature, each class in the system model is classified into one of these three layers. Figure 2.3 depicts the layout of stability model layers. The properties that characterize EBTs, BOs, and IOs are given in References [14,15].

EBTs are the classes that present the enduring and core knowledge of the underlying industry or business. Therefore, they are extremely stable and form the nucleus of the stability model. BOs are the classes that map the EBTs of the system into more concrete objects. BOs are semi-conceptual and externally stable, but they are internally adaptable. IOs are the classes that map the BOs of the system into physical objects. For instance, the BO "Agreement" can be mapped in real life as a physical "Contract," which is an IO.

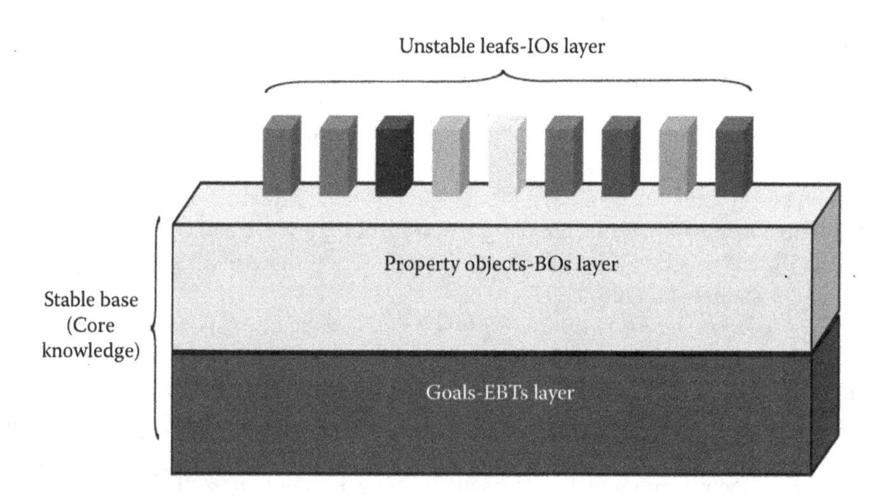

FIGURE 2.3 Stability model layers layout.

2.4.2 Stable Analysis Pattern Example

The purpose of this chapter is not to propose a new analysis pattern; however, it would be easier to illustrate the concept of stable analysis patterns through an example. We use the *Negotiation* analysis pattern as an example. *Negotiation* is a general concept that has many applications. In daily life, there are various situations where negotiation usually takes place. For instance, buying or selling properties usually involves some sort of negotiation (e.g., buying or selling a home or a car). In software systems, negotiation also appears frequently in the development of different applications. Developing software for online auctions and shopping might involve the negotiation of the price and/or the negotiation of different product aspects.

More technically, negotiation becomes an essential part in the development of next generation web-based devices and appliances. Today, devices that need to access the web diverge greatly in their capabilities, so negotiation algorithms between client agent and servers play a fundamental role in helping servers decide which representation of a document a device should be given. Therefore, having a stable pattern that can model the basic aspects of a negotiation problem would make it easier for the developer to build their system by reusing and extending this pattern. Figure 1.6 (Chapter 1) shows the stable object model of the *Negotiation* pattern. A short documentation for this pattern is provided in Appendix A.

As shown in Figure 1.6, the *Negotiation* pattern consists of the following participants:

- *Negotiation:* Represents the negotiation process itself. This class contains the behaviors and attributes that regulate the actual negotiation process.
- *AnyAgreement:* Represents the result of the negotiation. The ultimate goal of any negotiation is to reach an agreement. Thus, this object presents a core element in any negotiation. It is important to note that in many cases negotiation ends with no agreement and thus it is considered to be failed (the seller of the car did not agree on the price proposed by the buyer and vice versa), however, in this case, we expect that the agreement should provide this result by whatever mechanism. So, one can view the agreement object as the result of the negotiation, which is not necessarily a successful result.
- *AnyParty:* Represents the negotiation handlers. It models all the parties that are involved in the negotiation process. Party can be a person, organization, or a group with specific orientation.
- *AnyMedia:* Represents the media through which the negotiation will take place. For instance, one can negotiate the price of a good over the phone. Others might use an email or a mail to negotiate specific issues in their business.
- *Context:* Represents the matters to be negotiated. If we are buying a home, many issues could be negotiated. For instance, the price of the home, the payment procedure, etc. Defining the issue to be negotiated is an essential element of any negotiation process, otherwise, negotiation will have no meaning.

The prefix "any" that we used herein indicates that this is another pattern that provides an abstract model for the notion it precedes. For instance, AnyParty is a stand-alone stable pattern that models the party notation and, hence, can be used to model any party in any applications. The detailed structure of this pattern is beyond the scope of this chapter.

2.5 APPLYING STABLE ANALYSIS PATTERNS

In order to illustrate how stable analysis patterns can maintain both the generality and traceability quality factors, we have used the pattern to model two different applications: negotiation of buying a car, and content negotiation using composite capability/preference profile (CC/PP). Since the objective of these examples is to demonstrate the usage of the proposed pattern, stated examples do not

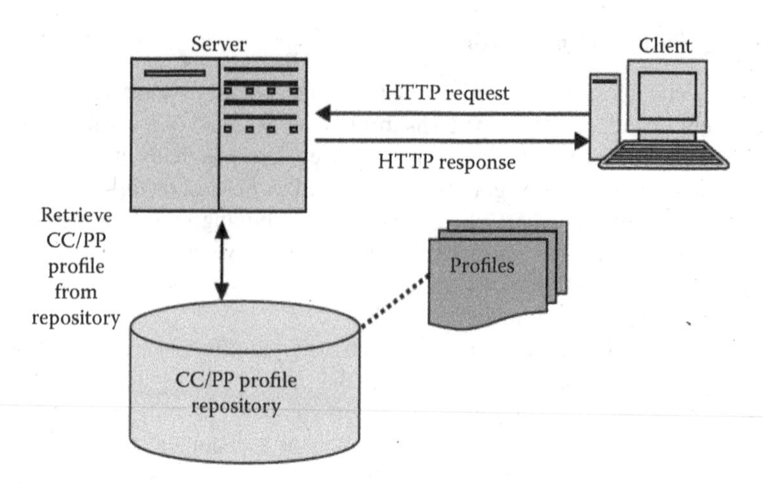

FIGURE 2.4 Possible scenario of content negotiation using CC/PP.

present the complete model for the problem, but focus on the part that involves the negotiation process. The full analysis (CRC cards, use case diagrams, use case descriptions, sequence diagrams, and state transition diagrams) of these two examples is given in Reference [8].

EXAMPLE 2.1: NEGOTIATION TO BUY A CAR

In buying a car, a negotiation concerning the car's price and warranty usually takes place. This example models the simple negotiation that might be involved in buying a car. Figure 1.7 (Chapter 1) shows the stability model of the negotiation used in buying a car. Classes that are not in the original *Negotiation* pattern are colored in gray.

EXAMPLE 2.2: CONTENT NEGOTIATION USING CC/PP

Today, specific heterogeneous devices are required to access the World Wide Web; yet, each device has its own set of capabilities. As a result, a negotiation between the client and the server should take place in order for the server to know the capabilities of these devices and provide the appropriate contents. One possible techniques of performing content negotiation is called *Composite Capability/Preference Profile* (CC/PP) [16]. A possible scenario of CC/PP content negotiation is given in Figure 2.4. Figure 1.8 (Chapter 1) shows the stability model of this example. Again, classes that are not in the original *Negotiation* pattern are colored in gray.

As shown in the two examples, the use of the *Negotiation* pattern is not achieved through an analogy. The pattern can be spotted easily and thus it is possible to trace it back in the developed system. On the other hand, the developed pattern does not lose its generality, as we were able to apply it to model the same problem in two different applications. Moreover, one can realize that each object in the *Negotiation* pattern has a clear role independent of the application the pattern will be used in. For instance, AnyMedia as an object still exists and it has the same role independent of the application; however, the type of media might vary based on the application.

2.6 SUMMARY

Stable analysis patterns introduce a new vision of developing and utilizing analysis patterns in building software systems. Although, current approaches of developing analysis patterns as templates and utilizing them through an analogy maintain pattern generality, it still scarifies its traceability. This makes the developed systems harder and more costly to maintain. Stable analysis patterns are developed and utilized so that they can preserve both the generality and traceability. In addition, stable analysis patterns guarantee the preservation of the classes' roles within the pattern;

thus, each class has the same role independent of the application that the pattern will be deployed in. Therefore, stable analysis patterns can form a more effective base for utilizing patterns in developing software systems.

REVIEW QUESTIONS

1. Explain and define Pattern.
2. Define generality. Why is it important?
3. Define traceability. Why is it important?
4. "Satisfying both generality and traceability is a factual challenge in current analysis patterns." Explain why.
5. Can generality be achieved if pattern as developed as templates? Justify your answer.
6. Why developing patterns as templates sacrifice their traceability?
7. What problems arise if analysis patterns are used through analogy? Explain.
8. Explain stability model.
9. What are the three layers of stability model? Explain with the help of a diagram.
10. Explain what negotiation is. Identify two applications of negotiation in day-to-day life.
11. Draw and explain negotiation stable model.
12. What are the participants in negotiation stable model? Explain.
13. Justify the existence of the following classes in negotiation stable model.
    a. AnyAgreement
    b. AnyMedia
    c. AnyContext
    d. AnyParty
14. Why is prefix "Any" used in front of the class names in negotiation stable pattern?
15. Apply the negotiation stable pattern to buy a car. Draw a class diagram for the same.
16. Explain content negotiation using CC/PP. Draw a class diagram for the same.
17. Explain the advantages of stability model over analogy approach.
18. Explain how traceability and generality is achieved using stability model.
19. "Each object in the negotiation pattern has a clear role, independent of the application the pattern will be used in." Explain and justify.
20. Compare stability approach with analogy approach.

EXERCISES

1. Model your dream vacation by using analogy approach and stable approach. Which one do you like more and why?
2. Model your dream house by using analogy approach and stable approach. Which one do you like more and why?
3. Model your car by using analogy approach and stable approach. Which one do you like more and why?
4. Try to identify the class in the stable design pattern for culture.
5. Try to connect those classes together to form a class diagram.
6. Try to find 2–3 scenarios where this pattern can be applicable.

PROJECTS

1. Consider a scenario where your laptop has stopped working and you take it to a repair center. The vendor asks for a huge sum of money, but you negotiate with him the price for repair. Create a class diagram for this using negotiation patterns as your base and draw a sequence diagram for negotiations.

2. Consider a scenario where you negotiate with a company over the salary and perks being offered for a position. Create a class diagram for this using negotiation patterns as your base and draw a sequence diagram for negotiations.
3. Consider a scenario where an online trade application is being developed. The two registered users can negotiate over the price of a product being traded. Create a class diagram for this application using negotiation patterns as your base and draw a sequence diagram for negotiations.

REFERENCES

1. D.C. Schmidt, M.E. Fayad, and R. Johnson, Software patterns, *Communications of the ACM*, 39(10), 1996, 37–39.
2. E. Gamma et al., *Design Patterns: Elements of Reusable Object-Oriented Software*, Addison-Wesley Professional Computing Series, New York, NY: Addison-Wesley, 1995.
3. M. Fowler, *Analysis Patterns—Reusable Object Models*, 1st ed., New York, NY: Addison-Wesley, 1997.
4. E.B. Fernandez, and X. Yuan, An analysis pattern for reservation and use of reusable entities, *Proceedings of Pattern Languages of Programs Conference (PLoP1999)*, Boca Raton, FL, 1999.
5. E.B. Fernandez, and X. Yuan, Semantic analysis patterns, *Proceedings of 19th International Conference on Conceptual Modeling (ER2000)*, Salt Lake City, UT, 2000, pp. 183–195.
6. E.B. Fernandez, Stock manager: An analysis pattern for inventories, *Proceedings of Pattern Languages of Programs Conference (PLoP2000)*, Monticello, IL, 2000.
7. R.T. Vaccare Braga et al., A confederation of patterns for business resource management, *Proceedings of Pattern Language of Programs' 98 (PLoP1998)*, Monticello, IL, September 1998.
8. H. Hamza, A foundation for building stable analysis patterns. *MS thesis*. Department of Computer Science, University of Nebraska-Lincoln, Lincoln, NE, 2002.
9. H. Hamza, and M.E. Fayad, Model-based software reuse using stable analysis patterns, ECOOP 2002, Workshop on Model-based Software Reuse, Malaga, Spain, June 2002.
10. H. Hamza, and M.E. Fayad, A Pattern language for building stable analysis patterns, *9th Conference on Pattern Language of Programs (PLoP2002)*, Monticello, IL, 2002.
11. M.E. Fayad, and A. Altman, Introduction to software stability, *Communications of the ACM*, 44(9), 2001, 95–98.
12. P. Coad, D. North, and M. Mayfield, *Object Models—Strategies, Patterns, & Applications*, 1st ed., Upper Saddle River, NJ: Prentice Hall, 1995.
13. M. Cline, and M. Girou, Enduring business themes, *Communications of the ACM*, 43(5), 2000, 101–106.
14. M.E. Fayad, Accomplishing software stability, *Communications of the ACM*, 45(1), 2002, 111–115.
15. M.E. Fayad, How to deal with software stability, *Communications of the ACM*, 45(4), 2002, 109–112.
16. F. Reynolds, J. Hjelm, S. Dawkins, and S. Singhal, Composite Capability/Preference Profiles (CC/PP): A user side framework for content negotiation, World Wide Web Consortium, Note NOTE-CCPP-19990727, July 1999.

3 A Pattern Language for Building Stable Analysis Patterns

All great truths are simple in final analysis, and easily understood; if they are not, they are not great truths.

Napoleon Hill

Existing analysis patterns are believed to play a vital role in reducing the cost and in condensing the time of software product lifecycles. However, existing analysis patterns are yet to realize their full potential. One of the most common problems with existing analysis patterns is the lack of stability. In many cases, analysis pattern that molds specific problems simply fail to model the same problem when it appears in a different context, thereby forcing software developers to start analyzing the problem from scratch. As a result, the reusability of the pattern will start diminishing. This chapter presents a pattern language for building stable analysis patterns. The main objective of this language is to propose a way for achieving stability in the process of constructing analysis patterns.

3.1 INTRODUCTION

In a nutshell, analysis patterns are conceptual models that model the core knowledge of the problem. It is expected that the pattern that models a specific problem should be easily and successfully reused to shape the same problem regardless of the context in which the problem appears. However, this might not be the case with existing analysis patterns because many of today's analysis patterns are known to tackle well-known problems that span across many domains. Nonetheless, using such patterns is not as easy they should appear to be even though one might still try to model similar problems under different scenarios. As a result, it is not too surprising to see software developers preferring to start their analysis from scratch. Stable and robust analysis patterns are extremely useful in designing highly effective and reusable patterns. Such patterns can be used to sculpt the problem, whenever it appears, independent of the context of the problem.

Software stability concepts that were introduced in [1,2] have demonstrated great promise in the area of software reuse and lifecycle improvement because they apply the novel concepts of "Enduring Business Themes" (EBTs) and "Business Objects" (BOs). Hereby, we propose the advanced concepts of stable analysis patterns by applying stability model concepts to the notion of analysis patterns. The main idea behind stable analysis patterns is to analyze the given problem under consideration in terms of its EBTs and the BOs and to promote the goal of increased stability and broader reuse. The pattern language presented in this chapter explains the main steps that are required for building stable analysis patterns.

The rest of this chapter is organized as follows: Section 3.2 provides an overview of the proposed pattern language, and displays the relations between the different patterns. A detailed description of each pattern is given is Section 3.3 while concluding remarks are presented in Section 3.4.

3.2 PATTERN LANGUAGE OVERVIEW

The proposed pattern language highlighted in this book contains eight important patterns. These patterns record different steps required for building stable analysis patterns. All eight patterns are categorized into three main levels with each level having its own main objective. Figure 3.1 shows three levels of the pattern language and their corresponding patterns under each level. Each pattern has two digits number where first digit shows the level of the pattern, while the second digit the pattern's number.

The first level is the "*Concept Patterns*" level. Patterns in this level provide main concepts that are required in order to understand the rest of the pattern language. The concept patterns level consists of two patterns: the *Efficient Usable Analysis Models* [1.1], which describes the main essential properties required for building efficient and usable analysis models, and the *Stability Model* [1.2], which provides the fundamental concepts of the stability model as a solution for achieving stability.

The second is the "*Problem Analysis Patterns*" level. Patterns in this level show how to analyze the problem that needs to be modeled. The analysis of the problem can be carried out by following four main steps, where each step is documented as a pattern. First, *Identify The Problem* [2.3]. Second, *Identify Enduring Business Themes* [2.4]. Third, *Identify Business Objects* [2.5]. Fourth, put these elements into the *Proper Abstraction Level* [2.6].

The third in the series is the "Building-Process Patterns" level. After analyzing the problem in the previous level, we will need to know how to *Build Stable Analysis Patterns* [3.7], and how to

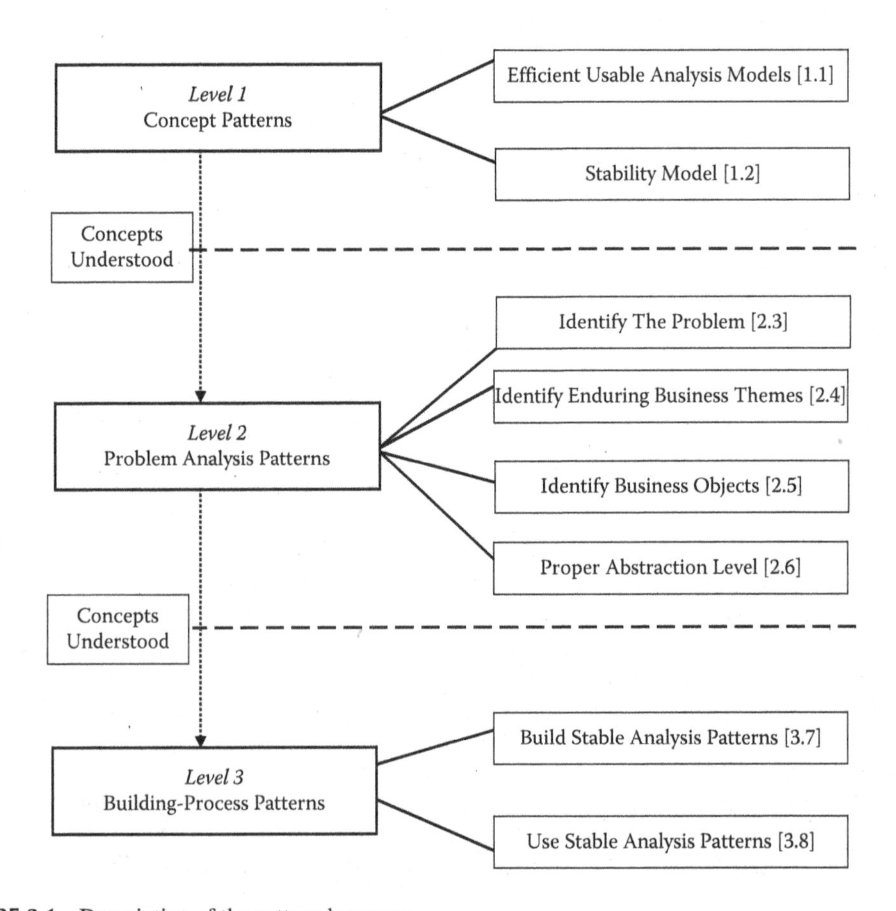

FIGURE 3.1 Description of the pattern language.

Use Stable Analysis Patterns [3.8]. Table 3.1 summarizes the patterns language for quick reference. The relationship among the language patterns is shown in Figure 3.2.

3.3 BUILDING STABLE ANALYSIS PATTERN LANGUAGE DESCRIPTION

> ***Level 1* Concept Patterns:** First, understand the main properties of the **Efficient Usable Analysis Models** [1.1]. Second, understand the basic concepts of **Stability Model** [1.2].

3.3.1 Pattern 1.1: Efficient Usable Analysis Models

3.3.1.1 Intent

Presents the essential properties of efficient and usable analysis models.

3.3.1.2 Context

Analysis is the first basic step in solving any problem. Some essential properties are needed for analysis models in order for them to be useful and usable. Keeping these properties in mind while building analysis models, will tremendously improve the quality of these models.

3.3.1.3 Problem

What are the essential properties that affect the usability and the effectiveness of analysis models?

3.3.1.4 Forces

- As models, analysis models must satisfy all six basic model properties that were introduced in [3]. These are, to be simple, complete and most likely accurate, testable, stable, to have visual representation, and to be easily understood. However, these six properties are insufficient because of the nature of analysis models.
- There are some misconceptions when we try to understand some of the models' essential properties. For instance, for the model to be "easy to understand" is not the same as to be "simple." Many analysis models are easy to understand; yet, most of them are not simple. Such confusions tremendously affect the resultant models.

TABLE 3.1
Summary of the Patterns Language

Pattern	Problem	Solution
Efficient Usable Analysis Models [1.1]	What are the essential properties that affect the usability and the effectiveness of analysis models?	Page 39
Stability Model [1.2]	How to accomplish software stability?	Page 42
Identify the Problem [2.3]	How to focus on a specific problem that the analysis pattern will model?	Page 45
Identify Enduring Business Themes [2.4]	How to identify EBTs of the problem?	Page 46
Identify Business Objects [2.5]	How to identify BOs of the problem?	Page 49
Proper Abstraction Level [2.6]	How to achieve the proper abstraction level?	Page 52
Build Stable Analysis Patterns [3.7]	How to assemble the problem model components to build the stable analysis pattern? How to define the relation between the identified EBTs and BOs?	Page 54
Use Stable Analysis Patterns [3.8]	How to use the contracted stable analysis pattern to model the problem within a specific context?	Page 58

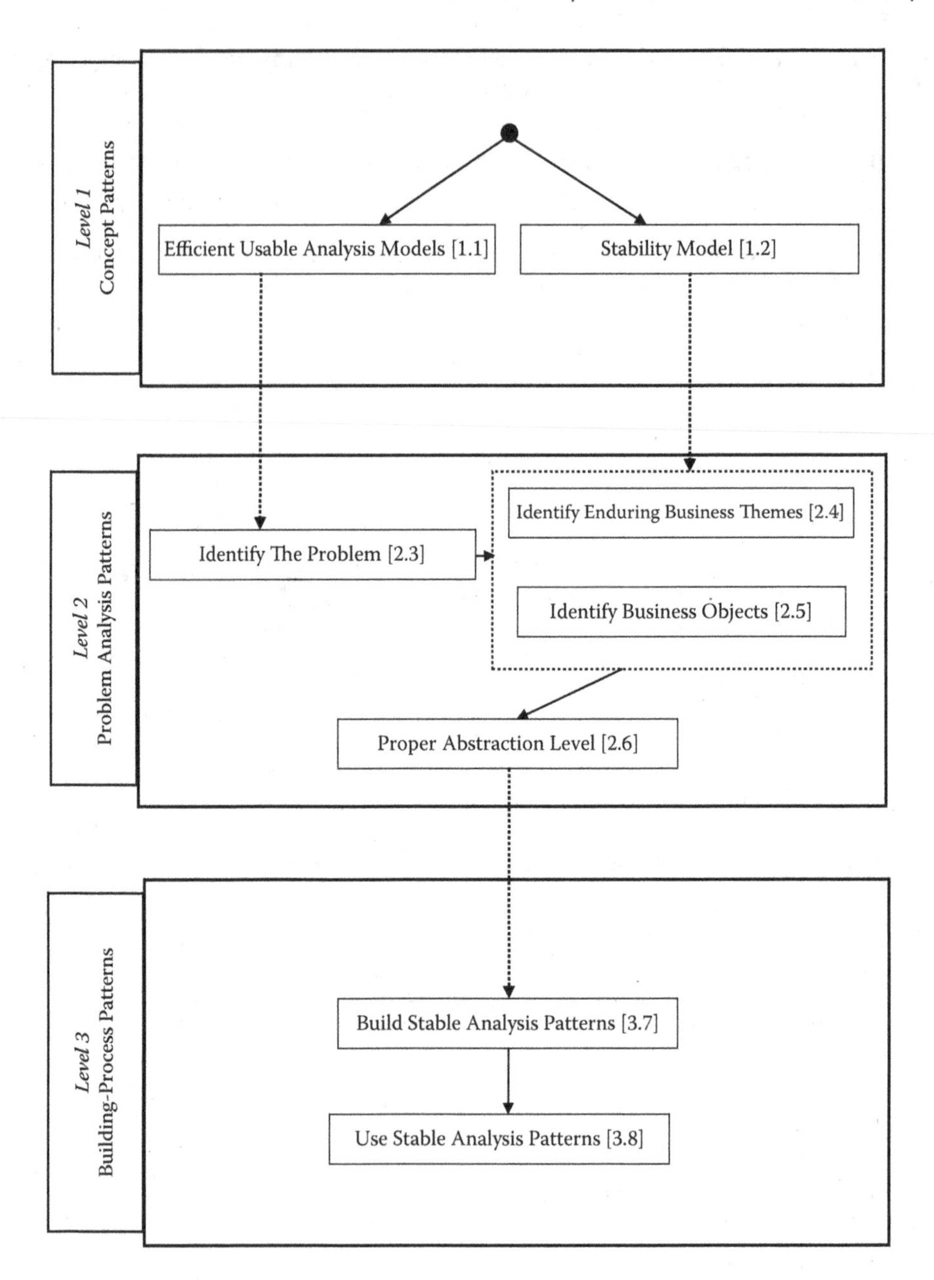

FIGURE 3.2 The pattern language chart.

3.3.1.5 Solution

We will introduce eight essential properties that cover the main qualities of any analysis model, although satisfying these eight properties does not guarantee an efficient, reusable model. However, in practice, a lack of any of these properties will conspicuously affect the reusability and the effectiveness of the model.

1. *Simple:* a pattern is not intended to represent a model for a complete system. Rather, it molds as a specific problem that commonly appears within larger systems. Systems, by

their nature, combine and present many problems. Thus, they are modeled by using a collection of analysis patterns. In fact, each analysis pattern should focus on one specific problem; otherwise, many problems might arise. Without decomposing a system into individual components, models become unreasonably complex, the generality of the patterns are adversely affected and they eventually become highly nonintuitive. If a pattern is used to represent an overly broad portion of a system, the generality of the resulting patterns is wholly sacrificed—the classic maxim holds here: "the probability of the occurrence of all the problems together is less than the probability of the occurrence of each problem individually." For instance, modeling the "payment" problem with "buying a car" is not effective since the "payment" problem may appear in innumerable number of problems. Here, pattern completeness is also sacrificed, when we attempt to create a system at an improper level of resolution. The analyst's attention is not focused on a specific problem and it is more likely that important features of the system and its subcomponents will be overlooked by the developer.

2. *Complete and most likely accurate:* closely related to the concept of simplicity, this property guarantees that all the required information is present within the system. In order to be considered complete, the model should not omit or ignore any component. The model must be able to express all the essential concepts of its properties. For example, trying to shape the whole rental system of any property may lead to missing some of the important components of this system. Renting a car will involve issues related to insurance, which is not the case, when someone is renting a book from a library. As a result, a pattern that forms the whole rental system will not be complete or accurate nor does it display the property of simplicity.
3. *Testable:* for the model to be testable, it must be specific and unambiguous. Thus, all the classes and the relationships of the model should be qualified and validated.
4. *Stable:* stability issue influences the reusability of a model. Stable models are easily adapted and modified without the risk of obsolescence.
5. *Graphical or visual:* conceptual models are difficult to visualize. Therefore, having a graphical representation for the model assists understanding.
6. *Easy to understand:* Conceptual models are usually complicated because they represent a high level of abstraction in its design. Therefore, they should be explained in a clear tone and in depth to assist better understanding of the system; otherwise, use of the pattern is neither attractive nor effective.
7. *General:* This property is crucial to ensure model reusability. Pattern models lacking generality become ineffectual, since analysts will try to build new models, rather than spend unnecessary time and effort to adapt an unruly pattern to fit into an application. Generality means a pattern that models a specific problem that can be easily used to reproduce the same problem independent of context. Pattern generality may be divided into two categories: Patterns that solve problems frequently occurring in different contexts (domain-less patterns), and patterns that solve problems frequently appearing within specific contexts (domain-specific patterns). In the latter, the pattern is considered to be general, even if it is only applicable in a certain domain; however, here we should ensure that the problem does not occur in other contexts.
8. *Easy to use and reuse:* analysis patterns should be presented in a clear way so that it could be easily reused. It is also critical to remember that patterns are consumed more in larger models. Patterns that are easy to use and designed for reuse stand a greater chance of actually being reused.

Next Pattern: After learning the essential properties that influence the usability and effectiveness of analysis models, we need now to understand the basic concepts of the "*Stability Model.*"

3.3.2 Pattern 1.2: Stability Model

3.3.2.1 Intent

The main goal is to describe the structure of stability model and its basic concepts ("EBTs," "BOs," and "IOs"). It also shows the relationships between these elements and how they work together to build stable models.

3.3.2.2 Context

Stability is a highly desired and mandatory feature for any working engineering system. In the domain of software engineering, stable analysis models, stable design models, stable software architectures, stable patterns, etc., will definitely reduce the cost and improve the overall quality of software engineering products.

3.3.2.3 Problem

How to accomplish software stability?

3.3.2.4 Forces

- It is usually very problematical to separate analysis, design, and implementation issues while shaping the problem. Usually, analysts evaluate the problem (problem domain) with some design issues in mind (solution space). Because, different solutions can exist for the same problem, mixing the modeling of a problem with its solution issues may adversely affect the reusability of this model. As a result, people who want to approach the same problem with different solutions will need to remodel the problem from scratch. As a result, many analysis models of the day will lack the factor of stability.
- Analysis models are required to capture the core knowledge of the problem they model.

3.3.2.5 Solution

Software stability concepts that were introduced in [1,2] have demonstrated vast potential in the area of software reuse and lifecycle improvement. Figure 3.1 shows the architecture of the stability model. In the stability model, the model of the system is viewed as three layers: EBTs layer, BOs layer, and IOs layer.

Each class in the system is classified into one of these three layers according to its nature. For instance, classes that present the enduring and basic concepts of the system are classified as EBTs. Since EBTs represent the enduring concepts of the system, they are extremely stable, and thus they form the nucleus of the stability model.

The classes that are tangible and externally stable, but are internally adaptable, are classified as BOs. For instance, human beings are BOs. They are externally stable; however, they can change internally (humans can get married, or become ill). The EBTs and the BOs form the stability model core.

The IOs layer contains the unstable classes, and thus they might be replaced, added, or even removed from the system without affecting the core of the system (Figure 3.3).

EXAMPLE

This example shows the method to apply software stability concepts in the modeling of a simple problem.

Problem

Model the business of a small shipping company. The company provides several different shipping services.

Solution

We will model the problem by using software stability concepts. First, we need to identify the EBTs, BOs, and IOs of this problem to build its stable model (Figure 3.4). Possible EBTs in this problem are:

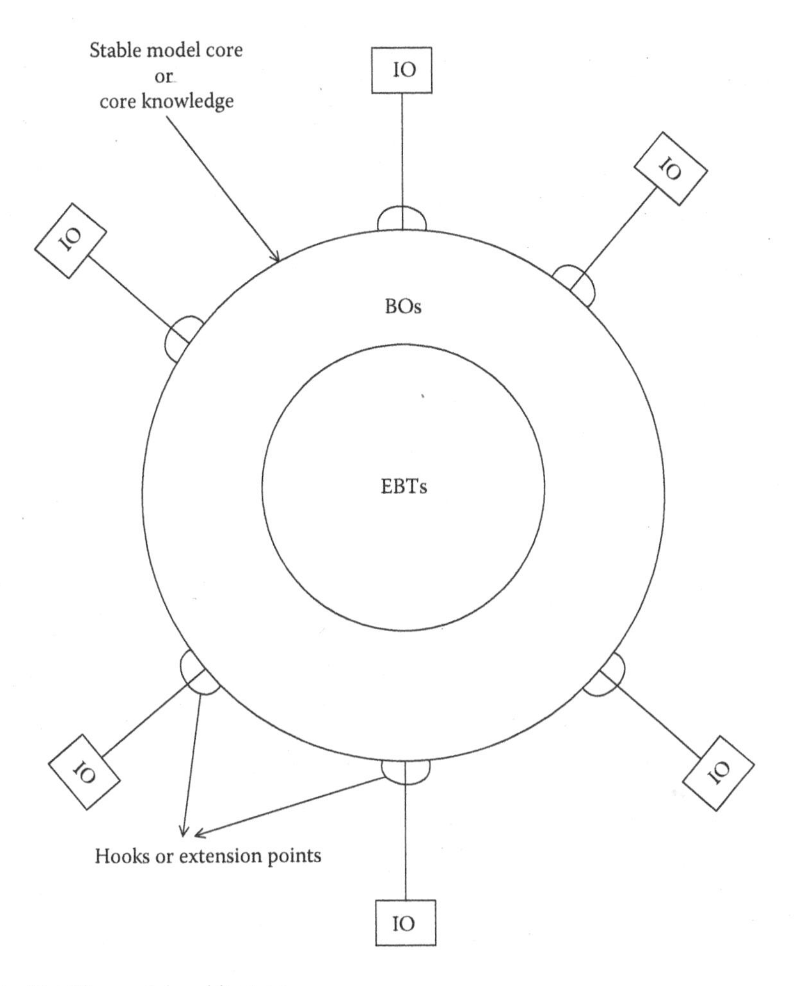

FIGURE 3.3 Stability model architecture.

- "Transportation," which represents the main purpose of this business.
- "Scheduling," which shows how to manage the shipping dates, destinations, etc.

These EBTs are vastly abstract. However, they describe the reasons for the existence of the system. To make these abstract and intangible objects more tangible, we will need to map them into more tangible objects. Identifying the BOs of the problem will do this mapping. One possible BO in this problem is "Schedule." This BO will map the EBT "Scheduling" into more tangible object: "Schedule." The BO is externally stable; however, it can change internally. For instance, schedule changes from one day to another; however, such changes will not affect the existence of the schedule as an object in the problem model.

Now, we need more concrete objects that can physically map the "BOs" into fully tangible objects. These objects are the IOs of the system. In this problem, we can think about different implementations for the BO "Schedule." For instance, we can make "Schedule" by using a piece of "Paper" or by using some sophisticated "Software" program, or even by using both of them. By modeling the problem using the stability concepts, such changes will never affect the core of our model.

Figure 3.5 shows how the EBT "Scheduling" can be mapped into the more tangible object (BO) "Schedule," and finally into concrete objects (IO) "Paper" and "Software."

Next Pattern: After understanding the required concepts, the next step is to analyze the problem. To analyze the problem we need to "*Identify the Problem*," "*Identify Enduring Business Themes*," "*Identify Business Objects*," and accomplish the "*Proper Abstraction Level*."

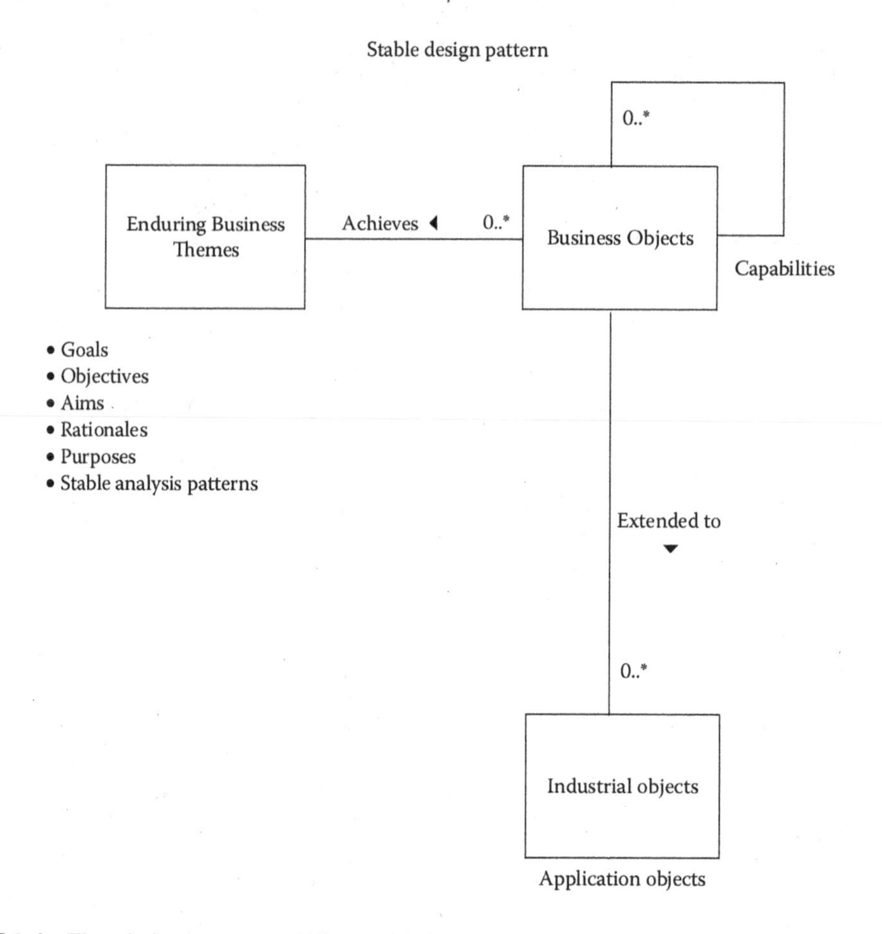

FIGURE 3.4 The relation between stability model elements.

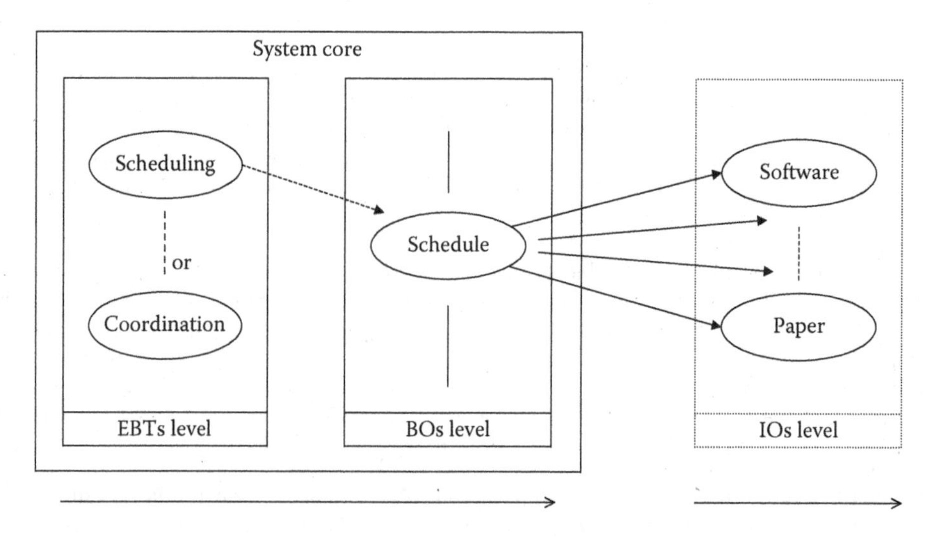

FIGURE 3.5 Example of the mapping between EBTs, BOs, and IOs.

Level 2 **Problem Analysis Patterns:** First, **Identify The Problem** [4]. Then, Identify **Enduring Business Themes** [5]. After that, **Identify Business Objects** [6]. Finally, place the identified EBTs and BOs in the **Proper Abstraction Level** [7].

3.3.3 Pattern 2.3: Identify the Problem

3.3.3.1 Intent

Shows how to focus on the problem we need to analyze.

3.3.3.2 Context

Analysis patterns reusability has a direct relationship with the number of problems that they model. If a pattern is used to model many problems, the generality of the resulting pattern is sacrificed, since the probability of the occurrence of all the problems together is less than the probability of the occurrence of each problem individually. Focusing on a specific problem is one of the key factors that help improve the reusability of the pattern.

3.3.3.3 Problem

How to focus on a specific problem that the analysis pattern will model.

3.3.3.4 Forces

- Many problems appear together frequently in many contexts. As a result, they will be modeled as one problem.
- Sometimes, it is quite hard to separate small problems from a bigger problem.
- In reality, not all of the small problems that we can separate are qualified to form practical stand-alone problems.

3.3.3.5 Solution

Before we start modeling the problem, we will need to check whether or not this problem can be further divided into smaller, real problems. The following questions help to do so: "What is the problem that we need to solve?" "Can we divide this problem further into a list of smaller problems?" "Are there any known possible scenarios, where these smaller problems can appear?"

If we can find practical scenarios for each of the smaller problems that we have separated, then we need to model each of them separately. If the smaller problems have no practical use, they should be modeled together.

EXAMPLE

We will choose a simple example to illustrate the idea of problem separation. We will consider the "account" problem. In fact, it was not too long ago when the word "account" was merely used to indicate banking and financial accounts. Today, the word "account" alone becomes a vague concept, if it is not allied with a word related to a certain context. For instance, apart from all of the traditional well-known business and banking accounts, we also have e-mail accounts, online shopping accounts, online learning accounts, subscription accounts, and many others.

One possible model for the account problem is the *Account pattern* provided by Fowler [8]. Figure 3.6 shows the class diagram of the *Account Pattern*. This pattern models two different problems at the same time. The first problem is the "account" problem and the second problem is the "entry" problem. These are two independent problems. Even though they appear together in many contexts, there is now a possibility of having entries without an account, or accounts without entries. Figure 3.7 shows some examples of accounts without entries, while Figure 3.8 gives

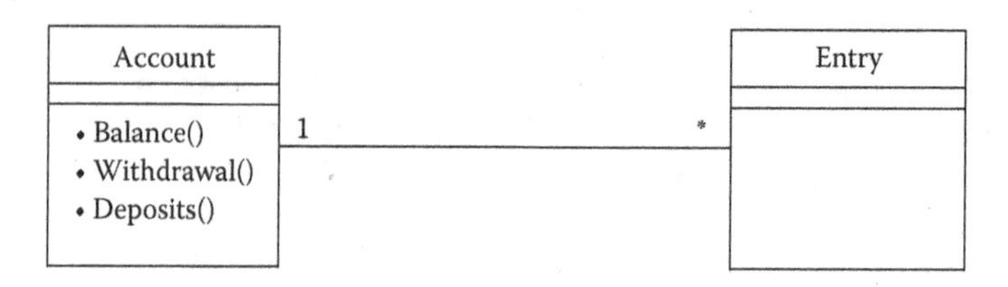

FIGURE 3.6 *Account* pattern provided by Martin [7].

(1) *Free online services account*: Now, many online companies provide free goods or services. For example, some companies provide learning software packages, or learning documents. In order to access these materials, these providers require you to create an account with the company. This account is simply a passport provided to enable you to access their service; you do not have anything in this account that can be considered to be your property. In fact, the only thing that you can do with this account are the limited functions prescribed by the company that issued the account.

(2) *Access account to the copy machine*: Suppose you have an account to access the copy machine in your school or work. This account is no more than a passport for you to use the copier. There are no entries in this case. (Note that in this example it is possible to use Martin's pattern by changing the names of the behaviors of his patterns. In fact, all the behaviors in Martin's pattern are not relevant in this case).

FIGURE 3.7 Examples of accounts without entries.

The following table contains information about class schedules, at the University of Nebraska-Lincoln, Spring 2002. In this table, each piece of information forms an entry to the table. Here, we do not need accounts to store this information.

Call #	Course Title	Course #	Cr Hrs	Sec.	Time	Day	Room
2850/2867	Computer Architecture	430/830	003	001	0230-0320p	M W F	Freg 112
2855/2873	Software Engineering	461/861	003	001	0930-1045p	T R	Freg 111

FIGURE 3.8 Example of entries without accounts.

an example of entries without accounts. As a result, the generality of the pattern is limited. These factors contribute negatively to the reusability of the pattern.

The stable model of the "Account" problem will be built step by step, throughout the remainder sections of the chapter. From this point forward, we will also show how each pattern contributes to the building of the stable analysis pattern that models the "Account" problem.

Next Pattern: After we have identified the specific problem to model, now we will need to ***"Identify Enduring Business Themes"*** of this problem.

3.3.4 Pattern 2.4: Identify EBTs

3.3.4.1 Intent

Shows how to identify EBTs of the problem.

3.3.4.2 Context

First, identify the core elements of the problem. These are the enduring concepts.

3.3.4.3 Problem

How to identify EBTs of the problem?

3.3.4.4 Forces

- EBTs should capture the core knowledge of the problem; however, some EBTs capture the core knowledge of the problem within a specific context. Such EBTs should be discarded from the model.
- Unfortunately, experience with the domain is not always an accurate generator for the relevant EBTs.
- Even though many of the selected EBTs appear strongly related to the problem at first, many of them in fact have nothing to do with the problem being modeled.
- EBTs of the problem should be as small as possible. Extracting the EBTs that have a real relation to the problem is usually very hard.
- Some of the EBTs might lack one or more of the EBTs essential properties. In this case, we should re-identify them as BOs or IOs.

3.3.4.5 Solution

One approach that helps to extract the appropriate EBTs of the problem is to follow these three essential steps:

Step 1 Create Initial EBTs List: In order to create the initial list of the EBTs of the problem, answer the question: "What is this 'problem' for?" In other words, "What are the reasons for the existence of the 'problem'?"

The output of this step is the list of the initial EBTs of the problem. These EBTs are still tentative and some of them are not as strongly related to the problem as they might appear.

Step 2 Filter the EBTs List: This step is to eliminate the redundant and irrelevant EBTs from the initial list. This step is also important due to the fact that people usually construct the initial EBTs list with specific context in mind, even if they do not intend to do so. The output of this step is a modified EBTs list, which is usually smaller than the initial list.

Step 3 Check the Main EBTs Properties: This step is to examine the EBTs obtained in previous steps against the main essential properties of the EBTs. The typical procedure is to answer the following questions for each EBT in the list, "The desired answer is written in **bold** beside each question":

- Can we replace this EBT with another one? **No.**
- Is this EBT stable internally and externally? In other words, does this EBT reflect the core knowledge of the problem we are trying to model? **Yes.**
- Does this EBT belong to a specific domain? **No.**
- Can we directly represent this EBT physically? **No.** It is important to note that the EBTs should not have direct physical representations (IO); otherwise they should be considered BOs. (Refer to stability model architecture shown in Figure 3.3). For example: "Agreement" is a concept and one can see it as an EBT. However, "Agreement" also has a direct physical representation (for instance "Contract"). Therefore, "Agreement" is not an EBT, it is a BO.

Any EBT that does not satisfy one of these properties should be eliminated from the list. Figure 3.9 summarizes the three steps needed to identify the EBTs of the problem.

EXAMPLE

This example shows the identification steps for the EBTs of the "account" problem:

Step 1 Create Initial EBTs List: We will need to answer the question: "What is the 'Account' for?"

The initial EBTs list might contain the following EBTs:

- Storage
- Ownership

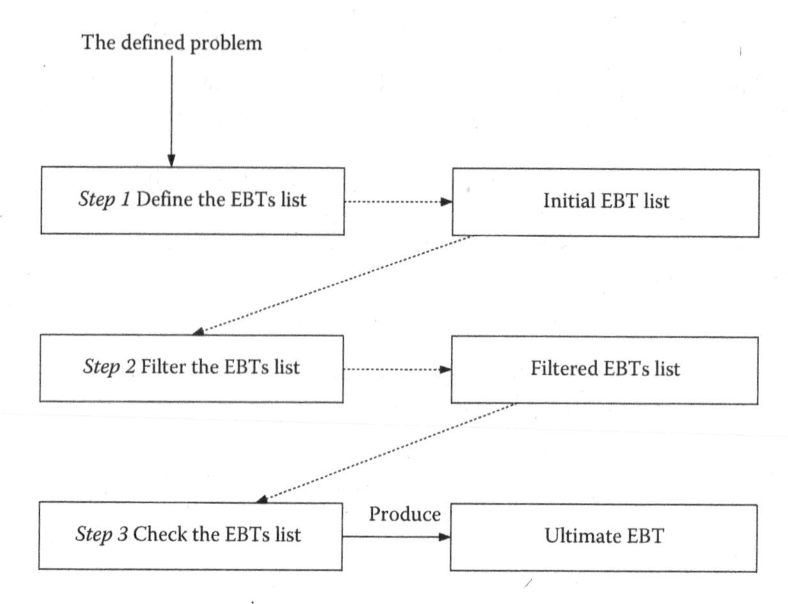

FIGURE 3.9 The steps for identifying the EBTs of the problem.

- Tractability
- Recording

Step 2 Filter EBTs List: In this step, we will need to extract the most appropriate EBTs for the "account" problem. To do so, we require examination of each EBT in the list and seeing whether or not it reflects the enduring concepts of the problem that we are trying to model.

Since our focus is on the modeling of the "account" problem alone, we have realized that most of the EBTs that we have defined are related to accounts that have entries. For instance, "Storage," "Tractability," and "Recording" are all concepts that are dependent upon the existence of entries within the account. If the account has no entries, "Storage," "Tractability," and "Recording" are unneeded. Therefore, we need to eliminate the EBTs "Storage," "Tractability," and "Recording" from the list. Now, we have only one EBT, "Ownership," instead of four.

Step 3 Check the Main EBTs Properties

- Can we replace the "Ownership" with other EBT? **No.**
- Is "Ownership" stable internally and externally? **Yes.**
- Does "Ownership" belong to a specific application or domain? **No.**
- Can we have direct physical representation for "Ownership"? **No.**

3.3.4.6 Discussion

1. There are different approaches that can be used to extract the EBTs of the problem; however, our experience shows that the shown approach is effective.
2. Debate and discussion are two important factors that definitely enhance the extracted EBTs.
3. For smaller problems, it is desired to narrow the number of EBTs to one or two. This usually makes the pattern more focused and more effective.
4. The final EBTs list obtained here is not necessarily the final EBTs that will appear in the pattern structure. These EBTs are usually enhanced and adapted in later steps.

Next Pattern: After we have identified the EBTs of the problem, we will now need to "***Identify Business Objects***" of the problem.

3.3.5 Pattern 2.5: Identify BOs

3.3.5.1 Intent

Shows how to identify BOs of the problem.

3.3.5.2 Context

After having used software stability concepts in the modeling of the problem to identify the EBTs, next important BOs are identified.

3.3.5.3 Problem

How to identify BOs of the problem.

3.3.5.4 Forces

- In some cases, it is not obvious whether the object is an EBT or BO. For instance, "Agreement" can be considered as an EBT, since it presents a concept. However, it is a BO.
- After the EBTs of the problem have been identified, the conceptualization becomes more involved, since the BOs of the problem must be based on the defined EBTs. This makes the extracting of the BOs difficult.
- Usually, there is no one-to-one mapping between the EBTs of the problem and its BOs. It is possible for EBTs to have no direct mapping to the BOs and for the BOs to have no direct mapping to the EBTs. Moreover, one EBT can be mapped into several BOs.
- Besides the main BOs that we can identify for the problem, it is also possible to think of some "hidden" BOs. These hidden BOs have no direct relationship with the defined EBTs. Instead, they are related to the main BOs and to the other hidden BOs in the problem.

3.3.5.5 Solution

One approach that helps to extract the appropriate BOs of the problem is to follow the following four steps:

Step 1 Identify the Main BO of the Problem: In this step, we will identify the main set of BOs that are directly related to each of the EBTs that exist in the problem. There could be one or more BOs corresponding to each EBT in the problem. Nonetheless, some of the EBTs may have no corresponding BOs.

The main set of BOs of the problem can be identified by answering one or more of the following questions for each EBT:

[Note: Some questions do not apply for some of the EBTs. This depends on the nature of each EBT.]

- How can we approach the goal that this EBT presents? (e.g., To achieve the goal of the EBT, "Scheduling" or "Organization," we can use, the BO "Schedule." Another example: for the EBT "Negotiation," we need the BOs: "NegotiationContext," and "NegotiationMedia" to perform the "Negotiation").
- What are the results of doing/using this EBT? (e.g., for the EBT "Negotiation," the eventual result is to reach an "Agreement" so this is one possible BO that maps this EBT).
- Who should do/use this EBT? (e.g., The BO "Party" uses/ does "Negotiation." This "Party" can be a person, a company, or an organization. Therefore, "Party" is one possible BO that maps the EBT "Negotiation").

Step 2 Filter the Main BOs List: This step is to purify the main BOs identified in the previous step. The objective of this step is to eliminate the redundant and irrelevant BOs from the initial list. One way to achieve this goal is to debate the listed BOs with a group.

Step 3 Identify the Hidden BOs of the Problem: This step is to identify the hidden BOs of the problem. The name hidden comes from the fact that these BOs have no direct relationships with any of the EBTs of the problem. Thus, we cannot extract them in the first two steps we have performed.

For example, assume that we need to model a simple transportation system that offers transportation services for different types of materials (gas, water, etc.). One possible EBT is "Transportation." One possible BO that maps this EBT is "Transport." A possible IO that can physically represent this BO is "Trucks." In this problem, one possible hidden BO is the BO "Materials." We do not have a direct EBT that the BO "Materials" can be mapped to; however, there is a clear relationship between the two BOs "Transport" and "Materials."

Before thinking about the hidden BOs in the problem, visualize a provisional scenario for each EBT and its corresponding BOs. Then answer the question, "What is still missing in the problem?" Usually, the answer to this question is the list of the hidden BOs of the problem. Some problems do not have any hidden BOs, especially in the case of the small-scale problems.

Step 4 Check the Characteristics of the BOs: This step is to ensure that the identified BOs satisfy the main BOs' characteristics.

The main BO characteristics are summarized as follows:

- BOs are partially tangible.
- BOs are externally stable, and they should remain stable throughout the life of the problem.
- BOs are adaptable; thus, they might change internally.
- BOs can have direct physical representation (IOs) (Figure 3.10).

EXAMPLE

In this example, we will need to identify the possible BOs for the "account" problem. In order to identify these BOs, we will apply the four steps we have proposed.

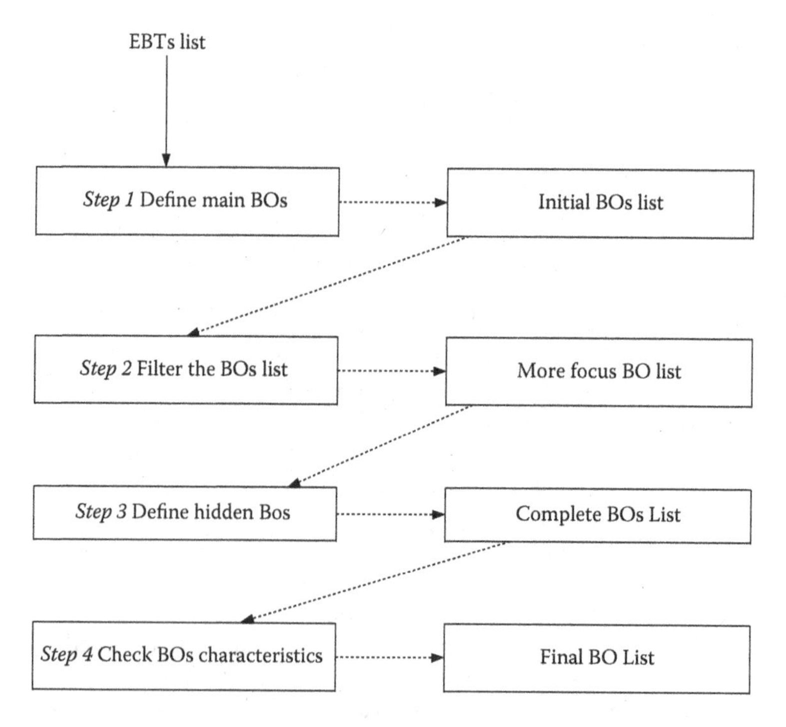

FIGURE 3.10 The steps for identifying the BOs of the problem.

The input of the first step is the EBTs list that contains one EBT, "Ownership."

Step 1 Identify the Main BO of the Problem:

- How can we approach the goal of "Ownership"?
  In the account problem, to achieve "Ownership," we should have something to own. The "Account" itself is what makes the meaning of "Ownership."
- What are the results of doing/using "Ownership"? By having "Ownership," we have "Privacy"; but this is not a BO, it is a redundant EBT, and so we exclude it. "Agreement" is a possible BO for the problem; when you own an account you will have to agree with its policy.
- Who should do/use "Ownership"? For the "Owner" to be able to the use the "Account," he or she should follow the responsibilities defined by the "Ownership" policy.
  BOs main list: "Owner," "Account," and "Agreement."

Step 2 Filter the Main BOs List: Now, the following BOs remain: "Owner," "Account," and "Agreement." While modeling this problem, we have debated the accuracy of using the BO "Agreement" in our model. "Agreement" is a general term that appears in many contexts. For instance, in the "negotiation" problem, we will need the BO "Agreement." Therefore, having the BO "Agreement" as part of the "Account" pattern will counter the simplicity property of the pattern. "Agreement" is a stand-alone problem that occurs in many contexts, and hence it is more appropriate to model the "Agreement" problem independent of the context of this problem. Therefore, we have excluded the BO "Agreement" from the main list.The filtered BOs list contains "Account" and "Owner."

Step 3 Identify the Hidden BOs of the Problem: After identifying and filtering the main BOs list, now answer the question, "Does the account problem need anything else to be complete?"

We have an "Account," its "Owner," and the concept of "Ownership" that regulates the responsibilities and benefits of the "Owner." This is all that is needed to model any basic account.

Step 4 Check the Characteristics of the BOs:

- BO should be partially tangible. "Owner" and "Account" are partially tangible.
- BOs are externally stable, and they should be so throughout the life of the problem. "Owner" and "Account" are always stable. In other words, we cannot have any account without having these two objects in its model.
- BOs are adaptable, thus, they might change internally. "Account" and "Owner" might change internally. For instance, you can add or remove some feature from your banking "Account" (e.g., adding the overdraft protection service); however, this is an internal change inside the account. Externally, there is nothing that has changed. Also, for the BO "Owner," the "Owner" may become ill; for example, however, he is still the owner of the account.
- BOs can have direct physical representations (IOs). There are different possible physical representations for the BO "Account" depending on its context. The "Account" could be, physically, a code as in the case of the copy machine. The BO "Owner" is physically the person who owns this account and uses it.

3.3.5.6 Discussion

1. Different approaches can be used to extract the EBTs of the problem; however, our experience shows that the displayed approach is quite effective.
2. For smaller problems, it is desirable to narrow down the number of BOs. Usually, for small size problems, we may have 2–4 focused BOs.
3. As for the EBTs, the final BOs list we have defined by using this pattern can be further enhanced and modified during the next stages.

Next Pattern: After we have identified the EBTs and the BOs of the problem, now we will need to accomplish the "***Proper Abstraction Level***."

3.3.6 Pattern 2.6: Proper Abstraction Level

3.3.6.1 Intent

Helps to accomplish an appropriate abstraction level for the analysis pattern.

3.3.6.2 Context

If the pattern lacks the desired abstraction level, the reusability of this pattern becomes doubtful. The ultimate goal of ensuring proper level of abstraction is to make the pattern as useful as possible, by covering all of the possible situations that might appear in the problems of the pattern models.

Identifying the EBTs and BOs of the problem was the first step toward establishing proper abstraction level. During the identification process of the EBTs and BOs, we have excluded the EBTs and the BOs that make the model adhere to a specific domain. However, this level of abstraction is not sufficient to meet our needs. Therefore, additional work is needed in order to enhance the abstraction level of our model.

3.3.6.3 Problem

How to achieve the proper abstraction level?

3.3.6.4 Forces

- Usually, the names of the EBTs and the BOs resulting from the previous stages are not accurate.
- Even though we can find an appropriate name for each EBT and BO in the problem, it is quite hard to define the attributes and operations for each class that can fit all the contexts that the problem might appear in.

3.3.6.5 Solution

After identifying the EBTs and the BOs of the problem, ensure that these elements display a proper level of abstraction. Our approach to achieving a proper abstraction level is summarized in the following points:

- We would prefer not to assign specific attributes or operations to any of the EBTs or the BOs, even though some developers may argue that by not doing so, we might encounter difficulties while using the pattern. We also believe that assigning specific attributes and operations can lead to confusion rather than help in understanding the model.
- We will also need to inspect both the names and the structure of each EBT and BO that we have identified in the problem. In this step, it is most likely that some BO names will need to be changed. One way of conducting such an inspection is examining the EBTs and the BOs of the problem against different situations that usually appear in the context of the problem we are trying to model. Also, thinking about exceptional situations and examining whether or not the identified EBTs and BOs will handle them will also assist us to further abstract the pattern and make it more general. This will be illustrated with an example, in the following section.
- At times, the problem that we try to solve may be too big or small which necessitates us to redefine to make them suit the given situation. Such problems usually arise when identified EBTs and BOs cannot be handled by simply changing their names or by modifying structure.

EXAMPLE

Consider the "account" problem again. So far, we have identified the following EBTs and BOs:

EBTs: "Ownership"
BOs: "Owner" and "AnyAccount."

At the first glance, the name of the BO "Owner" seems very apt and appropriate. However, when we think more about some of the possible situations that might occur in any account context, there are possibilities of witnessing a problematic scenario. What might happen when there are too many users who share the same account? For instance, in the case of credit card accounts, one individual could be the owner of the account, and he or she may allow other users to use the credit card account with specifically assigned privileges. In this case, using the BO "Owner" limits the pattern to the specific accounts, where there is only one possible user who can use the account: the owner of the account.

Now, it is obvious that the BO "Owner" lacks an appropriate level of abstraction that helps to handle the different situations of the problem. Therefore, we may need to redefine the BO "Owner" in such a way as to make it more general.

To suggest a solution to the problem, we have changed the name of the BO "Owner" to "Holder," which is more general in nature. Then, we will be able to change the inheritance structure to capture the different roles of the different account holders.

Figure 3.11 shows the detailed steps taken while contemplating the problem.

3.3.6.6 Discussion

The level of abstraction invariably determines how broadly one can use the pattern. However, too much abstraction can prove to be negative. There should be a tradeoff between flexibility and

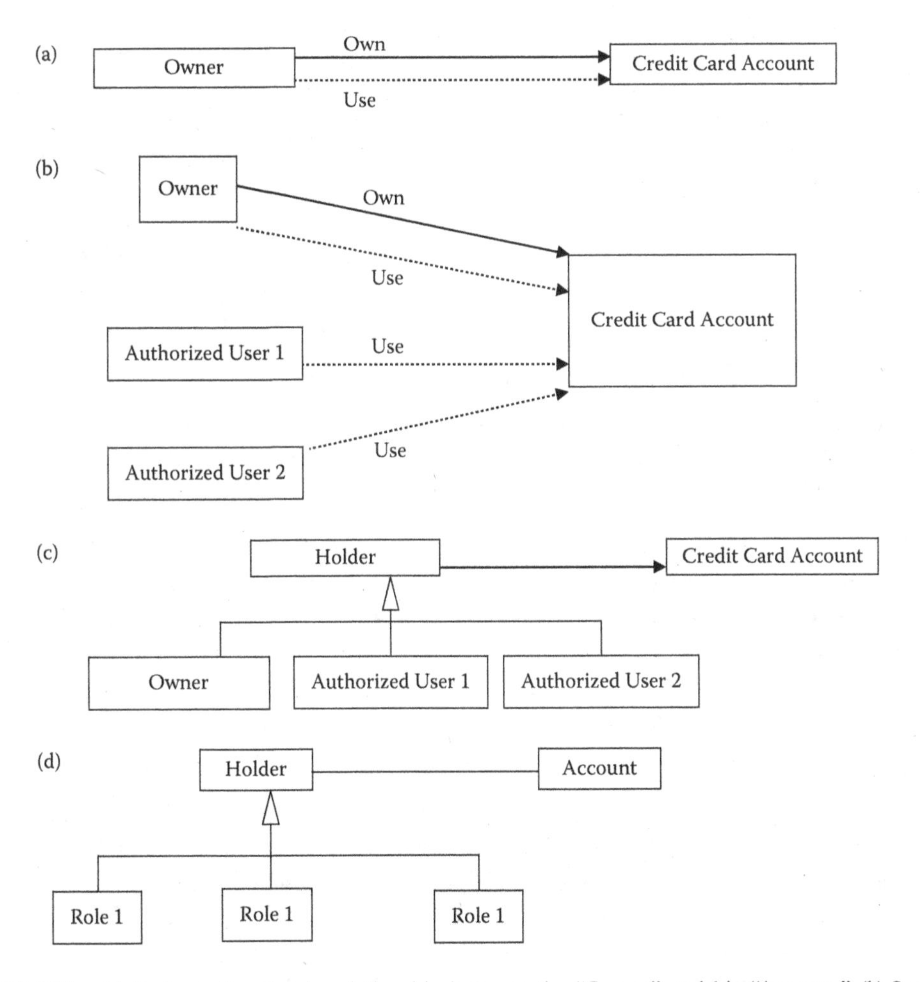

FIGURE 3.11 (a) It shows the classic relationship between the "Owner" and his "Account." (b) It shows the possible situation that cannot be handled with the current representation. (c) It shows possible solution to handle the problem. (d) It gives the final abstraction level that handles the problem.

reusability issues of the model on one hand, and complexity of understanding and reuse of the model on the other. The level of abstraction plays a vital role in this equation. A low level of abstraction will reduce the reusability factor of the model; re-adapting it again will no longer be an easy task. However, the resulting model will be relatively simple, when compared to a similar model designed with a higher level of abstraction. Conversely, too much abstraction will lead to models that are too complex and intricate. Therefore, the optimum solution is to create a conceptual model that is simple enough to be easily used and comprehend, but with a level of abstraction high enough to make it more general and hence more reusable. Several different approaches currently exist to handle the abstraction problem. Some of these views are illustrated below:

- Developers, who extract analysis patterns from several projects, believe that it is unsafe to further abstract patterns that are generated within certain projects in order to make them reusable in other contexts. The *Account* pattern shown in Figure 3.6 shows an example of this approach. Since no one can be an expert in all fields, domain expert analysts often extract domain-specific patterns, even if the problem they model occurs in many other contexts. For example, by following this approach, we might end up synthesizing a list of patterns that model the account problem within different contexts; for instance, account patterns that model banking accounts, account patterns that model web applications accounts, and so on. It will be far more efficient, if we have one pattern called "*AnyAccount*" that can capture the core structure of the different account types. Hence, we can use this pattern whenever we need to model the account problem regardless of the context of this account.
- Analogy is another view of abstraction problem. As per this approach, patterns that model complete systems in one context are reused by making an analogy between the pattern and the new application. Thus, by using analogy, they change the names of the pattern's classes to be relevant to the new application. Sometimes, you might have to add/remove a few classes to adapt the pattern to the new application. This approach to the task of accomplishing abstraction levels results in the building of templates instead of patterns. Figure 3.12 shows an example of this group' pattern [9]. The objective of the shown pattern is to provide a model that can be reused to model the problem of renting any resource. Figure 3.13 shows an example of using this pattern in the context of a library service [9].

Next Pattern: Once, we have identified the underlying problem, its EBTs, BOs, and the appropriate abstraction level, the next step is to "***Build Stable Analysis Patterns***," and to "***Use Stable Analysis Patterns***."

Level 3 **Building-Process Patterns:** In this level, first, **Build Stable Analysis Patterns** [8], and then **Use Stable Analysis Patterns** [10].

3.3.7 Pattern 3.7: Build Stable Analysis Patterns

3.3.7.1 Intent

After defining the problem, its EBTs, BOs, and their proper abstraction level, the next step is to patch up everything together to design and build a stable analysis pattern, which we need to glue things together in order to build the stable analysis pattern that shapes the problem.

3.3.7.2 Context

After focusing on a specific problem and defining different stability model elements of this problem, we will now need to understand the process of identifying the relationship between these elements.

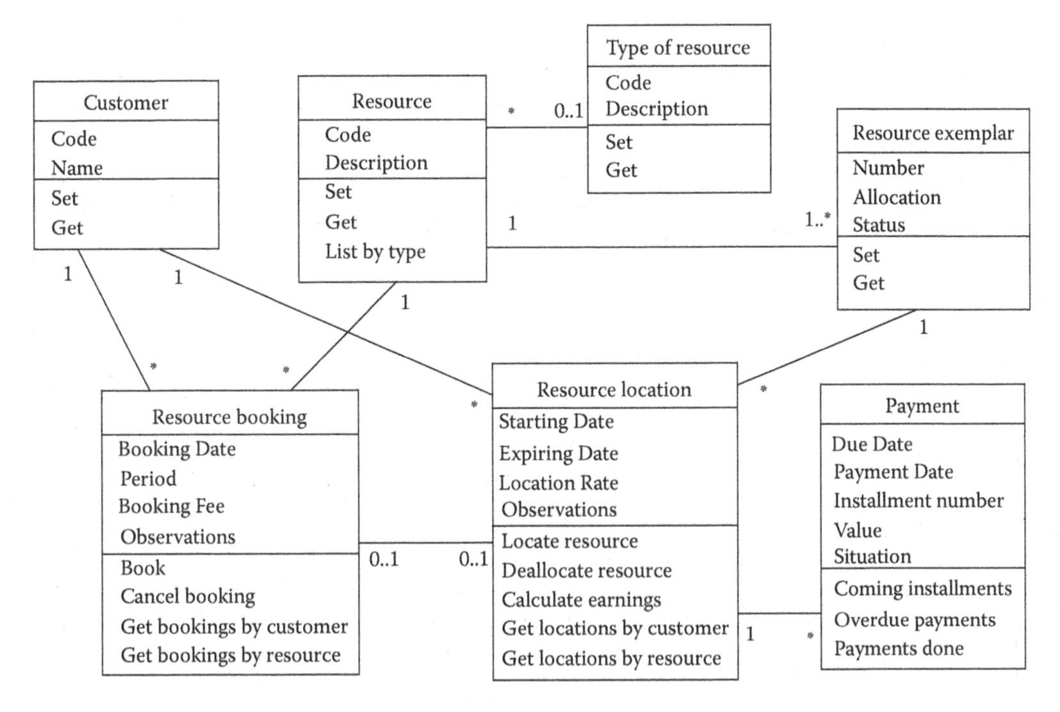

FIGURE 3.12 Resource rental pattern. (R.T. Vaccare Braga et al., *Pattern Language of Programs*, Monticello, IL, 1998.)

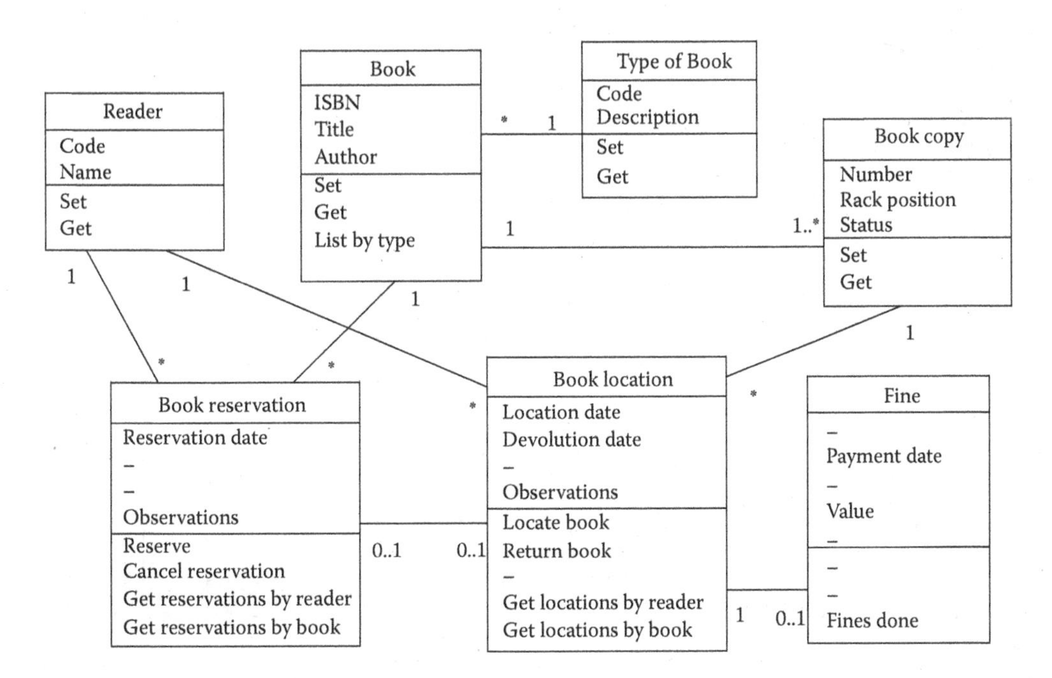

FIGURE 3.13 Instantiation of resource rental pattern for a library service. (R.T. Vaccare Braga et al., *Pattern Language of Programs*, Monticello, IL, 1998.)

3.3.7.3 Problem

How to assemble the problem model components to build a stable analysis pattern? How to define the relations between the identified EBTs and BOs?

3.3.7.4 Forces

- Abstraction level of the pattern elements imposes difficulty in defining the different relationships between these elements.
- Some EBTs may not have corresponding BOs.
- We will also need to identify relations between the main and hidden BOs of the problem.

3.3.7.5 Solution

One known way of identifying different relationships among EBTs and the BOs of the problem is to put the pattern into a context; this makes visualization of these relationships easier. Nonetheless, in order to ensure the generality of the pattern, we will need to visualize the pattern under different contexts. With this approach, we can increase the accuracy of defining the relationships and multiplicities between the elements of the problem.

Another possible way for identifying the relationships between different EBTs and BOs of the problem is as follows:

First, join each EBT in the EBTs list with its corresponding BOs in the main BOs list. Second, define the relationships between the EBTs of the problem. Finally, define the relationships between the main BOs of the problem, and among the main BOs and "Hidden" BOs of the problem. Figure 3.14 shows the relationship between the EBTs and the BOs of the problem.

3.3.7.6 Examples

EXAMPLE 1: BUILDING "ANYACCOUNT" PATTERN

This example shows the process to join EBTs and the BOs of the "account" problem to build a stable analysis pattern *."AnyAccount."* Given below is the summary of the information in hand, related to the "account" problem (Table 3.2):

Define the relationships between the EBTs and the corresponding main BOs:
What is the relationship between the EBT "Ownership" and the BO "Holder"?
What is the relationship between EBT "Ownership" and BO "AnyAccount"?
"Ownership" presents the policy of owning the account, and it can range from "zero" to "many" accounts. For instance, in banking accounts, there is a predefined policy that regulates all

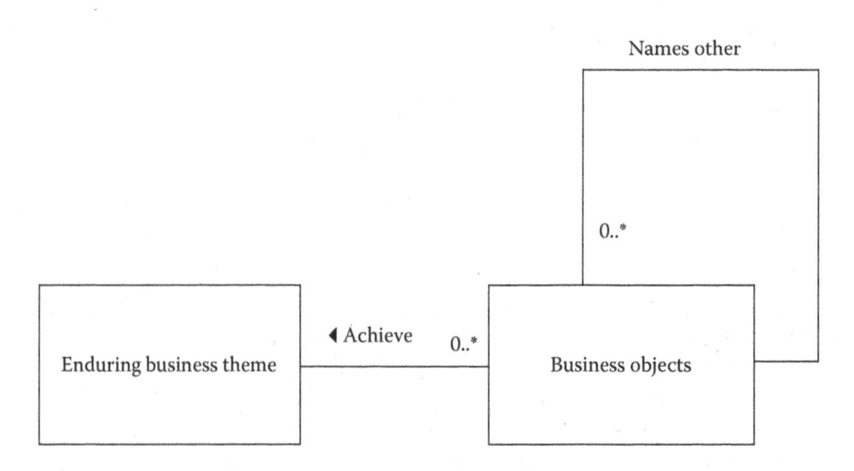

FIGURE 3.14 The relationship between the EBTs and the BOs of the problem.

TABLE 3.2
AnyAccount Information

Element	Description
Problem definition	Modeling any account for any context
EBTs list	"Ownership"
Main BOs list	"Holder" and "AnyAccount"
Hidden BOs list	None

checking accounts, and another one that controls all saving accounts, and so on. Since parts of this policy regulate the responsibilities and the benefits of the account "Holder," we also have an association between the EBT "Ownership" and the BO "Holder."

Define the relationships between the EBTs of the problem:
As there are no other EBTs in this problem, we will move to the last step.

Define the relationships between the BOs of the problem:
What is the apparent relationship between the BO "Holder" and tBO "AnyAccount"? The relationship is quite clear; the "Holder" uses the "AnyAccount," and hence, an association between the two classes should exist. Since the "Holder" can have as many "AnyAccounts" as it wants, the multiplicity is from "zero" to "many." Also, the "AnyAccount" can belong to one or more "Holder," keeping in mind the fact that the "Holder" of the "AnyAccount" could be the owner only, or could be the owner and any other authorized holders, as in the case of credit cards accounts. Figure 3.15 gives the class diagram of the *"AnyAccount"* pattern.

EXAMPLE 2: BUILDING "ANYENTRY" PATTERN

By following similar steps like in the case *"AnyAccount"* pattern, we have constructed another pattern called *"AnyEntry"* pattern. This pattern captures the core element of any entry independent of the context of the problem. Figure 3.16 gives the structure of the *"AnyEntry"* pattern.

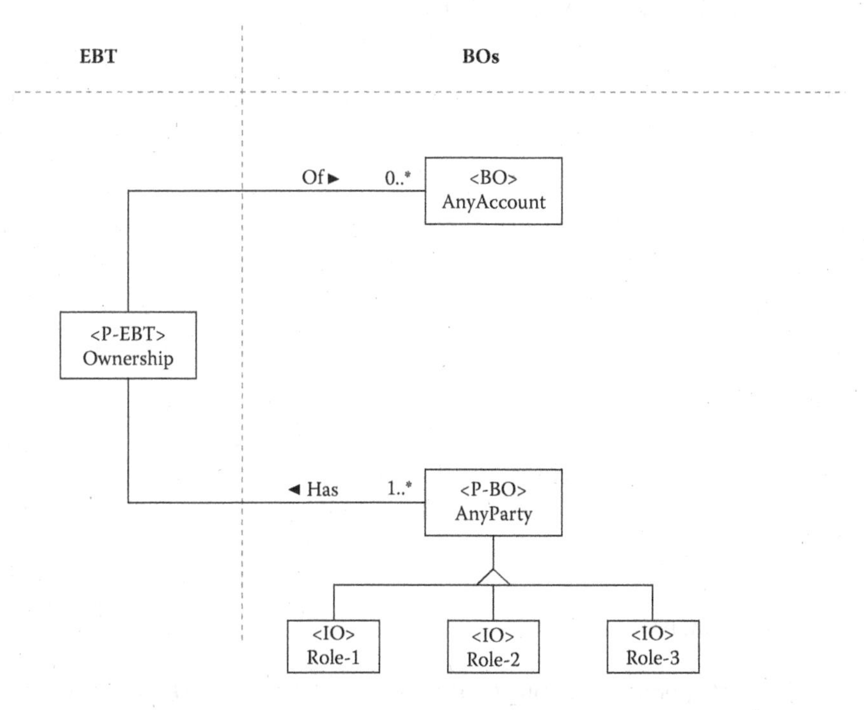

FIGURE 3.15 Any account pattern class diagram.

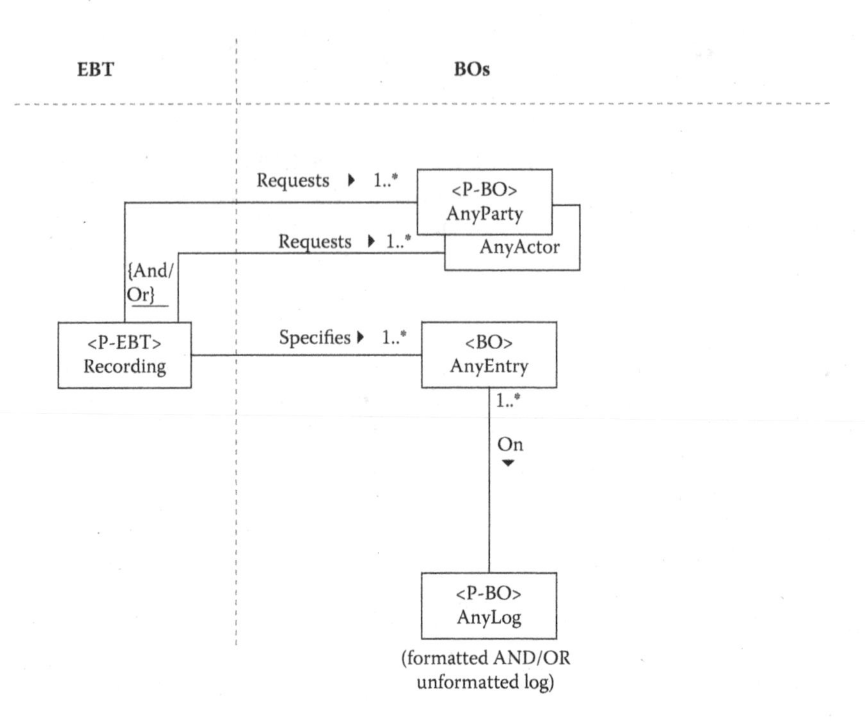

FIGURE 3.16 Any entry pattern class diagram.

Next Pattern: Once we have built the stable analysis pattern, now we will need to see how to ***Use Stable Analysis Patterns***.

3.3.8 Pattern 3.8: Use Stable Analysis Patterns

3.3.8.1 Intent

After "Build Stable Analysis Patterns" patterns are designed and built, they should to be applied in context.

3.3.8.2 Context

Subsequent to the construction of the stable analysis pattern and once refinement to the proper abstraction level is made, this pattern needs to be used to model the problem in context.

3.3.8.3 Problem

How to use the contracted stable analysis pattern to model the problem within a specific context?

3.3.8.4 Forces

- Analysis patterns usually have very high levels of abstraction, and hence, they will need to be properly instantiated in order to be used within a specific context.
- In many situations, multiple stable analysis patterns should be integrated to model bigger problems.

3.3.8.5 Solution

To utilize these stable analysis patterns:

1. Choose the most appropriate attributes and operations for each EBT and BO of the problem, based on their contexts.

2. Identify the IOs of the system based on the identified BOs. Both the main and hidden BOs can have corresponding IOs that are known to physically represent them. However, in some cases there is no one-to-one mapping between the BOs and the IOs of the problem.
3. Identify all EBTs, BOs, and IOs for the rest of the problem or the system being modeled.
4. In the case of using more than one pattern together, we will need to define the level at which the different patterns are to be connected. (Connection shall take place at the EBT level, the BO level, and/or the IOs level, depending on the problem nature.)

3.3.8.6 Comment

The names that we have chosen for each EBT and BO in the pattern will remain the same. However, sometimes we might need to modify some of the BOs' names for clarity of purpose only. The names of the EBTs will never change.

3.3.8.7 Example

In this section, two examples are provided to show how to use the patterns that have been developed (the "*AnyAccount*" and "*AnyEntry*" patterns) in specific contexts.

EXAMPLE 1: MODELING COPY MACHINE ACCOUNT

This simple problem displays how to use the *"AnyAccount"* pattern in the modeling of a simple copy machine account in one of the universities. Each student in the university has an account that he or she can use to access a central copy machine.

Figure 3.17 gives the object diagram of the *Copy Machine Account*. Possible IOs that map the BOs of the problem are identified. For the BO "Student," the ***"AnyAccount"*** pattern without

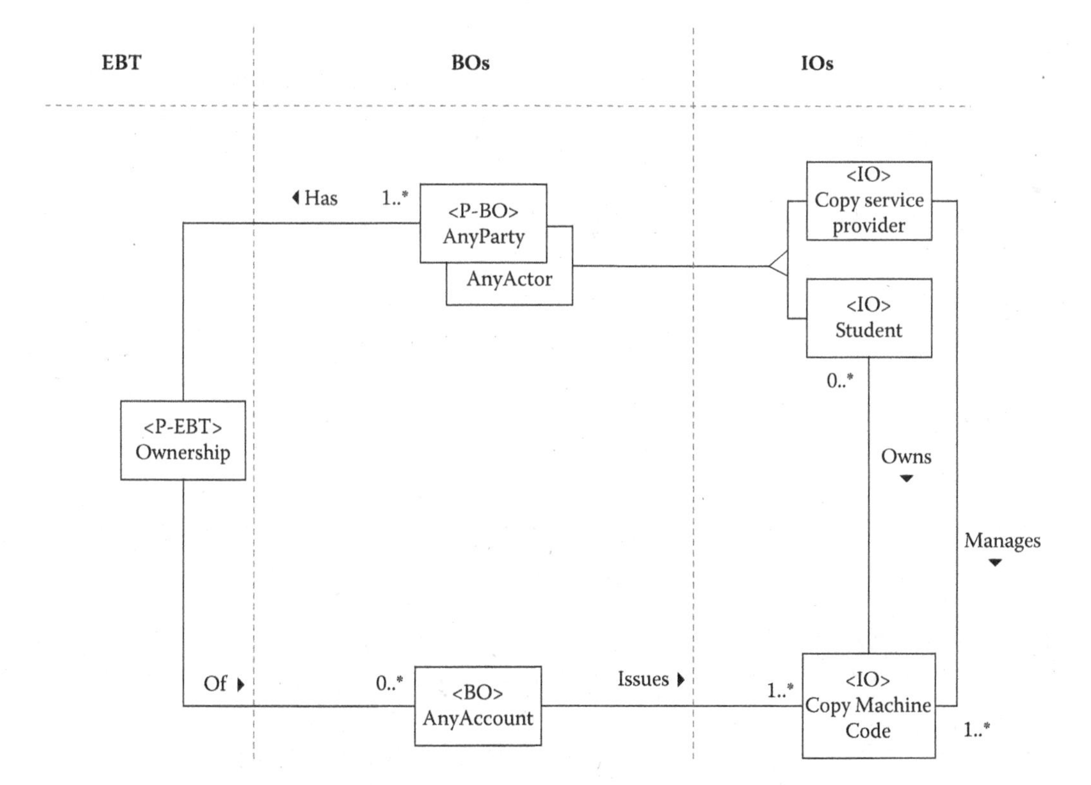

FIGURE 3.17 *Copy Machine Account* object diagram.

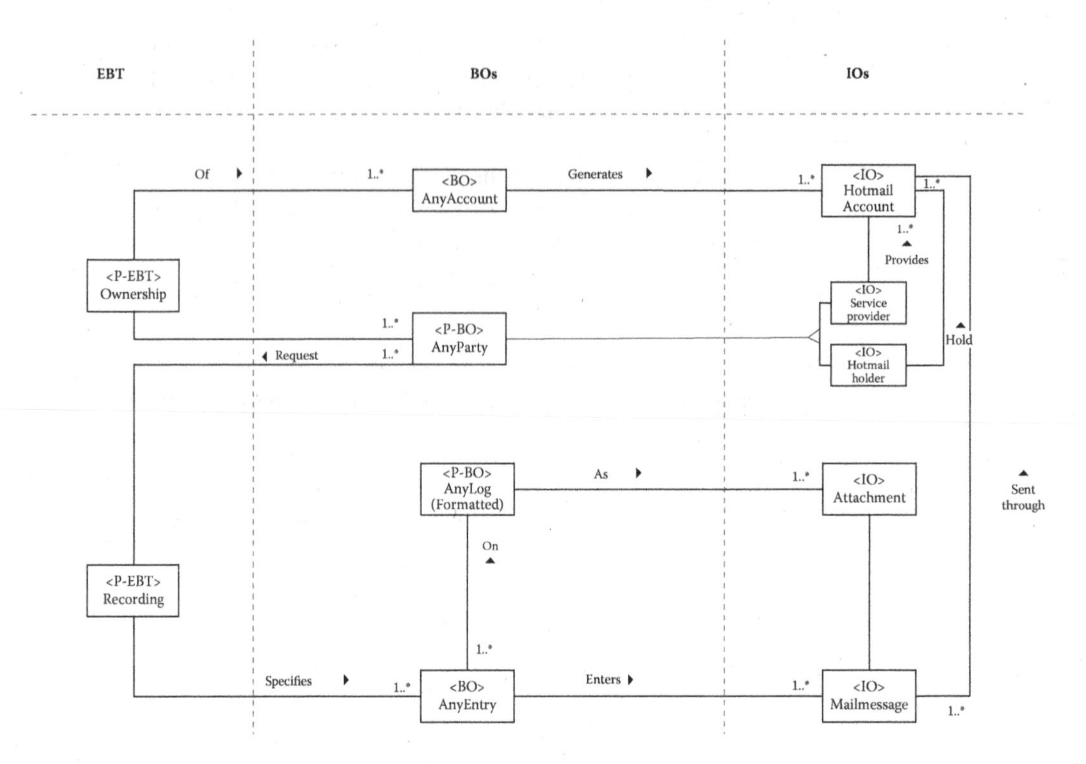

FIGURE 3.18 Hotmail account object diagram.

the inheritance part is used, since for each account there should be only one holder. For the BO "Account," one possible physical representation is the IO "Code." Each student has a "Code" in order to use the copy machine. If there should be any other physical representations for the BO "Account," all that would need to be done is to remove the current IO "Code" and insert the new IO into the model without affecting the core. In this problem, no extra EBTs, BOs, or IOs are needed.

EXAMPLE 2: MODELING HOTMAIL ACCOUNT

This example shows how to integrate more than one pattern to model bigger problems. The goal of the problem is to utilize two constructed patterns: the *"AnyAccount,"* and the "AnyEntry" patterns, in the modeling of a simple Hotmail Account. For simplicity, only the object diagram of the problem model is shown. Also, it is important to note that this model is not complete; it is merely for demonstration purpose. Figure 3.18 gives the object diagram of the *Hotmail Account.*

For each BO, one possible physical representation "IO" is displayed. For instance, "Hotmail" is one possible physical presentation for the "Host," however, for the purpose of generality this IO can change anytime in order to represent any other hosts, without affecting the core of the model. Notice that in this example, all of the connections between the two patterns are made in the IOs level only.

3.4 SUMMARY

The proposed pattern language in this book provides a novel approach for achieving stability while building analysis patterns. Thus, we can increase the effectiveness, and improve the reusability of analysis patterns. Future work and research includes demonstrating the effectiveness of this pattern language by applying it to many problems in different contexts [11].

3.5 OPEN AND RESEARCH ISSUE

1. Catalog all stable analysis patterns. Implement them and show how to utilize in any application. Analysis patterns will support the creation of the conceptual model of the problem space of the applications (requirements). An analysis pattern also provides a reusable, well-proven requirement model representation of a recurrent sub-problem (sub-requirement) and facilitates the proper transformation of actual requirements into an analysis model of an application. Traditionally, existing analysis patterns specify the static components of the corresponding model of the requirements of an application and the stable model, which is called stable analysis patterns and it illustrates both the static and dynamic requirements of an application.
2. Show the roles of stable analysis patterns for creating a complete formation of software architectures on demand. The rapid growth of technology, coupled with the tightened development time and production cost constraints, have imposed tremendous pressure and an intense desire for software enterprises to create new and innovative designs, which could respond to rapidly changing business and operating environments. Enterprises must invest in building *stable architectures*, that is, based on conceptual knowledge more than application context. Any system based on a stable architecture is built in a way that captures the enduring goals (stable analysis patterns) that the software is supposed to meet. We will refer to these emerging trends of architectures as *Architectures on Demand* as they can be *adaptable*, *customizable*, *extensible*, *personalize-able*, *self-configurable*, and *self-manageable*, according to the future requirements and changes in the operating environments. In a nutshell, *Adaptability* refers to the extent software system architecture can accommodate changes in its environment, that is constrained by the hardware and software. *Customizability* refers to the ability of the architecture to be managed and customized by an agent, its users, and benefiting applications, etc. *Extensibility* means that the architecture is designed to include mechanisms for expanding/enhancing the system with new capabilities, without having to make major changes to the architecture and the underlying infrastructure. A good architecture provides the design principles to ensure this—a roadmap for that portion of the road that is yet to be built. Self-configurable and self-manageable architectures refer to the architectures of systems that can manage and "self-heal" its properties dynamically at runtime at the level of components, connectors, and the underlying infrastructure.
3. Show a method to extract stable analysis patterns from user's requirements and support the formation and the creation of the requirements specifications.
4. Show how to utilize stable analysis patterns in domain analysis. Domain analysis is a process that utilizes domain knowledge of the software systems users' requirements to identify, capture, and organize with the purpose of making it domain-less and reusable stable analysis patterns when creating any new software and any system.

REVIEW QUESTIONS

1. Explain main disadvantages of starting analysis from scratch for each project.
2. What does a pattern represent—a complete system or a specific commonly occurring problem?
3. List eight essential properties that influence usability and effectiveness analysis models. Give brief description for each property.
4. What are the three foundation concepts of stability model?
5. Give a description for three foundation concepts in stability model.
6. Describe three main steps to identifying EBTs.
7. (T/F) EBTs should have a direct physical representation.

8. Where do you start developing BOs for any particular EBT? What questions do you need to ask?
9. What criteria do you use to filter the list of BOs?
10. What are the four characteristics that a BO must satisfy to fully qualify as a BO in any particular pattern?
11. What characteristics do you check for each BO?
12. How do you apply a stable analysis pattern to a particular context?

EXERCISES

1. Use the existing pattern in this chapter to model a student email account.
   a. Draw a class diagram of using AnyAccount pattern.
   b. Generate a significant use case for this context.
   c. Map the use case to a sequence diagram.
2. Use the AnyEntry pattern to model entries in a spreadsheet style application.
   a. Draw a class diagram of using AnyEntry pattern.
   b. Generate a significant use case for this context.
   c. Map the use case to a sequence diagram.

PROJECTS

You're charged with creating a software system for organizing collaborations between team members in a project group. Each member has his or her own account and there are different levels of permission given to each account (admin, creator, read-only). These access levels determine what kind of entries each member can enter into the system. Some users have the authority to manage entries while others can only read or comment on them. Use the patterns in this chapter to develop the system diagram.

REFERENCES

1. A. Geyer-Schulz, and M. Hahsler, Software Engineering with Analysis Patterns, Technical Reports 01/2001, Institute of Information Processing and Economics, University of Economics Wine, Augasse 2–6, 1090 Wine, 2001.
2. E.B. Fernandez, and X. Yuan, An analysis pattern for reservation and use of reusable entities, *the Pattern Languages of Programs Conference*, Monticello, IL, 1999. http://st-www.cs.uiuc.edu/~plop/plop99
3. M.E. Fayad, and M. Laitinen. *Transition to Object-Oriented Software Developments*, New York, NY: Wiley, 1998.
4. E. Gamma et al., *Design Patterns: Elements of Reusable Object-Oriented Software*, 1st ed., New York, NY: Addison-Wesley, 1995.
5. F. Buschmann et al., *Pattern-Oriented Software Architecture, a System of Patterns*, Chichester, WA: Wiley, 1996.
6. H. Hamza, and M.E. Fayad, Model-base software reuse using stable analysis patterns, *16th European Conference on Object-Oriented Programming*, Malaga, Spain, 2002.
7. M.E. Fayad, and A. Altman, Introduction to software stability, *Communications of the ACM*, 44(9), 2001, 95–98.
8. M. Fowler, *Analysis Patterns: Reusable Object Models*, Boston, MA: Addison-Wesley, 1997.
9. R.T. Vaccare Braga et al., A confederation of patterns for business resource management, *Pattern Language of Programs*, Monticello, IL, 1998.
10. M.E. Fayad, and H. Hamza, Introduction to stable analysis patterns, *(Error) Communications of the ACM, Thinking Objectively*, 45(9), 2002.
11. M.E. Fayad, and H. Hamza, Comparative study of analysis patterns, *(Error) Communications of the ACM, Thinking Objectively*, 45(11), 2002.

4 Model-Based Software Reuse Using Stable Analysis Patterns

Nothing lasts forever, not even the best machines. And everything can be reused.

Hephaestus
Rick Riordan, The Lost Hero

The intricate challenge of building efficient reusable software artifacts is the key focus of several schools of thought of software engineering. The existing analysis patterns claimed to be recurring and reusable models. However, there are several deficiencies with existing analysis patterns. These deficiencies make it is not recurring and difficult to use analysis patterns as efficient reusable artifacts. This chapter proposes nine essential properties (or metrics) to measure pattern reusability. In addition, the concept of stability analysis patterns is also introduced. This chapter also distinguishes stable analysis patterns with some analysis patterns by using the proposed metrics.

4.1 INTRODUCTION

Since the origin of object-oriented concepts, researchers and practitioners alike, have stuck to their belief that reuse vastly improves the ultimate quality of software products, while simultaneously reducing cost and condensing lifecycles. Many reuse software communities have evolved in recent years, including aspect-oriented programming (AOP), component-based software engineering community, and many others.

Analysis patterns, as reusable artifacts, have been widely recognized by global software engineering community as a major technological advance over conventional reuse techniques, and it is rightly so. Unfortunately, analysis patterns have not realized their fullest potential. Analysis patterns are still insufficiently mature to be considered as a strong base for building robust and reusable software models. Understanding the cause of this immaturity is the first step in achieving real reuse of analysis patterns.

As software models, analysis patterns must satisfy all the six basic model properties introduced in Reference [1]. That is, to be simple, complete, and most probably accurate, testable, stable, to have visual representation, and to be easily understood. In addition to these six properties, reusable artifacts must also satisfy three additional metrics: first to be general, second to be easily and actually reused, and third to be a full representation of the core aspects. Thus, a pattern that models a specific problem should be constructed, so that it is easily reused whenever the problem occurs, and independent of the context in which the problem appears.

Software stability concepts introduced in References [2–5] have demonstrated great promise in the area of software reuse and lifecycle improvement. Stability models apply the concepts of "Enduring Business Themes" (EBTs) and "Business Objects" (BOs). These concepts have been shown to produce models that are both stable over time, and stable across various paradigm shifts within a domain or application context. By applying stability model concepts to the notion of analysis patterns, we propose and recommend the concept of stable analysis patterns. The important idea behind the stable analysis patterns is to analyze the problem under consideration in terms of its EBTs and the BOs, along with the goal of increased stability and broader reuse.

In the remainder of this chapter, we will examine the nine essential properties of analysis patterns (Section 4.2), and study different methodologies for building analysis patterns (Section 4.3). We will also study some example patterns reflecting each of the aforementioned methodologies (Section 4.4), and compare these approaches (Section 4.5). Conclusions are presented in Section 4.6.

4.2 ESSENTIAL PROPERTIES OF ANALYSIS PATTERNS

Here we examine all nine properties (or metrics) of efficient reusable models. Satisfying these nine properties may not guarantee an efficient reusable model; however, in practice, without any of these properties the reusability of the model might be affected in a significant manner. These nine essential properties are

1. *Simple:* A pattern is not intended to represent a model for a complete system; rather it models a specific problem that commonly appears within larger systems. Systems, by their nature, combine many problems and bottlenecks. Thus, they are modeled by using a collection of analysis patterns. In fact, each analysis pattern should focus on one specific problem; otherwise, many problems would arise in the future. Without decomposing a system into individual components, models become unreasonably complex, the generality of the patterns would be adversely affected, and the entire model becomes highly nonintuitive. If a pattern is used to model an overly broad portion of a system, the generality of resulting patterns is sacrificed overall—the maxim eventually holds here: the probability of the occurrence of all the problems together is less than the probability of the occurrence of each problem individually. For example, modeling the "payment" problem with "buying a car" is not really effective, since the "payment" problem may appear in unlimited problems. Pattern completeness is also sacrificed, when we model a system at an improper level of resolution, because the analyst's focus is not on a specific problem, it is likely that important features of the system and its subcomponents will be overlooked.
2. *Complete and most likely accurate:* Closely related to the concept of simplicity, this property guarantees that all the required information is present. In order to be considered complete, the model should not omit any component. The model must be able to express the essential concepts of its properties. For example, trying to model the whole rental system of any property will force us to miss some parts of this system. Renting a car will involve something related to its insurance; however, renting a book from a library has nothing to do with the insurance problem. As a result, pattern that models the rental system, besides lacking the simplicity property, will not be complete nor accurate.
3. *Testable:* For the model to be testable, it must be specific, doubtless, and unambiguous. Thus, all the classes and the relationships of the model could be qualified and validated.
4. *Stable:* Stability influences the reusability of a model. Stable models are easily adapted and modified without the risk of obsolescence.
5. *Graphical or visual:* Conceptual models are difficult to visualize. Therefore, having a graphical representation for the model helps in deeper understanding.
6. *Easy to understand:* Conceptual models are extremely complex, as they represent a high level of abstraction. Therefore, it is required for analysis patterns to be well described in such a way that they assist in communicating a deeper and comprehensive understanding of the system. Otherwise, use of the pattern is neither attractive nor effective.
7. *Applicability:* This property is essential to ensure model reusability. A pattern model lacking generality become useless and redundant, since analysts tend to build new models, rather than spending time and effort to adapt an unruly pattern to fit into an application. Generality means that a pattern that models a specific problem is easily used to model the same problem that is independent of context. Pattern generality may be divided into

two categories: Patterns that solve problems that frequently appear in different contexts (domain-less patterns), and patterns that solve problems that frequently appear within specific contexts (domain-specific patterns). In the latter sense, the pattern is still considered to be general, even if it is only applicable in a certain domain, but in this case, we should make sure that the problem that this pattern models does not occur in other contexts.

8. *Easy to use and reuse (usability and reusability):* Analysis patterns should be presented in a clear and easily comprehensible manner that makes them easily reused. It is important to remember that patterns are consumed in larger models. Patterns that are easy to use and designed for reuse stand a greater chance of actually being reused.
9. *Presentation of the core aspects:* Analysis patterns should expose the core aspects of the problem. If the analysis pattern is not able to uncover the issues underneath the surface of the problem, then using such a pattern will result in inadequate and/or defected analysis.

4.3 CLASSIFICATION OF ANALYSIS PATTERN METHODOLOGIES

One possible classification for analysis patterns is based on the methodology of construction. Generally, different building methodologies categorize analysis patterns into three groups:

Group I: Direct approach: The methodological approach used to build analysis patterns is experiential. Simply, patterns are produced during the course of specific projects. Since, no one can be an expert in all fields, domain experts often produce domain-specific patterns, even if the problem tackled occurs in many other contexts. Proponents of this category believe that it is unsafe to further abstract patterns generated within certain projects in order to make them reusable in other contexts. They also argue that the patterns result from extended debate and that the patterns are tested and validated in the project. Therefore, there is no guarantee that these patterns will be successfully reused in other contexts.

Group II: Analogy approach: The methodological approach used to build analysis patterns is the analogy. According to this group, patterns that model complete systems in one context are reused by making an analogy between the pattern and the new application. Thus, by analogy, they change the names of the pattern's classes to be relevant to the new application. Even though this group believes that analysis patterns should be built in a way the makes them reused to model the same problem regardless of its context, it usually ends up in building templates rather than building patterns.

Group III: Stability approach: Which is our group, the methodology to build the analysis patterns is exclusively based on the software stability concepts [2–5]. By analyzing the problem in terms of its EBTs and the BOs, the resultant pattern touches the core knowledge of the problem. The goal of this methodology is stability. As a result, these stable patterns could be used to model the same problem regardless of its context.

4.4 EVALUATION OF ANALYSIS PATTERNS GROUPS

In order to objectively compare the effectiveness of the patterns that are generated by the three different groups' methodologies as described in the previous section, a pattern that reflects the methodology of each group will be examined against the essential properties mentioned earlier in Section 4.2.

4.4.1 Group I: Direct Approach

The *Account Pattern* provided by Fowler [6] is the true representative of this group. Figure 3.6 shows the class diagram of the *Account Pattern*. The purpose of this pattern is to provide a stable model for the "account" problem; thus, we can use this pattern to create banking account. In fact, it was not long ago when the word *account* was merely used to indicate banking and financial accounts. However, the word *account* alone becomes a vague concept, if it is not allied with a word related to a certain context. For instance, apart from traditionally well-known business and banking accounts, we also have email accounts, online shopping accounts, online learning accounts, subscription account, and many other similar accounts. Consequently, using words such as balance and withdrawal while modeling the account, makes the use of the pattern to model accounts in different contexts, time and effort-consuming, if not, impossible. For instance, suppose we want to model an email account using Martin's pattern, perhaps the most obvious changes are all the classes' behaviors, which are completely irrelevant to the email application.

From the simplicity point of view, Fowler's *Account Pattern* is not considered simple in the sense that it models two different problems at the same time. The first problem is the "account" problem and the second problem is the "entry" problem. In fact, these are two independent problems even though they appear together in many contexts and situations; however, there is a possibility of having entries without an account, or having an account without entries. As a result, the generality of the pattern is rather limited. These factors eventually contribute negatively to the reusability of the pattern.

Fowler's pattern is also not complete because it lacks some of the "basic" concepts that appear frequently in banking accounts. For instance, let us assume that we want to use this pattern to design a banking account. In the case of banking accounts, it is possible that two or more persons may be holders of the same account. Perhaps, there is a primary holder who has the full authorization to manage and control the account, while each of the secondary account holders will have specific privileges for using the same account. Such a situation cannot be handled when we are using Martin's account. Thus, Martin's pattern is not applicable to some of the usual financial and banking account situations. Since significant effort is required to adapt this pattern to other circumstances, the stability of the system is quite limited and the pattern structure is unstable over time.

This pattern is graphical too in the sense that it has a graphical model as a class diagram. Having such a graphical representation will make the pattern visually testable.

4.4.2 Group II: Analogy Approach

Figure 2.1 provides the class diagram of the *Resource Rental Pattern* [7]. The objective of the pattern is to provide a model that could be reused to model the problem of renting any resource. Figure 2.2 provides an example of the *Resource Rental Pattern* applied in the context of a library service [7]. Many examples for applying this pattern to different applications are suggested in Reference [7].

This pattern reproduces a complete resource rental system; thus, it also models a collection of problems, whereas each of these problems could be modeled individually. For example, the "payment" process is a stand-alone problem, which could appear in many other contexts. Therefore, having a pattern that represents the "payment" problem alone will be more effective, since such pattern will be reused in many other applications.

The resource rental pattern lacks the factor of simplicity. In addition, it is not too general, because it is not applicable for any resource rental. One of the most basic steps in the automobile rental process is the question of insurance. There is nothing in this model that can be used to address insurance issues.

The verification process is very essential in many of the rental issues. However, incidentally this issue is not addressed in this pattern. In many cases, renting a resource might require a membership (as in the case of the universities library) or other identification (such as the driver license in the case of renting a car). There is a link between the completeness of the model and its stability. Reuse is extremely challenging, when applying incomplete patterns, because many new classes are needed to complete the model.

In our example, let us assume that we want to model a car rental system using this pattern. Now, we have reached the conclusion that we should add the verification process classes and the insurance process classes. Substantial analysis is needed to complete this new model, hence the analyst may be inclined to build a new model from scratch. Therefore, the pattern will not be reused. As in the first group, the existence of a class diagram to represent the model makes it more graphical. As a result, the pattern is readily tested.

4.4.3 Group III: Stability Approach

Our proposed solution chooses this approach for building reusable analysis models as a model-based software reuse and to avoid most of the common problems that we found in the other groups. To make the evaluation of this group patterns more interesting and enticing, the pattern example chosen here is the stability version of the *Account* pattern that was pioneered by Fowler.

From the stability point of view, the model that focuses on the account problem has nothing to do with the entry problem. Thus, it is required to have different patterns for each individual problem. In this manner, the simplicity of our model is guaranteed because each pattern is intended to focus on a specific problem.

Stability is enhanced by supplementing with other classes that do not exist in Fowler's model. Figure 4.1 shows the proposed analysis pattern "*AnyAccount*" that gives a stable model for the "account" problem. The new classes that appear in the stability model help us in handling those circumstances that Fowler's model fails to cover; thus, the model becomes more accurate and complete. For instance, the use of the EBT of "Ownership" and the BO of "Holder" in the modeling of the account assists in contexts, where there is a difference between the account owner, and those who are authorized to use the account under certain rules, should be made clear and precise.

"Ownership" is an enduring concept, which will never change independent of context. On the other hand, the "Holder" addressed here is externally stable and it never changes with time, although the holders of the account could internally change (e.g., holder may get ill); however, they are still the holder of the account. As we can see in the pattern class diagram, the inherited objects from the "Holder" object represents different roles of the various levels of the usability of that account.

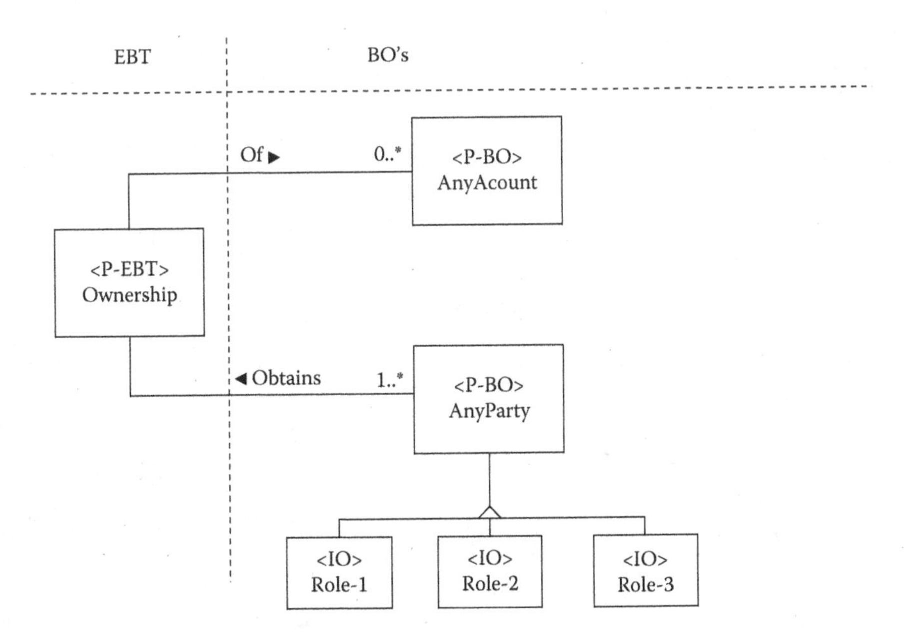

FIGURE 4.1 AnyAccount stable design pattern—class diagram.

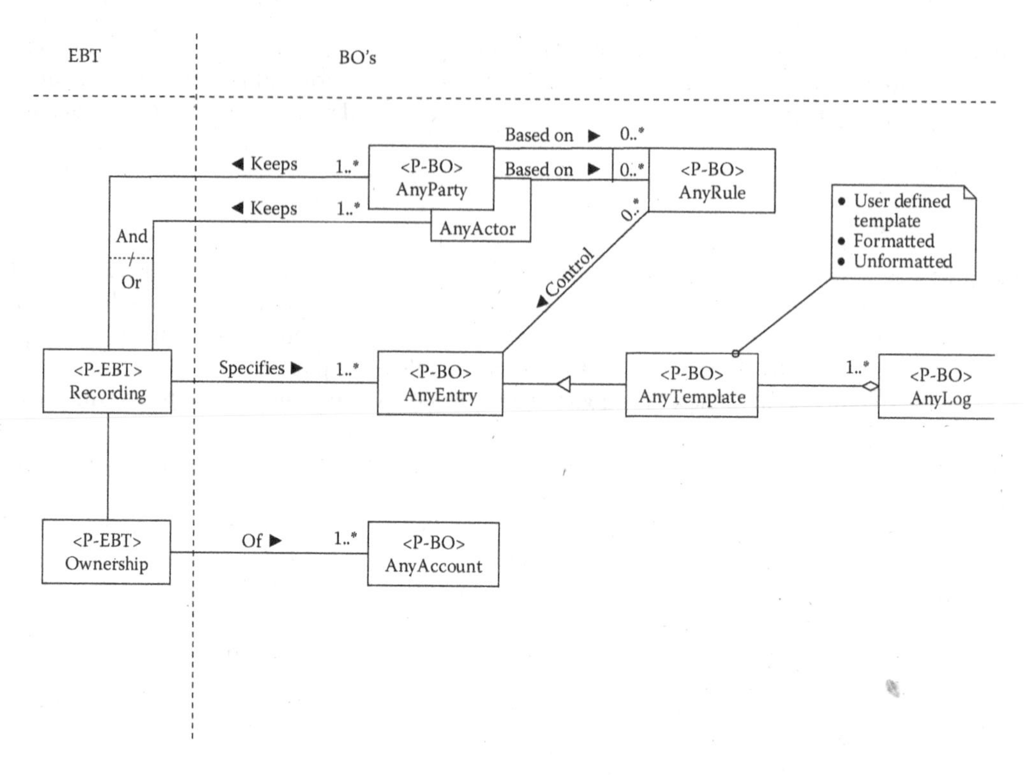

FIGURE 4.2 AnyAccount with entry stable architecture pattern—class diagram.

This pattern's structure is stable over time and general enough to handle different applications that involve accounts and different situations within the same application.

On the other hand, discussing about "entry" as a stand-alone problem forces us to build a pattern that represents any entry regardless of the context. Using stability concepts, we were able to create a stable pattern that reproduces any entry for any application. This pattern is called "*AnyEntry*" pattern and it is shown as part of Figure 4.2.

By combining the pattern that models the account problem (the "*AnyAccount*" pattern) with that which models the entry problem (the "*AnyEntry*" pattern), we can easily demonstrate the ease of reusing stability models to construct comprehensive models. Figure 4.2 displays the class diagram for this third pattern. The "*AccountWithEntry*" pattern could be used to develop any account that has entries associated with it, as in the case of banking accounts and email accounts for instance.

4.5 COMPARISON OF ANALYSIS PATTERNS GROUPS

Based on the essential properties discussed in Section 4.2, Table 4.1 summarizes the results of the three analysis pattern groups.

4.6 SUMMARY

Analysis patterns could form a solid and robust foundation for building reusable software assets. However, evaluation of some analysis patterns in the earlier section shows that these patterns lack many essential properties. As a result, their reusability might diminish. Software stability has been proposed as a solution for removing numerous deficiencies that are encountered in analysis patterns. *Stable analysis patterns* demonstrate significant improvements over the software analysis

TABLE 4.1
Comparison of Analysis Pattern Groups

Properties	Evaluation		
	Group I	**Group II**	**Our Group**
Simple	No. It models two different problems.	No. It models an entire system.	Yes. Each model focuses on one problem.
Complete and most likely accurate	No. It does not cover all the circumstances that might occur in the application.	No. It is not sufficiently general to address the requirements of different renting applications.	Yes. Because each model focuses on a specific problem, all situations within the problem can be easily covered.
Stable	No. This pattern cannot model all types of today's accounts. Thus, we will always need to do major changes to reuse this pattern for different applications.	No. Using this pattern to model different application will need major changes. For instance, adding the verification process to the model will need a lot of changes.	Yes. The patterns in this group are built with stability in mind. The use of EBTs and BOs, ensure stability in the model.
Testable	Yes. Since the pattern can be visualized; thus, we can, at least, visually test it.	Yes. Since the pattern can be visualized; thus, we can, at least, visually test it.	Yes. Since the pattern can be visualized; thus, we can, at least, visually test it.
Easy to understand	Yes. Generally speaking, despite the accuracy of the pattern, it is easy to understand its structure.	Yes. Despite the accuracy of the pattern, it is easy to understand its structure.	Yes. Since the used EBTs and the BOs reflect the concepts that we are familiar with, it is easy to understand the model structure.
Graphical or visual	Yes. The pattern has a graphical presentation, which is the class diagram.	Yes. The pattern has a graphical presentation, which is the class diagram.	Yes. The pattern has a graphical presentation, which is the class diagram.
General	No. We cannot use it to model the account in other contexts other than monetary application. Also, it does not cover the cases of having accounts without entries and vice versa. Moreover, the pattern does not cover some of the situations such as having more than one holder for the same account.	No. This pattern cannot be used to model the rental of some resources. For instance car rental; since there is nothing in the model that covers the insurance issues, which is an essential part of any cat rental process.	Yes. Because of the stability concept, our models focus on a specific problem trying to flush the core knowledge underneath the surface of the problem. Since the core knowledge of any problem is constant, regardless of the context that this problem might appear in, the model of the problem is general and can apply to the problem whenever it occur.
Easy to use and reuse	No. Using the patterns of this group in different application that they originally built for, if possible it is not an easy task.	No. As we mentioned before, we will need to do major changes to use this pattern in different applications such as for car renting.	Yes. Using the pattern by itself or the integration of few patterns are both easy to be done. This property is demonstrated by introducing the third pattern shown in Figure 4.2.
Presentation of the core aspects	No. The pattern does not illustrate the core aspects of itself.	No. The pattern ignores the core aspects of itself.	Yes. The pattern is unified and models the core aspects.

patterns that were examined earlier. Therefore, the application of stable analysis patterns is an important approach that needs further research and extended relook by reuse and development communities.

4.7 OPEN AND RESEARCH ISSUES

1. Stable architecture patterns (SArchPs) can be mapped smoothly to any type of well-known software architectures, such as model-driven architecture (MDA) [8,9], pattern-oriented architecture [10,11], or system of patterns [12], aspect-oriented architectures [13,14], etc. We will explore the idea of creating any type of architecture using EBTs, BOs, or a combination of EBTs and BOs.
2. EBTs + BOs = SArchPs form an MDA as shown in Sidebar 4.1. The rapid growth of emerging technology, coupled with the tightened or constricted software development time and production cost constraints, have imposed and exerted tremendous pressure, and an intense desire for software enterprises and firms to design and create new and innovative designs, to respond to a rapidly changing business environment. Enterprises must heavily invest in building stable architectures that are readily adapted in many different ways to meet the new challenges and risks. These kinds of architectures are called *architectures on demand,* as they can be adapted accordingly to meet the future requirements and changes in the system. The primary focus of this chapter is to show how software stability concepts are used to develop on-demand architectures. This issue also focuses on three key aspects: (1) EBTs, or business goals and transformations, which we call stable analysis patterns; (2) BOs, or business process design, which we call stable design patterns; and (3) IOs, or application objects. Both EBTs and BOs form a stable core, and, thus provide architectures on demand for any domain. We will call these architectures as "stable architectural patterns." Data must be collected in relation to how often and how many architectures on demand can be generated per knowledge map. EBTs and BOs are *stable software patterns* and a combination of EBTs and BOs form the *core knowledge* for a given domain. The core knowledge for any domain is called a *Stable Architectural Pattern* that you can extend and adapt through the application of hooks. The quality of stable architectural patterns creates competitive advantages through differentiation and productivity. It will also integrate partners in order to increase adaptive capabilities [15].
3. EBTs + BOs = unified software engine (USE) [16] creates software architectures on demand and you will develop and test, and deploy unlimited applications right away and it will speed the process of software development in a matter of hours, days, and a few weeks. USE for Any Domain leads to a very highly reusable and USEs technology for developing unlimited service and/or production systems, which are called service engines and production engines. USE's for any domain is an open research issue and topic, because building such engines is not an easy exercise, specifically, when several conflicting factors can undermine or impede their quality success, such as cost, time, and lack of systematic approaches. The main difference between software developments (business as usual), application and enterprise frameworks, and the USE's also needs further research and development in a comprehensive manner.
4. EBTs + BOs = Understanding the requirements .They are considered the best way to come up with the problem space (analysis) and the ultimate solution space (design and architecture) of any system.

REVIEW QUESTIONS

1. List and explain essential properties of analysis patterns.
2. "A pattern is not intended to represent a model for a complete system." Explain.

3. Why is it important that each pattern must focus on one specific problem?
4. Why is it important to decompose a system into individual components?
5. "If a pattern is used to model an overly broad portion of a system, the generality of resulting pattern is sacrificed." Explain with the help of an example.
6. Explain why a pattern representing rental system will not be complete and accurate.
7. What are the key characteristics of a model that makes it easily testable?
8. List advantages of stable models.
9. Why is graphical representation of a model important?
10. How is applicability of a pattern related to pattern reusability?
11. What are the two categories of pattern generality? Explain.
12. Why is it important for a pattern to present the core aspects of a system?
13. Explain the classification of analysis patterns based on the methodology of construction. Briefly describe each category.
14. What is the concept behind constructing an analysis pattern by using direct approach?
15. What are the drawbacks of constructing analysis patterns using direct approach?
16. What is the concept behind constructing analysis pattern using analogy approach?
17. "Analogy approach makes people end up building templates rather than building patterns." Explain.
18. Explain the concept behind constructing analysis pattern by using stability approach.
19. What are the advantages of using stability approach over direct and analogy approach?
20. What is Fowler's account pattern? Explain with the help of a diagram.
21. Examine Fowler's account pattern on following characteristics:
    a. Simplicity
    b. Generality
    c. Completeness
    d. Stability
    e. Visual and graphical model
    f. Testability
22. Draw a class diagram for resource rental pattern and its installation for library service by using analogy approach.
23. Examine resource rental pattern on following characteristics:
    a. Simplicity
    b. Generality
    c. Reusability
    d. Completeness
    e. Stability
    f. Visual and graphical model
    g. Testability
24. Why is it required to have two different patterns for account and entry?
25. Draw a stable class diagram for account pattern.
26. Explain and justify the existence of ownership and holder class in account pattern.
27. Draw a class diagram for account with entry pattern.
28. Compare account stable pattern with account pattern developed by using direct approach.
29. What are the advantages of separating account and entry patterns?
30. Compare and contrast three approaches of constructing analysis pattern.

EXERCISES

1. Write two challenges of AnyAccount stability pattern.
2. Write two constraints of AnyAccount stability pattern.
3. Write two challenges of AnyEntry stability pattern.

4. Write two constraints of AnyEntry stability pattern.
5. Write four scenarios of AnyAccount stability pattern.
6. Write four scenarios of AnyEntry stability pattern.
7. Draw class diagram for following using AnyAccount stability pattern and write its description:
   a. Email account
   b. Bank account
   c. Chatting account
   d. Online shopping account
8. Draw class diagram for following using AnyEntry stability pattern and write its description:
   a. Email entry
   b. Bank entry
   c. Chatting entry
   d. Online shopping entry

PROJECTS

1. You are a part of a team that creates videos of cooking and uploads it on the YouTube. The team has its own account on it and people can watch videos and subscribe to the channel. They can also like or prefer the video. The team uploads one video clip every day. Use the account and entry stability pattern to draw a class diagram for this application.
2. Google has a facility called AdSense, where you can open your account and make your website AdSense enabled, so that Google can put advertisements on your website. Google pays you, if a person clicks on the advertisement displayed on your website. Use the account and entry stability pattern to draw a class diagram for this application.
3. There are numerous auditions regularly held for a reality show called ABC Idol. It is a show, where talented singer gets a chance to show his or her talent. Anyone can open an account on ABC Idol's website and post video of self-performance. Later, ABC Idol would shortlist the list of final participants by evaluating all uploaded videos. Use the account and entry stability pattern to draw a class diagram for this application.
4. When a person rents a flat, he or she has to open an account with a company that provides electricity in the area. The company then bills the person based on the monthly electricity usage. The company either sends out a paper bill or can email the bill to the person. Then, the person can either pay the bill by check or can apply for a direct debit or can even pay over phone. A bill is generated at the end of each month. Use the account and entry stability pattern to draw a class diagram for this application.
5. You are part of a project team developing a software application for your local community art fair. Art work is accepted from all members of the community. The event occurs yearly and active community members contribute their art work to the yearly event. Art pieces may also be submitted anonymously as a one-time occurrence or over multiple submissions by the same anonymous artist. Use the account and entry stability patterns to develop a class diagram for this application.

REFERENCES

1. M.E. Fayad, and M. Laitinen, *Transition to Object-Oriented Software Developments*, 1st ed., New York, NY: Wiley, 1998.
2. M.E. Fayad, and A. Altman, Introduction to software stability, *Communications of the ACM*, 44(9), 2001, 95–98.
3. M.E. Fayad, Accomplishing software stability, *Communications of the ACM*, 45(1), 2002, 111–115.
4. M.E. Fayad, How to deal with software stability, *Communications of the ACM*, 45(4), 2002, 109–112.

5. M.E. Fayad, and S. Wu, Merging multiple conventional models into one stable model, *Communications of the ACM*, 45(9), 2002, 102–106.
6. M. Fowler, *Analysis Patterns: Reusable Object Models*, Boston, MA: Addison-Wesley, 1997.
7. R.T.V. Braga et al., *A Confederation of Patterns for Business Resource Management*, Monticello, IL: Pattern Language of Programs, 1998.
8. A. Kleppe, *MDA Explained, The Model Driven Architecture: Practice and Promise*, Boston, MA: Addison-Wesley, 2003.
9. C. Raistrick, C. Carter, I. Wilkie, and P. Francis, *John Wright Model Driven Architecture with Executable UML*, Cambridge, MA: Cambridge University Press, 2004.
10. S.M. Yacoub, and H.H. Ammar, Pattern-oriented analysis and design (POAD): A structural composition approach to glue design patterns, *Proceedings of TOOLS '00 Proceedings of the Technology of Object-Oriented Languages and Systems*, IEEE Computer Society, Washington, DC, p. 273.
11. S.M. Yacoub, and H.H. Ammar, UML support for designing software systems as a composition of design patterns, *UML*, 2185, 2001, 149–165.
12. F. Buschmann, R. Meunier, H. Rohnert, P. Sommerlad, and M. Stal, *Pattern-Oriented Software Architecture: A System of Patterns*, vol. 1. New York, NY: Wiley, 1996.
13. P. Tarr, H. Ossher, W. Harrison, and S.M. Sutton Jr., N degrees of separation: Multi-dimensional separation of concerns, in *Proceedings of the 21st International Conference on Software Engineering (ICSE 1999)*, Los Angeles, CA, IEEE Computer Society Press, 1999, 107–119.
14. G.C. Murphy, R.J. Walker, E.L.A. Baniassad, M.P. Robillard, A. Lai, and M.A. Kersten, Does aspect-oriented programming work? *Communications of the ACM*, 44(10), 2001, 75–77.
15. M.E. Fayad, H. A. Sanchez, S.G.K. Hegde, A. Basia, and A. Vakil, *Software Patterns, Knowledge Maps, and Domain Analysis*, Boca Raton, FL: Auerbach Publications, 2014, p. 422.
16. M.E. Fayad, *Unified software engine (USE)*. Technical Report No. 1, San Jose, CA, San Jose State University, January 2010.

5 Stable Patterns' Documentation

Templates, UML Forms, Rules, and Heuristics

Don't document the program; program the document.

The pattern documentation template addresses many, major key questions and issues of each stable pattern in a knowledge map while each section of the template answers key questions about the stable pattern involved. The template is a very consistent oratorical structure and stylistic format that answers the most common questions about individual stable patterns in a very logical manner. The template is suitable for documenting stable atomic patterns and stable architectural patterns where

1. Stable atomic patterns are stable analysis patterns (SAPs (also called EBT goals, rationales, aims, purposes, and objectives),
2. Stable design patterns (SDPs) (also called BOs or capabilities used to achieve each of the goals), and
3. Stable architecture patterns SArchPs (a combination of SAPs and SDPs).

A knowledge map is a collection of interrelated patterns consisting of EBTs and BOs.

5.1 INTRODUCTION

There are three types of templates for documenting stable patterns:

1. *Full or detailed template:* Each stable pattern should have a full template that gives a detailed documentation for that pattern. The full template consists of 20 fields that will be discussed in detail in the next section. The full template fields are
   a. Pattern name
   b. Known as (optional)
   c. Context with two or more scenarios
   d. Requirements (problems)
      i. Functional requirements
      ii. Nonfunctional requirements
   e. Challenges
   f. Constraints
   g. Solution
      i. Pattern structure
      ii. Participants
      iii. Class responsibility and collaborations (CRC) cards
   h. Consequences
   i. Applicability
      i. Five scenarios in a table

        ii. Two case studies, each consists of
            A. Case study title
            B. Case study description
            C. Class diagram
            D. Significant use case with test cases
            E. Sequence diagram
    j. Related pattern and evaluation
    k. Measurability and evaluation
        i. Quantitative measurement
        ii. Qualitative measurement
    l. Modeling issues (optional)
    m. Analysis issues (optional)
    n. Design and implementation issues
    o. Testability (optional)
    p. Formalization (optional)
    q. Business issues (optional)
    r. Known usage
    s. Tips and heuristics
    t. References
2. *Mid-size template:* a mid-size template is a medium-size version of the full template. This template is usually used to document the pattern, while this pattern can be used as a short paper or a short documentation. Mid-size template consists of only nine fields. The mid-size template fields are
    a. Pattern name
    b. Known as (optional)
    c. Context with two scenarios
    d. Requirements (problems)
        i. Functional requirements
        ii. Nonfunctional requirements
    e. Challenges
    f. Constraints
    g. Solution
        i. Pattern structure
    h. Applicability
        i. Five scenarios in a table
        ii. One case study consists of
            A. Case study title
            B. Case study description
            C. Class diagram
            D. Significant use case with test cases
            E. Sequence diagram
    i. References
3. *Short template:* A short template is a reduced size version of the full template. This template is usually used to document the pattern, when this pattern appears within the main pattern that we are interested in documenting. Short template consists of only six fields. The short-size template fields are
    a. Pattern name
    b. Context with two scenarios
    c. Requirements (problems)
        i. Functional requirements
        ii. Nonfunctional requirements

   d. Solution
      i. Pattern structure
   e. Applicability
      i. Five scenarios in a table
   f. References

This chapter illustrates the template to be used in documenting stable patterns (analysis, design, and architectural). It also defines rules and provides some useful heuristics for filling this template.

5.2 PATTERNS' DOCUMENTATION TEMPLATES

5.2.1 Current Stable Models' Templates and Their Problems

Many documentation templates are used for documenting patterns, while [1] furnished a complete survey of pattern documentation templates, such as Alexandrian Form [2,3] and Gang-of-Pattern [4]. However, a number of issues exist with the existing documentation templates:

1. Alexandrian form is limited to a few entries that ignore the essentials of software patterns.
2. Limited coverage—limited and tied down to a micro-architecture level.
3. Missing the classification and scope.
4. Lack of requirements analysis or specifications.
5. Lack of implementation.
6. Lack of applicability.
7. Missing the CRC cards of the base classes of the pattern and use cases of the application.
8. Lack of testability.
9. Lack of measurements.
10. Missing the story teller (contexts).
11. Lack of quality assurance.
12. Lack of verification and validation (V&V).
13. Lack of deployment.

Major differences exist between the existing documentation templates and Fayad's stable pattern documentation templates:

1. Stable software patterns cover all the software development stages: requirements analysis, software architecture, detailed design, implementation, testing, and deployment. Stable software patterns are small frameworks, stable frame-lets, etc.
2. It utilizes software stability as a proven technique for classification and scope.
3. It covers the requirement analysis or specifications of each pattern using an EBT or a goal.
4. It covers BOs of each pattern that represents the design aspect of the pattern and we call it "Capabilities" that are used to achieve the goal of the documented pattern.
5. IOs or application objects or implementation.
6. It shows several examples of applicability "exemplars" of the pattern and show that we can generate many applications for a single pattern.
7. It provides CRC cards of the base classes of the pattern and use cases of the application.
8. It provides a scenario-like or story teller of each application.
9. It covers testability aspects of the documented pattern.
10. It provides a detailed discussion of measurability issues and metrics.
11. It consists of many quality factors such as stability, extensibility, customizability, maintainability, scalability, adaptability, etc.
12. It covers verification and validation (V&V).
13. It indirectly covers many issues related to deployment.

5.2.2 Full Stable Patterns' Template Description with Rules and Heuristics

The following are the fields of the full template.

5.2.2.1 Pattern Name

This field presents the name of the presented pattern. It also focuses on a precise name for the pattern to be described. The pattern name is a descriptive, natural, and singular noun or a noun phrase. The pattern name uniquely identifies it. Patterns' names will form the foundation for a stable pattern vocabulary for useful discussions, as a base for abstractions, and publications, software developers and software engineers recognize and use them in actual practice. Nicknames or aliases are allowed optionally and it will be commonly used too. The pattern name is an indication of the classification and scope.

The prefix *Any* should be attached to the beginning of the design pattern name and only the design pattern name; thus, the pattern name *Party* becomes *AnyParty*. Each of the analysis and design patterns represents an atomic notion that means the pattern is a domain independent (like *AnyParty*, this can appear in the model of any application and in any domain). *Pattern name rules and heuristics are presented in* Tables 5.1 and 5.2.

5.2.2.2 Known As

This field lists all the terms that are similar to the name of the pattern. Two possible sources can be used to fill this field.

1. Similar patterns that are proposed in the literature.
2. Other names that you may find relative to the developed pattern.

In some cases, using several names might be logical, so you can keep a list of just a few of them under this field. List all the terms that are similar to the name of the pattern. Two possible sources can be used to fill this section: 1—similar patterns that are proposed in the literature; 2—other

TABLE 5.1
Pattern Name Rules

Rules	Pattern Name

- Why have you chosen that specific name for the pattern?
- Pattern is a class.
- Justify the given name (such as why use "*Any*…" as a prefix).
- Names of all design patterns that model atomic notion should be preceded by the word *Any*.
- Pattern names should be *bold and italic*, whenever they appear in the text.
- It is a single word (singular noun) or a short phrase (singular noun phrase).
- Length is a few lines.

TABLE 5.2
Pattern Name Heuristics

Heuristic	Pattern Name

- The given name has to be meaningful, descriptive, and natural.
- The given name must refer to the atomic concept and the core knowledge of the pattern.
- The given name has to be unique and descriptive, refer only to the pattern it describes.
- Multiple names for the same pattern might cause confusion, but it cannot be avoided. In this case, the ***aliases*** must be stated (if applicable).
- Pattern names must imply the characteristics of being EBTs; for stable analysis patterns, BOs; for stable design patterns, or architectural patterns.

TABLE 5.3
Known as Rules

Rules Known As

- List all the similar names for the documented pattern.
- Just list similar names and state why?

TABLE 5.4
Known as Heuristics

Heuristic Known As

1. Names match exactly the same meaning of the pattern name.
2. Names match the same commonalty with elements of the documented patterns.
3. Names indicate a similar behavior and workflow.
4. Names may differ in some corner cases and must be stated or described as cautions or indicate consequences.

names that you may find relative to the developed pattern. In some cases, several names will be prudent, so you can keep a list of few of these names under this section. *Known as rules and heuristics are presented in* Tables 5.3 and 5.4.

5.2.2.3 Context

Possible scenarios for the situations in which the pattern may recur (Sidebar 5.1) are provided in this field. It contains a scenario template for documenting scenarios. *This must be discussed thoroughly to avoid critic.* It is important in this section that you motivate the problem you solve in an attractive way. For example, if I were writing a pattern about trust, I would flush the trust in the context of e-commerce, for example. Keep this section short, yet exciting (This section somewhat serves as an introduction in conventional paper). *Context/scenarios' rules and heuristics are presented in* Tables 5.5 and 5.6.

TABLE 5.5
Context/Scenarios' Rules

Rules Context

- Context must be clearly described.
- Describe the boundaries.
- List all the preconditions for which the solution is valuable.

TABLE 5.6
Context/Scenarios' Heuristics

Heuristic Context

- Show by examples (scenarios), where the pattern can be applied.
- Scenarios have to be very specific and detailed.
- The scenario title reflects the intent of the scenario and the user needs to be able to accomplish through the scenario.
- For example "account"… would have ownership and handler context, can be applied to banking … etc.
- Within any scenario,
  1. Map each BO to 1..* IOs except actor and/or party.
  2. Map each actor and/or party to 2..* Roles (IOs).

5.2.2.4 Problem

To build a stable analysis pattern, we should first understand the core purpose of the pattern and the essence of the pattern, and focus on what the actual requirements of the pattern, and how they should be achieved. This is one of the hardest parts in pattern writing. Do not try to finalize writing it in the first iteration itself, most probably; you will not be able to! The problem should focus on the core purpose of the pattern and should be able to answer the question: will I benefit from the pattern? Always try to keep this section as short as possible, otherwise the reader may be confused.

The problem describes the *intent of the pattern*: the goal that occurs within the context, challenges, and constraints. To summarize, what are the requirements (as exist in the problem space) to solve or design (as exist in the solution space) by using this pattern? There are two types of pattern requirements: functional requirements and nonfunctional requirements (Sidebar 5.2). *Functional requirements' rules and heuristics are given in Tables 5.7 and 5.8 and nonfunctional requirements' rules and heuristics are presented in Tables 5.9 and 5.10.*

5.2.2.5 Challenges and Constraints

The challenges and constraints that the pattern needs to resolve are illustrated in this field. In particular, in this section one may try to say, this is not a trivial problem, and trivial solution may not

TABLE 5.7
Functional Requirements' Rules

Rules	Functional Requirements

- For an EBT as a SAP = List of BOs that achieve the goal.
- For a BO as an SDP = an ultimate goal that is suitable for all possible contexts of the SDP + the rest of the BOs.
- Describe the EBT and BOs and highlight the generic and possible behavior of each one of them to satisfy all contexts.
- Avoid corner cases (context).

You may create a list of the subgoals for requirements of the pattern.

TABLE 5.8
Functional Requirements' Heuristics

Heuristic	Functional Requirements

1. Make it concise; otherwise, it will confuse the reader
2. This is the hardest part of the document
3. Each SAP has 1 EBT and average 5 BOs

TABLE 5.9
Nonfunctional Requirements' Rules

Rules	Nonfunctional Requirements

- Each pattern (SAP, SDP, or SArchP) has a number of nonfunctional requirements.
- Nonfunctional requirements—limited 2 ultimate 3 nonfunctional requirements.
- Nonfunctional requirements are quality factors that the functional requirements must comply.

TABLE 5.10
Nonfunctional Requirements' Heuristics

Heuristic	Nonfunctional Requirements

1. Nonfunctional requirements' essential properties are (1) quality factors, (2) must be enduring, (3) positive, and (4) control functional requirements.
2. Each of the nonfunctional requirements is a SAP.

work at all. Be clear and brief. One major mistake in writing this section is that one may mix the problem statement with the challenges and constraints themselves. After writing this section, try to read the problem statement again and make sure that they are not the same! It always happens! It also discusses the main issue of the problem and the trade-off between these issues. Reading these constraints and challenges will help a reader in realizing the extent of difficulty of the problem while at the same time it helps in appreciating the importance of having a solution to this problem. It must discuss problem's constraints. Challenges and constraints may contradict with one another. Sidebar 5.3 illustrates a template for documenting each of the pattern challenges. *Challenges and constraints' rules and heuristics are presented in* Tables 5.11 and 5.12.

5.2.2.6 Solution

There are three parts:

1. *Pattern structure and participants:* Gives the class diagram of the pattern (EBT or BO) as shown in Sidebar 5.4. It also introduces each class briefly and its role. Associations, aggregations, dependencies, and specializations should be included in the class diagram. Association classes, constraints, interfaces, tagged values, and notes must be included in the class diagram. Include the hooks (show each of the BOs' connections to IOs) too. A *full description of the class diagram should be included with the final submission.*

TABLE 5.11
Challenges and Constraints' Rules

Rules	Challenges and Constraints

- Forces represent the "Why?"
- Describe some of the challenge that must be overcome by the pattern.
- Forces also represent the problem's constraints.
- Forces expose the particulars of the problem.
- Forces must be listed based on their priorities.
- Describe the constraints that are related to the pattern such as multiplicities, limits, and range.

TABLE 5.12
Heuristic Challenges and Constraints

- Pattern is a free form and a number of challenges or forces each of the patterns face, when you apply.
- List challenges by relating each BO or a number of BOs with as many IOs as you can. Examine the challenge every time when you increment the IOs, until such a time you cannot proceed further.
- It is important to understand and document the challenges to find out what can and cannot be done with the pattern.

The participants of any pattern can be classified into two main categories: classes and patterns. We usually present the participants as follows: *1. Classes.* List all the classes in the pattern and a short description for each. *2. Patterns.* List all the patterns in the main pattern and a short description for each. *Pattern structure rules and heuristics are presented in* Tables 5.13 and 5.14.

2. *CRC cards:* Summarizes the responsibility and collaboration of each participant (class). Each participant should have only one well-defined responsibility in its CRC card. Participants with more than one responsibility should be presented with more than one CRC card, when each CRC card will handle one of these responsibilities. Refer to CRC

TABLE 5.13
Pattern Structure Rules

Rules Pattern Structure

- Describe the constraints related to the pattern such as multiplicities, limits, and range.
- Describe some of the challenge that must be overcome by the pattern.
- Note: Not ALL IO and BO may have inheritance.
- Follow UML Class Diagram.
- Use tagged value as shown in Table 5.15.

TABLE 5.14
Pattern Structure Heuristics

Heuristic Pattern Structure

- EBTs or pattern-EBT's multiplicities is usually one or a fixed number of instances.
- Avoid many-to-many relationship.
- Use the software stability, model template as shown in Sidebar 5.4 for creating the pattern structure.
- Each **Class** in the main pattern structure should have a stereotype (EBT, BO, or IO).
- Each **Pattern** in the main pattern structure should have tag value.
  (**Pattern-EBT** = ***Stable Analysis Pattern***, or **Pattern-BO** = ***Stable Design Pattern***)
- EBTs usually do not have multiplicity around them (always considered to be 1, a fixed number, or a range if you are not sure).
- Between any two components in the pattern-class diagram, the multiplicity, roles, and relationship role should be added.
- Follow the modeling tip, while creating the pattern structure (Table 5.16).

TABLE 5.15
Tag Value used in the Pattern Structure

Tag Name	Applied to	Value Type	Description
<EBT>	Class		
<BO>	Class		
<IO>	Class		
<P-EBT>	Pattern		Represents stable analysis pattern
<P-BO>	Pattern		Represents stable design pattern

TABLE 5.16
Modeling Tip

Heuristic	Modeling Tip

Doing EBTs such as: reading, searching, and others are usually modeled as follows:

<<BO>> AnyParty — <<EBT>> Searching — <<BO>> SearchEngine

One common issue is to connect the ***AnyParty*** to the ***SearchEngine***, which is not true. Because of the relationship between ***AnyParty*** to the ***SearchEngine*** is only through the ***Searching.*** Therefore, if the ***AnyParty*** does not do ***Searching***, then at this moment there is no relationship between the ***AnyParty*** and the ***SearchEngine***.

TABLE 5.17
CRC Cards' Rules

Rules	CRC Cards

- A role must be specified per the card's class.
- Each CRC card should contain one and only one responsibility. This responsibility should be unique and it must be specified within context.
- The clients' names are listed under each other without using bullets or dashes.
- Server part should contain all the services provided by the current class/pattern listed without bullets. Each sever will be written as an operation starting with small letter and has two curve parenthesizes at the end.
- Classes with multiple roles, such as actors, must have a CRC card per role.

TABLE 5.18
CRC Cards' Heuristics

Heuristic	CRC-Cards

- CRC card filling process: (1) identify the CRC card's class name, (2) name its role, (3) name its responsibility under responsibility section, (4) list card class's services (operations) under server, and (5) identify clients in the client's section. The clients must have direct or indirect relationships with the card class.
- CRC cards should map correctly of the pattern structure.
- Create a CRC card for each EBT, BOs, P-EBT, and P-BO.
- Make sure to use tag values for EBTs and BOs.
- You must have two or more clients section of the collaboration.

card Layout (Sidebar 5.5). *CRC cards rules and heuristics are presented in* Tables 5.17 and 5.18.

3. *Behavior model* (whenever is possible): If the abstraction of the pattern prevents you from writing an appropriate behavior model, then you can flush the dynamics of the pattern later.

5.2.2.7 Consequences

How does the pattern (EBT or BO) support its objectives? What is the trade-offs and results of using the pattern? Nothing is perfect, and each problem has many issues that are usually hard, if not impossible, to be satisfied at once. This section lists the affect expected by using this pattern: good results (benefits), bad (drawbacks), ugly (which might cause major problems in specific situations), and the trade-offs (which shows what you gain, vs., what you lose from using this pattern).

5.2.2.8 Applicability

The applicability of five scenarios in a table as shown in below.

EBT	BOs	A1	A2	A3	A4	A5
EBT Name	BO1	IO11	IO21			
	BO2	IO12	IO22			
	BOn	IO1x	IO2y			

5.2.2.9 Applicability with Illustrated Examples

Offers clear and detailed twin case studies for applying the pattern in different contexts. The following sub-elements represent the required details in one case:

1. *Case studies:* Shows the scenario of two case studies from different contexts.
2. *Class diagram:* Presents the EBTs, BOs, and IOs.
3. *Use case template:* Gives detailed description for a complete use case. Include test cases for the EBT and all the BOs—abstraction of actors, roles, and classes, classes' type, such as EBT, BOs, IOs, attributes, and operations. Refer to Sidebar 5.6—use case template.
4. *Behavior diagram:* Map the above use case into a sequence diagram.
   Applicability rules and heuristics are presented in Tables 5.19 and 5.20.

5.2.2.10 Related Patterns and Measurability

This shows other patterns that usually interact with the described pattern, and those that are included within the described pattern. Related patterns can be classified as *related analysis* or/and *related design patterns.* Related patterns usually share common forces and rationale. In addition, it is possible that you might give some insights of other patterns that can or need be used with the proposed patterns; for example, in the case of AnyAccount pattern, we may point out to the AnyEntry pattern as a complementary pattern. There are rooms for contrasting and comparing the existing

TABLE 5.19
Applicability Rules

Rules Applicability

- Select relevant and significant case studies.
- Create class diagram for each of the case studies by using the stability model application template, which consists of EBT, BOs, and IOs.
- Create a significant use case with two or more actors.
- Map the use case to sequence diagram as a behavior model.

TABLE 5.20
Applicability Heuristics

Heuristic Applicability

- Mapping the pattern to the application objects or IOs. Each BO is connected to one or more IOs, except that AnyParty and/or AnyActor are connected to two or more roles (IOs).
- Provide test cases within each step of the use case.

patterns with the documented pattern. This section also gives a few metrics for measuring several things related to the pattern structure, such as complexity and size, cyclomatic complexity, lack of cohesion, coupling between object classes, etc. This section is divided into two parts

1. *Related patterns: rules and evaluations* are presented in Tables 5.21 and 5.22.
2. *Measurability: rules and evaluations* are presented in Tables 5.24 and 5.25.

TABLE 5.21
Related Patterns' Rules

Rules	Related Patterns

- Two different approaches:
- Search for an existing traditional pattern on the same topic. Compare with traditional existing pattern's model with reference to ours.
- If existing patterns do not exist, select a single definition of the name of our pattern, develop a traditional model class diagram and describe it briefly.

TABLE 5.22
Related Patterns' Evaluation

Evaluation	Related Patterns

- Use the model essential properties (Sidebar 5.7) and the template showing in Table 5.23. Compare this pattern with its counterparts or tradition model with similar context.
- Use 5 to 6 adequacies (Sidebar 5.8) and the template shown in Table 5.23 to compare both the pattern with its similar traditional pattern model.

TABLE 5.23
Comparison Template

Comparison		Evaluation Template				
#	Adequacy/Property Name	W.	Traditional Pattern	W1	Stable Pattern	W2
1						
2						
3						
4						
5						
6						
Result		100				

TABLE 5.24
Measurability Rules

Rules	Measurability

- Measurability will compare our pattern with other models on issues like number of behaviors and number of classes. Justification of why the numbers of behavior or classes are so high or low.
- You may compare and comments on other quality factors, such as reuse, extensibility, integration, scalability, applicability, etc.

TABLE 5.25
Measurability Evaluation

Evaluation	Measurability

- Two approaches: Compare the traditional with stability models in two of the following approaches:

(a) Quantitative measurability like
+ Number of behaviors or operations per class
+ Number of attributes per class
+ Number of associations
+ Number of inheritance
+ Number of aggregations
+ Number of interactions per class
+ Number of EBTs versus number of requirements classes in TM
+ Number of classes
+ Documentation—number of pages
+ Number of IOs
+ Number of applications
+ Estimation metrics
+ Measurement metrics
+ Others

(b) Qualitative measurability
-- Scalability
-- Maintainability
-- Documentation
-- Expressiveness
-- Adaptability
-- Configurability
-- Reuse
-- Extensibility
-- Arrangement & re-arrangement
-- Others

5.2.2.11 Modeling Issues, Criteria, and Constraints

One may need to address many modeling issues, criteria, and constraints that you need to address, to explain them by making sure that the model satisfies all the modeling criteria and constraints. Table 5.26 discusses modeling issues and rules.

5.2.2.12 Design and Implementation Issues

For each EBT, discuss and explain the important issues that are required for linking the analysis phase to the design phase, and for each BO, discuss the important issues required for linking the design phase to the implementation phase: for example, hooks. Describe all design issues (EBT): for example, hooking issues. Alternatively, discuss the implementation issues (BO): for example, why we are using relationship rather than inheritance, hooking, hot spots problems. Show segments of code here. Table 5.27 lists design and implementation issues and rules.

5.2.2.13 Testability

Describes the test cases, test scenarios, testing patterns, etc. (this is a very important point, but sometimes it is very hard to write them for an isolated pattern, we are not sure the best way to write this part!) You can use three ways to document testability: (1) test procedures and test cases within classes' members of the patterns, (2) propose testing patterns that are useful for this pattern and other existing patterns, and (3) check if the pattern will fit with as many

TABLE 5.26
Modeling Rules

Rules Modeling

Modeling issues are

- *Abstraction:* describe the abstraction process of this pattern, list, and discuss briefly the abstractions within this pattern.
  + Show the abstractions that are required for the patterns (EBT, BOs, and IOs)
  + Elaborate on the abstraction of why EBTs and BOs are selected?
  + Show examples of unselected EBTs, and Why?
  + Show examples of unselected BOs, and Why?
- *Static models:* Illustrate and describe one or two of the static models of this pattern, and list and discuss briefly the complete story of the pattern's model by using actual objects.
  + Determine the sample model that you are planning to use: CRC cards, class diagram, component diagram, etc.
  – Show the model
  + Highlight the entire story of the pattern's model by using objects
  + Repeat it with other objects
- *Dynamic models:* Illustrate and describe one or two of the dynamic models of this pattern, and list and discuss briefly the behavior of the pattern through selected dynamic models.
  + Determine the sample model that you are planning to use: interaction diagram or state transition diagram—show the model
- *Modeling essentials:* Examine the pattern by using the modeling essentials, and list and discuss briefly the outcome of this examination.
  + List or reference the model essentials, and use them as criteria to examine the pattern
  + Elaborate on methods to examine the model of the pattern by using the model essential criteria
  + Briefly describe the outcome
- *Concurrent development:* Show the role of the concurrent development of developing this pattern. Describe.
  + Describe and exhibit with illustration the concurrent development of this pattern
- *Modeling heuristics:* Examine the pattern by using the modeling heuristics, and list and discuss briefly the outcome of this examination.
  + List or index the modeling heuristics, and use them as criteria to examine the pattern
  + Explain methods to examine the model of the pattern by using the modeling heuristics
  + Briefly describe the outcome
  + Modeling heuristics: such as
    + No dangling
    + No star
    + No tree
    + No sequence
    + General enough to be reused in different applications
    + Others

scenarios as possible, without changing the core design. Table 5.28 shows testability issues and rules.

5.2.2.14 Formalization using Z++, Object Z, or Object-Constraints Language (OCL) (Optional)

Describes the pattern structure by using the formal language (Z++ or Object Z), BNF, EBNF, and/or XML. Table 5.29 lists formalization rules.

5.2.2.15 Business Issues

Cover one or more of the following issues.

- *Business rules:* Describe and document business rules, and how you can extend them in the mentioned context and scenarios that are listed.

TABLE 5.27
Design and Implementation Issues

Rules Design and Implementation

Here is a list of analysis issues

- Divide and conquer
- Understanding
- Simplicity
- One unique base that suitable to many applications
- Goals
- Fitting with business modeling
- Requirements specifications models
- Packaging
- Components
- Type (TOP) (A)
- Actors/roles
- Responsibility and collaborations
- Generic and reusable models
- Others

Design issues (EBT)

- For example hooking issues
- Implementation issues (BO)
- For example, why using aggregation or delegation rather than inheritance
- For example, hooking, hot spots problems
- Can show code here

Here is a sample list of design and implementation issues

- Framework models (D)
- Classes (TOP) (D)
- Collaborations (D)
- Refinement (D)
- Generic and reusable designs (D)
- Precision (I)
- Hooks (I)
- Pluggable parts (I)
- Navigation (I)
- Object identity (I)
- Object state (I)
- Associations/aggregations (I)
- Collections (I)
- Static invariants (I)
- Boolean operators (I)
- Collection operators (I)
- Dictionary (D) (I)
- Behavior models (D) (I)
- Pre-post-conditions specify actions (I)
- Joint actions (use Cases) (D)
- Localized actions (I)
- Action parameters (I)
- Actions and effects (I)
- Concurrent actions (I)
- Collaborations (I)
- Interaction diagrams (D)

(Continued)

TABLE 5.27 (*Continued*)
Design and Implementation Issues

Rules Design and Implementation

- Sequence diagrams with actions (D) (I)
- Pattern 1: Continuity
- Pattern 2: Performance
- Pattern 3: Reuse
- Pattern 4: Flexibility
- Pattern 5: Orthogonal abstractions
- Pattern 6: Refinement
- Pattern 7: Deliverables
- Pattern 8: Recursive refinement
- Package (D) (I)

Here is a list of Java patterns

- Fundamental design patterns
  1. Delegation (when not to use inheritance)
  2. Proxy
- Creational patterns
  1. Abstract factory
  2. Builder
  3. Factory method
  4. Object pool
  5. Prototype
  6. Singleton
- Partitioning patterns
  1. Composite
  2. Filter
  3. Layered initialization
- Structural patterns
  1. Adaptor
  2. Bridge
  3. Cache management
  4. Decorator
  5. Dynamic linkage
  6. Façade
  7. Flyweight
  8. Iterator
  9. Virtual proxy
- Behavioral patterns
  1. Chain of responsibility
  2. Command
  3. Little language/interpreter
  4. Mediator
  5. Null object
  6. Observer
  7. Snapshot
  8. State
  9. Strategy
  10. Template method
  11. Visitor

TABLE 5.28
Design and Implementation Issues

Rules Testability

- Try to find scenarios within the context that cannot work with this pattern.
- Show how to test the requirements and the design artifacts within use cases.
- Can also use exhaustive testing of behaviors (may require more pages) by using testing patterns.

TABLE 5.29
Formalization Rules

Rules Formalization

- Describes the pattern structure by using a formal language (OCL, Z++, Object Z, or other).
- Generates a formal code of any of compliable formal language.
- Generates a grammatical template or syntax by using EBNF to be used by advance programmers to link this pattern and program it with their own codes.

- Define the business rules, business policies, business facts, in relation to the pattern.
- Illustrate the business rules that derived from the pattern.

Check the following links:

- http://en.wikipedia.org/wiki/Business_rules
- http://www.businessrulesgroup.org/bra.shtml
- http://www.businessrulesgroup.org/first_paper/br01c0.htm—pdf format file.
- http://www.businessrulesgroup.org/brmanifesto.htm

Define all business rules that relate to the pattern.
Illustrate all business rules that are derived from the pattern.

- *Business models:* Issues
- Business model design and innovation
- Business models' samples http://en.wikipedia.org/wiki/Business_model:
  - Subscription business model
  - Razor and blades business model (bait and hook)
  - Pyramid scheme business model
  - Multi-level marketing business model
  - Network effects business model
  - Monopolistic business model
  - Cutting out the middleman model
  - Auction business model
  - Online auction business model
  - Bricks and clicks business model
  - Loyalty business models
  - Collective business models
  - Industrialization of services business model
  - Servitization of products business model

    - Low-cost carrier business model
    - Online content business model
    - Premium business model
    - Direct sales model
    - Professional open-source model
    - Various distribution business models
- Describe the pattern. If it is part of or whether it is a business model
- Describe the direct impact of the pattern on the business model or
- Describe the indirect impacts of the pattern on the business model

Describe the same for the following business issues:

- **Business standards**
    - Vertical standards versus horizontal standards
- **Business integration**
    - Data integration
    - People integration
    - Tools integration
- **Business processes or workflow**: Here are some of the business processes issues:
    - Business process management (BPM) is a systematic approach to improving those processes
    - Business process modeling and design
    - Business process improvement
    - Continuous business process improvement
    - Business process categories: management processes, operational processes, and supporting processes
    - Business process ROI
    - Business process rules
    - Business process mapping
- **e-Business**
    - **e-Commerce**
    - **e-Business models** http://en.wikipedia.org/wiki/E-Business
        - E-shops
        - E-commerce
        - E-procurement
        - E-malls
        - E-auctions
        - Virtual communities
        - Collaboration platforms
        - Third-party marketplaces
        - Value-chain integrators
        - Value-chain service providers
        - Information brokerage
        - Telecommunication
    - **E-Business categories** http://en.wikipedia.org/wiki/E-Business
        - business-to-business (B2B)
        - business-to-consumer (B2C)
        - business-to-employee (B2E)
        - business-to-government (B2G)
        - government-to-business (G2B)
        - government-to-government (G2G)

- government-to-citizen (G2C)
- consumer-to-consumer (C2C)
- consumer-to-business (C2B)

- **Web applications**
- **Business patterns**
  - Business modeling with UML
  - Business knowledge map
- **Business strategies**
  - Business strategy modeling
  - Business strategy frameworks
  - Strategic management
  - Strategic analysis
  - Strategy implementation
  - Strategy global business
- **Business performance management**
  - Methodologies
  - Business performance management framework
  - Business performance management knowledge map
  - Assessment and indication
- **Business transformation**
- **Enduring business themes**
- **Security and privacy**

5.2.2.16 Known Usage

Give examples of the use of the pattern within existing systems or examples of known applications that may benefit from the proposed pattern. Mention some projects that use it.

5.2.2.17 Tips and Heuristics

List and briefly describe all the lessons learnt, tips, and heuristics from the utilization of this pattern, if any. Table 5.30 lists tips on how to find heuristics.

5.3 SUMMARY

This chapter is documentary in nature and it communicates a series of easily understandable materials that describe, instruct, and explain the rationale behind the stable analysis patterns and their design and implementation. It also highlights various methods, processes, and procedures to create various components of a stable analysis pattern. Rules and heuristics are highlighted throughout the chapter, while information on documentation templates provides a pattern developer necessary tools and resources required to create a new pattern based on stability principles. In fact, creating a stable pattern is a challenging exercise, and a developer needs necessary expertise and knowledge that is

TABLE 5.30
Tips and Heuristics

Rules	Tips and Heuristics
	• What did you discover? • Why did you include or exclude different classes? • Are there any tips on usage such as scaling, adaptability, flexibility?

usually required to uncover accidental problems and pitfalls that might crop up during designing stage.

REVIEW QUESTIONS

1. List three different types of templates for documenting stable patterns.
2. List some problems that are associated with current templates and provide brief explanation about each one of them.
3. What are the major differences that exist between the existing documentation templates and Fayad's stable pattern documentation templates?
4. What is pattern name and provide details on it?
5. Mention some of the pattern name rules.
6. What are "Known as" rules?
7. What is Context?
8. Mention some rules that are associated with it.
9. What is a "Problem"?
10. What are functional requirements?
11. Mention some functional requirements rules.
12. What are nonfunctional requirements?
13. List some nonfunctional requirement rules.
14. List some common challenges and constraints that you come across while designing a stable pattern.
15. What is pattern structure and who are the participants in it?
16. What are CRC cards?
17. What is a behavior model?
18. Why is applicability factor very important while designing a stable pattern?
19. What are measurability rules?
20. List some modeling rules.
21. List some design and implementation rules.

EXERCISES

1. Create a complete stable pattern by using the tips and heuristics provided in this chapter.
2. Write down some common case studies to highlight the importance of stable analysis patterns.
3. List all constraints, problems, benefits, and advantages of creating a stable pattern by using pattern structure and CRC cards.

Write down CRC cards for the patterns that you want to create.

PROJECT

1. Use the detailed template to document the following patterns:
   - Passion
   - Violence
   - Intelligence
2. Use the mid-size template to document the following patterns:
   - Equality
   - Liberty
   - Security
3. Use the short template to document the following patterns:
   - Slavery
   - Torture
   - Freedom

SIDEBARS

SIDEBAR 5.1 CONTEXT/SCENARIO TEMPLATE

Scenario ID: Give each use scenario a unique numeric identifier.
Scenario title (context name/domain): State a concise and descriptive name for the scenario. This reflects the intent of the scenario and the user needs to be able to accomplish through the scenario.
Scenario elements: Provide a list of BOs and corresponding IOs)
Scenario description: Provides a brief description of a complete scenario.
[Hint: describe scenario with mapping each BO to IOs]
Rules:

1. Map each BO to 1..* IOs except actor and/or party.
2. Map each actor and/or party to 2..* Roles (IOs).

SIDEBAR 5.2 NONFUNCTIONAL REQUIREMENTS

Nonfunctional requirements represent the quality factors of an EBT or a BO, or the systems of patterns. Other terms for nonfunctional requirements are "qualities," "quality goals," "quality of service requirements," "constraints" and "nonbehavioral requirements" [1] Informally, these are sometimes called the "ilities," from attributes like stability and portability.

Nonfunctional requirements have five properties [2]:

1. Correspond to quality factors or quality criteria.
2. Positive criteria within context (always).
3. Most likely enduring or goals.
4. Control the behavior of the EBT, or BO, or system of patterns.
5. Considered to be the essential properties of an EBT or a BO.

Functional requirements define what a system is supposed to *do* and nonfunctional requirements define how a system is supposed to *be* [1–3].

References

1. A. Stellman, and J. Greene, *Applied Software Project Management*, Sebastopol, CA: O'Reilly Media, 2005, p. 113, ISBN 978-0-596-00948-9.
2. M.E. Fayad, H. A. Sanchez, S.G.K. Hegde, A. Basia, and A. Vakil, *Software Patterns, Knowledge Maps, and Domain Analysis*, Boca Raton, FL: Auerbach Publications, 2014, 422 p.
3. K.E. Wiegers, *Software Requirements*, 2nd ed., Redmond, WA: Microsoft Press, 2003, ISBN 0-7356-1879-8.

SIDEBAR 5.3 CHALLENGES TEMPLATE

Challenge ID	Give each of the challenges a unique numeric identifier in a simple form: 01, 02, 03, and so on.
Challenge Title	State a concise description of the challenge. These reflect the tasks the user needs to be able to accomplish using the system.
Context	Rules should be predefined to avoid confusions.
Description	In competition say soccer game, there are AnyRule() such as scoring mechanisms, are not predefined by (game organization) AnyParty(). Hence, players of the team (AnyParty()) are confused as what to use as AnyMechanism() to win the game. It leads to confusion in the AnyParty() [teams] involved in the game.
Solution	Rules and scoring mechanisms should be properly defined and communicated to AnyParty() to avoid chaos.

SIDEBAR 5.4 FAYAD'S SOFTWARE STABLE MODEL TEMPLATES

Fayad's Stable Analysis/Design/Architectural Patterns (Layout)

Stable Analysis Pattern or Stable Design Pattern Layout (One EBT and 2-14 BOs)

EBT	BOs

Stable Architectural Patterns Layout (2-5 EBTs where 3 is the most common)

EBTs	BOs
EBT1	
EBT2	
....	

Fayad's Stable Analysis/Design Pattern Applications Layout

Stable Analysis Pattern or Stable Design Pattern Applications Layout

EBT	BOs	IOs

Stable Architectural Pattern Applications Layout

EBTs	BOs	IOs
EBT1		
EBT2		
....		

SIDEBAR 5.5 FAYAD'S CRC CARD

CRC cards are index cards that are utilized for mapping candidate's classes in predefined design scenarios; for example, use case scenarios. The objectives of CRC cards are to facilitate the design process, while insuring an active participation of involved designers.

Fayad's CRC card version includes a clear role for each class. This class role will also be useful when defining the class responsibility. Each class will be allowed to have only one unique responsibility. If more than one responsibility is identified, additional classes should be formed. Limiting responsibilities will help prevent low cohesion and high coupling as well as reduce the possibility of macho classes. Finally, the revised CRC card will include the services offered by each class. This will help verify the validity of the class responsibility as well as ensure that overlapping functionality is avoided [1,2]. The Fayad's CRC card format is shown in Table SB5.5.1.

TABLE SB5.5.1
Fayad's CRC-Card Format

Class/Pattern Name (Class/Pattern Role) (Class Type)		
	Collaboration	
Responsibility	Clients	Server
• A single responsibility for this class/pattern should be listed here briefly • Unique • Within context	• A list of all the classes/patterns that have a relationship with the named class/pattern • 2 or more clients	• A list of all the servers that named class/pattern provides • 5–12 operations (services)
Attributes: Class/Pattern 7 or more attributes		

References

1. M.E. Fayad, H.S. Hamza, and H.A. Sánchez, A pattern for an effective class responsibility collaborator (CRC) cards, The 2003 IEEE International Conference on Information Reuse and Integration, Las Vegas, NV, October 2003.
2. M.E. Fayad, H.A. Sánchez, and H.S. Hamza. A pattern language for CRC cards, Proceedings of Pattern Language of Programs' 2004 (PLOP'04), Monticello, IL, September 2004.

SIDEBAR 5.6 FAYAD'S USE CASE TEMPLATE

Gathering Requirements with Use Cases

The utilization of scenarios (or use cases) to understand and comprehend requirements is not an innovative idea; however, these scenarios were treated quite informally until Jacobson's work raised the notion of use case [1]. To date, a use case represents a typical interaction between users and a software system. Recently, use cases have been included in the UML-based unified process methodology [UML 2–4], as a way to capture the high-level user functional requirements of a system. A simple use case recipe can be defined using the following steps [5]:

Step 1. Identify who will use the system directly and treat them as actors.
Step 2. Pick one of those actors.
Step 3. Define what that actor wants to do with the system. Each of the things that the actor wants to do with the system will become the use case.
Step 4. For each of those use cases, find out the common course of action when that actor is using the system. What normally happens? This will be the basic course.
Step 5. Describe that basic course in the description for the use case.
Step 6. Once you have found out the basic course, consider possible alternates and integrate them as extending use cases.
Step 7. Review each use case descriptions against the descriptions of the other use cases. Is there any glaring commonality? Soon, extract those out.
Step 8. Repeat steps 2–7 for each actor.

Although this naive procedure looks straightforward to achieve, a developer involved in middle- or large-scale projects will know that the picture is more complicated. At the beginning, one can capture the use case set by interacting with your typical users and by discussing various things, they might want from their systems. Besides that, there are other issues to consider too:

- Are we sure that we have captured all of the user's viewpoints about the system?
- What if some of the actors do not really understand the true business needs?
- How can we avoid an explosion of use cases?
- What is the smoothest way to accomplish the transition between use cases based on analysis and system design?

We believe that the technique proposed (and further applied in many projects, such as the Philips New York Project, and NASA QRAS (quantitative risk analysis system)) offers some primary guidance in a sense, and provides a well-defined set of practice to alleviate development efforts during requirement analysis. Our use case template is a simple use case [5] when compared to the following use cases by many object-oriented proponents, like [6–8].

Fayad's Use Case Template [5]

The first step would be the development of use cases. According to [9], good software engineering practice is not use-case driven. However, use cases do provide one of the easiest methods to start developing system requirements. The use case model is about describing what our system will do at a higher level and with a user focus for scoping the project, and giving the application some focused structure. Use cases are not a functional decomposition model, and they do not require how a system will accomplish its functionality. In this project, the use cases were developed by using a group approach. Our use case template delineates the different parts of the use case, thereby making it easier to see different parts of each use case (see Figure SB5.6.1).

<table>
<tr><td colspan="2">Use Case Id</td><td colspan="2">Give each use case a unique numeric identifier in a hierarchical form: X.Y. Related use cases can be grouped in the hierarchy. Functional requirements can be traced back to a labeled use case</td></tr>
<tr><td colspan="2">Use Case Title</td><td colspan="2">State a concise and descriptive use case name. These reflect the tasks the user needs to be able to accomplish by using the system. Insert a use case name that starts with a verb</td></tr>
<tr><td colspan="2">AnyActor and/or Any/Party</td><td colspan="2">Roles</td></tr>
<tr><td colspan="2">List AnyActor's or AnyParty's Types* for traditional model (OR)
List AnyActor or AnyParty for stability model</td><td colspan="2">Insert corresponding roles</td></tr>
<tr><td colspan="4">• AnyActor has four different types: human, hardware, software, and creatures such as animals, trees, and animated characters and,
• AnyParty has four types as well: human, organizations, countries, and political parties.</td></tr>
<tr><td>Class Name</td><td>Type*</td><td>Attributes</td><td>Operations</td></tr>
<tr><td></td><td></td><td></td><td></td></tr>
<tr><td></td><td></td><td></td><td></td></tr>
<tr><td></td><td></td><td></td><td></td></tr>
<tr><td></td><td></td><td></td><td></td></tr>
<tr><td colspan="4">• Each class is classified as an actor, role, or a system class, if you are using traditional model
• Each class is classified as an EBT, BO, or IO, if you are using stability model.

Use Case Description:</td></tr>
<tr><td colspan="4">1. Describes the first step in the scenario
2. Describes the second step in the scenario.
3. Other steps
• The use case should have at least five or more steps, and it should be written with the factor of stability in mind, with test cases of EBTs and BOs names, attributes, and operations only.
• Use case description must be numbered.
• Each use case may contain 6–12+ steps.</td></tr>
<tr><td colspan="4">Alternatives:</td></tr>
<tr><td colspan="4">2. (As an example) Insert an alternative scenario for step 2 in the original sequence.
Repeat the alternative, as many time as required, for any of the use cases steps</td></tr>
</table>

FIGURE SB5.6.1 Fayad's use case template.

References

1. I. Jacobson, M. Christerson, P. Jonsson, and G. Övergaard, *Object-Oriented Software Engineering—A Use Case Driven Approach*, Reading, MA: Addison-Wesley, 1992.
2. M.J. Chonoles, and J.A. Schardt. *UML 2 for Dummies*, New York, NY: Wiley Publishing, ISBN 0-7645-2614-6.
3. I. Jacobson, B. Grady, and J. Rumbaugh *The Unified Software Development Process*, Reading, MA: Addison-Wesley Longman, ISBN 0-201-57169-2.
4. M. Penker, and E. Hans-Erik, *Business Modeling with UML*, New York, NY: John Wiley & Sons, 2000, ISBN 0-471-29551-5.
5. M.E. Fayad, H.A. Sanchez, S.G.K. Hegde, A. Basia, and A. Vakil, *Software Patterns, Knowledge Maps, and Domain Analysis*, Boca Raton, FL: Auerbach Publications, December 2014.
6. A. Cockburn, *Writing Effective Use Cases*, Reading, MA: Addison-Wesley, 2001.
7. M. Fowler, *UML Distilled*, 3rd ed., Reading, MA: Addison-Wesley, 2004.
8. I. Jacobsonr, I. Spence, and K. Bittner, *Use Case 2.0: The Guide to Succeeding with Use Cases*, SA: Ivar Jacobson International, 2011.
9. T. Korson, Constructing useful use cases, *Component Strategies (formerly Object Magazine) Column*, March 1999.

SIDEBAR 5.7 MODEL ESSENTIALS PROPERTIES [1]

The model is an abstraction of reality and lets you see the relationship between different parts and the whole, such as text, diagram, pictures, frames, images, etc. Modeling is a well-established human activity. All models are descriptions of something (i.e., a representation that is not the real thing) that allow us to answer questions about the real thing, which capture only those features deemed essential by the modeler, and that can be validated by experimenting with physical things or by quizzing experts. A single thing can be represented by a large number of different models. There are two types of modeling: intangible modeling that includes static and dynamic models (e.g., logical models, behavior models, and object models) and tangible modeling (e.g., physical models).

The logical model, that is the intangible model, represents the key abstractions and mechanisms that define the system's architecture. The logical model also describes the system behavior and defines the roles and responsibilities of the objects that carry out the system behavior. Logical modeling frees analysts from the consideration of limitations inherent in real-world mechanisms. UML uses object diagrams and class diagrams to illustrate the key static abstractions in systems and sequence diagrams, interaction diagrams, activity diagram, and state transition diagrams to model the behavior of the system [2–4].

The physical model of the system, that is the tangible model, uses symbols to represent things that must be physically present. In the software sense, the physical model is used to describe either the system's context or implementation. System context diagrams are physical model of the system [5,6]. There are six essential properties to any good model, where they can be used to examine different types of models.

1. *Simple*—This property covers those attributes of the object-oriented model that present modeling aspects of the problem domain in the most understandable manner. This property measures the model and technique complexity in terms of number of process steps, notational aspects, constraints, and modeling and design rules.
2. *Complete (most likely to be correct)*—This property ensures that model artifacts are free of conflicting information and all the required information is present.

For example, component names within the model should be uniform and incomplete sections of the model should not exist. This property determines whether the object-oriented model provides internal consistency and completeness of the model's artifacts. The model must be able to convey the essential concepts of its properties.

3. *Stable to technological change*—Unfortunately, object-oriented models are fuzzy, due to the absence of quantitative heuristics, and most OO models are built upon false assumptions.
4. *Testable*—To be testable, the model must be specific, unambiguous, and quantitative wherever possible. For our purposes, we define simulation as an imitation of the actual model. This definition leads us to validate the characteristics of the model against the user's requirements.
5. *Easy to understand*—In addition to the familiarity of the modeling notations, the notational aspects, design constraints, and analysis and design rules of the model should be simple and easy to understand by the customers, users, and domain experts.
6. *Visual or graphical*—A picture worth a thousand words. As a user, you can visualize and describe the model. The graphical model is essential for visualization and simulation.

References

1. Fayad, M.E., and M. Laitinen, *Transition to Object-Oriented Software Development*, New York, NY: John Wiley & Sons, August 1998, pp. 10–11, ISBN 0-471-24529-1.
2. M.J. Chonoles, and J. A. Schardt, *UML 2 for Dummies*, New York, NY: Wiley Publishing, 2003. ISBN 0-7645-2614-6.
3. I. Jacobson, B. Grady, and J. Rumbaugh. *The Unified Software Development Process*, Boston, MA: Addison Wesley Longman, 1999, ISBN 0-201-57169-2.
4. M. Penker, and E. Hans-Erik, *Business Modeling with UML*, New York, NY: John Wiley & Sons, 2000, ISBN 0-471-29551-5.
5. A. Kossiakoff, and W.N. Sweet *Systems Engineering: Principles and Practices*, Hoboken, NJ: John Wiley & Sons, 2011, p. 266.
6. S. Robertson, and J.C. Robertson, *Mastering the Requirements Process*, London, England; Pearson Education, 17 mrt. 2006.

SIDEBAR 5.8 MODEL ADEQUACIES

Model adequacies examine whether the model is sufficient to satisfy a requirement or meet a need. It will also examine whether the quality of the model is adequate or not. Seventeen adequacies examine the quality of a good model:

1. Descriptive adequacy

   Descriptive adequacy refers to the ability to visualize objects in the models. Every defined object should be browse-able, allowing the user to view the structure of an object and its state at a particular point in time. This requires understanding and extracting metadata about objects that will be used to build a visual model of objects and their configurations. This visual model is domain-dependent—that is, based on domain data and objects' metadata. Descriptive adequacy requires that all of the knowledge representation is visual, as follows:

   - Visual models are structured to reflect natural structure of objects and their configurations.

- All the visual knowledge (data and operations) in the visual model is localized.
- Relationships among objects in the visual model are well defined.
- Interactions among objects in the visual model are limited and concise.
- The visual model must transcend objects, and instead highlight crosscutting aspects.

2. Logical adequacy
   Logical adequacy refers to the logical representation of the model that describes the model components' behavior, users, roles, and responsibilities. Logical adequacy indicates whether the model makes any sense or not.
3. Synthesis adequacy
   Synthesis adequacy refers to an integrated problem resolution methodology, or built-in trouble-shooting tools. Built-in trouble-shooting tools are very important in managing complex distributed systems, to avoid or reduce potential points of failure.
4. Analysis adequacy
   Analysis adequacy refers to integrated validation and verification tools. With built-in validation and verification, the process of maintenance and regression testing can be streamlined, and the cost of validation and verification minimized. This adequacy also examines any model, if it meets the needs or the state of quality analytically.
5. Blueprint adequacy
   Blueprint adequacy refers to various modeling features that provide integrated system specifications. Integrated system specifications are very important because they facilitate and ensure the extensibility of the system. An integrated blueprint for an enterprise framework should clearly identify both hot spots and frozen spots in the model. The adequacy examines the state of design of the model.
6. Epistemological adequacy
   Epistemological adequacy refers to tools for representing objects in the real world. There are two ways to view the world based on simplicity: (1) a perfect, but simple view—the world is represented in this view as an ideal environment and (2) as-is, but complex and detailed view—the world is represented as an ultimate reality. Additionally, there are two more ways to view an organization: (1) flat and single view and (2) layered and multiple views. It is obvious that most of modeling techniques, such as UML [1,2,3] model the world as an ideal environment and flat or single view of itself. Nevertheless, successful enterprise frameworks have made great leaps in representing objects in the real world and in providing the necessary tools to alter these objects as required by the business.
7. Notational adequacy
   Notational adequacy refers to the presentation constructs the impact the presentation tools have on the operation of the system.
8. Procedural adequacy
   Procedural adequacy refers to recognition, and search capabilities.
9. Contractual adequacy
   Contractual adequacy refers to the client tools for representing the system behavior.
10. Scalable adequacy
    Scalable adequacy relates to the constructs and tools supporting partitioning, composition, security, and access control. Can your models be scalable in all directions?
11. Administrative adequacy
    Administrative adequacy refers to tools for modeling the deployed system's performance, reliability and administrative characteristics, and to the actual tools for administering the system. Administrative adequacy also considers the availability of install set builders, start and stop procedures or scripts, integrated database management capabilities, archiving, fail-over mechanisms, etc.

12. Understanding adequacy
    Understanding adequacy relates to, be easy to understand.
13. Simplicity adequacy
    Simplicity adequacy relates to how simple your models will be.
14. Extensibility adequacy
    Extensibility adequacy relates to the degree of extensibility, adaptability, customizability, and configurability of any models.
15. Systematic adequacy
    Systematic adequacy relates to the systematic approaches that are utilized in modeling, such as bottom-up, top-down, and middle-out approaches. It also includes functional decomposition techniques, and the selection of the correct level of abstraction at each stage of analysis.
16. Behavioral adequacy
    Behavioral adequacy relates to the behavioral models that concentrate on behavioral aspects of the system that you model. It also includes accuracy of behavioral models, such as scenarios, activity diagrams, sequence diagrams, interaction diagrams, state transition diagrams, etc.
17. Analytical thinking adequacy
    Analytical thinking adequacy relates to the analytical thinking approaches and tools that concentrate on analytical aspects of the system. It also includes the utilization of analysis patterns.

References

1. M.S. Chonoles, and J.A. Schardt, *UML 2 for Dummies*, New York, NY: Wiley Publishing, 2003, ISBN 0-7645-2614-6.
2. I. Jacobson, B. Grady Booch, and J. Rumbaugh, *The Unified Software Development Process*. Boston, MA: Addison Wesley Longman, 1999, ISBN 0-201-57169-2.
3. M. Penker, and H.-E. Eriksson, *Business Modeling with UML*, New York, NY: John Wiley & Sons, 2000, ISBN 0-471-29551-5.

REFERENCES

1. W.J. Brown, R.C. Malveau, H.W. McCormick, and T.J. Mowbray, *AntiPatterns: Refactoring Software, Architectures, and Projects in Crisis*, New York, NY: John Wiley & Sons, Inc., 1998.
2. C. Alexander, *A Pattern Language: Towns, Buildings, Construction*, New York, NY: Oxford University Press, 1977, p. 1216.
3. C. Alexander, *The Timeless Way of Building*. New York, NY: Oxford University Press, 1979.
4. G. Erich, H. Richard, J. Ralph, and V. John, *Design Patterns: Elements of Reusable Object-Oriented Software*, Boston, MA: Addison Wesley, 1994.

Part II

SAPs

Detailed Documentation Templates

Part II contains five chapters listed below:

Chapter 6 titled "Competition Stable Analysis Pattern"
Chapter 7 titled "Corruption Stable Analysis Pattern"
Chapter 8 titled "Dignity Stable Analysis Pattern"
Chapter 9 titled "Trust Stable Analysis Pattern"
Chapter 10 titled "Accessibility Stable Analysis Pattern"

6 Competition Stable Analysis Pattern

Competition is the heart of engineering innovations.

M.E. Fayad

The English word competition is a widely used term. It relates to business domain apart from its diverse use in biology and evolution. In a widely used context, competition means rivalry between two entities that could be both plants and animals. Any competition is known to test the factor of winning ability of a person or group of persons. Competition is a common phenomenon that is witnessed in our daily lives. Everyone competes for something whether it is food, air, or living space. In essence, competition could be good or bad, fair or unfair, fierce or stale, or legal or illegal. In nutshell, competition is a vital tool to develop one's ability to live in a world that is full of competition. As Michael Jordan explained, "you have competition every day because you set such high standards for yourself that you have to go out every day and live up to that." In other words, competition brings out the best in a person or an entity [x].

In the domain of software stability model [1–4], Knowledge Maps [5], and Stable analysis patterns [6], competition is perhaps the most important variable that has a wider connotation and application. While traditional methods of creating patterns can work only on one application at a time, their usage is severely restricted and limited. On the contrary, software stability patterns are widely reusable, extendable, and practically applicable to any number of applications and scenarios. This chapter provides a deep insight on how the word *competition* could be used to design and create a wide array of stable patterns that can withstand the test of time and differing application scenarios.

6.1 INTRODUCTION

Competition is a phenomenon, which represents a sense of feeling or striving of gaining an upper hand over the rival in issues, which are of interest to both the parties and it involves a feeling of compulsion to prove one's self or to prove one's superiority.

The good part of competition is its ability to bring out the best in any striving party, as it tries its best and attempts to innovate new methods or techniques to beat its rival in a level-playing field. For example, playing chess with a superior or higher ranking player will help the player to devise new strategies or sharpen playing skills to get an upper hand.

The bad face of competition can be the loss of morality or ethics in order to beat a competitor. History is replete with examples, where a party or an individual has acquired illegal, immoral, or unethical means to beat their opponents in a given scenario. An example can be breach of a contract signed between two parties prior to taking part in competition. A very famous example can be given by the case of Mike Tyson biting off a part of ear of Lennox Lewis, his competitor, in the World Heavyweight Boxing Championship match. Such an attack is prohibited by the rules of boxing and can prove the bad effect of competition, where pressure to prove one's superiority makes a party or an individual to stoop down to high levels of immorality, which is completely uncalled and undesired for.

The ugliest face of competition could be a bloody war or any similar incidents, which lead to the huge loss of human lives. In fact, it is the most undesired form of competition and it invariably leads to numerous bad effects and debilitating shocks. The world has already witnessed the untold and retold horrors of World War I and II, and what can be worse than them? The historical facts available with us justify that Hitler persisted with the Battle of Stalingrad initiative for years, despite numerous warning that the German troops were suffering badly in a faraway land that was icy and inhospitable. It was clearly an ego clash between Stalin and Hitler, and neither of them wanted to lose by retreating or surrendering. In nutshell, a sense of competition that hinges on flashes of ego, false sense of superiority, and badly nourished ambition is not good either for an individual or for the entire world.

6.2 COMPETITION ANALYSIS PATTERN DOCUMENT

6.2.1 Pattern Name: Competition Stable Analysis Pattern

Competition can be defined as a rivalry between two (or more) parties. It can be seen as a positive way of achieving success through a negative thing. Competition creates an environment of work that eventually leads to success. Two parties try their best to work better than the other. This is also a main ingredient for winning. In business terms competition is

> The effort of two of more parties acting independently to secure the business of a third party by offering the most favorable terms.
>
> *Merriam-Webster*

Competition that occurs within a team results in better understanding of a problem statement and it results in better brainstorming. The team tries to achieve greater success, when it competes against the other. Everyone in the world tries to win, which is best advantage of competition. Competition is an important, emerging trend in today's world. Competition is being used in various fields, has its own context and rationale behind using the available resources to achieve the cherished goal that one is looking for. Beyond the common goal of the team, there is a great diversity in the procedures adopted by the teams to reach the final destination. The general idea of competition stresses on working harder and smarter to reach a common goal post, which many parties are trying to reach. This process is common to all the parties.

6.2.1.1 Known As

Competition is a term that is close in analogy to the term *race*. The word "race" best describes the process of the competition. The race does not only mean the domain of sports, but it also relates to a race to achieve a common goal by two or more different parties. In business, it can be a race in which two parties compete to gain more numbers of customers. The real meaning of any race or competition is to declare the winner among two or more contestants. The race is to prove that a party is equal in quality or ability to win over its competitors. But, there are some differences between the words *competition* and *race*. The word *race* may also have different issues attached like ethnic racism, running race, classifying a human being based on some parameters, etc. Still, the word *race* best describes the term competition in the best possible way.

6.2.1.2 Context

A number of parties compete against each other to reach a similar goal post. So, competition is used in the context where two or more parties are racing for something that is common. Competition arises in many different contexts and scenarios. The most advantageous outcome of competition is the best possible result for a particular aspect of race.

Competition can be applied to various applications, which include competition for winning a soccer game, for acquiring small-scale companies by bigger companies, competition by two animals for a piece of meat, two companies trying to conquer the market, competition to acquire grades in a class, etc. The objective of competition can differ from application to application.

As illustrated above, competition analysis pattern can be applied to numerous applications in diverse disciplines. In the later sections, application scenarios to illustrate the applicability of competition are shown in detail. Competition concept is illustrated in two distinct scenarios like competition for winning in a soccer game and competition for taking over a small steel company by a bigger scale company and how competition can be abstracted in different applications.

1. *To get an admission in a renowned university:* Everyone (AnyParty) dreams about going to renowned universities (AnyParty). The current trend is to go to medical profession or to become an engineer (AnyDomain). The first step toward applying to universities (AnyParty) involves taking necessary examinations (AnyMechanisms) like SAT(AnyEntity) and score good points in pre-requisite courses (AnyRule). Application regulations (AnyRule) differ from one university to the other. After the application deadline, university authorities (AnyParty) will evaluate (AnyMechanism) each profile and declare the students, who score the highest, for (AnyGain) the admission to the university.
2. *Baseball competition:* Baseball is an example of outdoor sports (AnyDomain), where two teams (AnyParty) compete (AnyMechanism) against each other to determine, who can score the highest points. When no set is rewarded for the winning team, many players gain a sense of pride. Competitive sports are governed by codified rules (AnyRule) that are agreed upon by the competitors (AnyParty). Violating these rules will be considered as unfair competition. Participating teams try to perform their best. Every game (AnyEntity) has measurable criteria and standards, and appointed judge give points to both teams. The team that scores more points becomes the winner (AnyGain) of the baseball match.
3. *Election:* One can find competition in politics or government (AnyEntity) too. In growing democracies (AnyDomain), an election (AnyMechanism) is a competition for an elected office. To attain a position (AnyGain) of power, two or more candidates (AnyParty) compete against one another. But, they should follow prescribed election regulations. (AnyRule). The seat of the elected office will be assumed for a set term by the winner.

6.2.1.3 Problem

6.2.1.3.1 Functional Requirements

Competition may exist in different forms, levels, sizes, and numbers of competitors. The most significant problem is the number of competitors and difficult in understanding the process of competition in a particular domain. In any domain, whether it is law, business, sports, science, economics, politics, education, ecology, etc., one needs to understand the best possible way to succeed in the competition. There are various aspects in the competition pattern like the mechanism used, entities considered in that mechanism, the domain to which it is related, rules that binds the competition, and the parties that are acting as competitors. All these aspects make the competition a complete process.

The main driving force in the analysis of any competitive environment also encompasses a struggle between egos. Sometimes, competitors have nothing to gain, but they still compete for the sake of preserving their so-called "image" or "ego," where loosing may not hurt their materialistic goals a lot, but it will be a big setback for their ego.

So, there is a category of forces that have an effect on the competition and they are not easily calculable. The best example relates to Hitler's destructive policy during the World War II, where his faulty persistence with the "Battle of Stalingrad" eventually resulted in heavy losses to the Nazi Forces. This was due to the fact that in that particular instance of "Competition for Supremacy," his withdrawal of troops from Stalingrad would have served the best purposes to the Nazi Troops.

However, the "ego" played a negative part in it and it refused to be treated as a secondary choice in strategy building.

Competition: A contest between organisms, animals, individuals, groups, etc., for territory, a niche, or a location of resources, for resources and goods, mates, for prestige, recognition, awards, or group or social status, for leadership. The ultimate goal of AnyParty is to win the competition. It could be getting an admission in university or winning a lottery or even buying milk from a store. Hence, competition is EBT. AnyActor, like students and competitors are involved too, which could belong to AnyParty like universities and schools. Examinations are AnyMechanism that conducts the necessary competition and subsequent winners receive prizes such as cash or car (AnyEntity).

AnyMechanism: Competition has a variety of different mechanisms for determining the best-suited group; politically, economically, and ecologically. Positively, competition may serve as a form of recreation or a challenge provided that it is nonhostile.

AnyParty: AnyParty is the legal user of the system. AnyParty is classified into four types: person, country, political party, and organization. There will be at least two or more AnyParty that will be directly involved in the competition. AnyParty has a name, unique ID, contact information, and it plays at least two or more roles at any given time. Competition has become an integral part for AnyParty, such as government officials or private organizations. They pass rules, set up laws, practices or express dominance of power to maintain discipline in the competition. AnyParty has a name, unique ID, contact information, and should be two or more to participate in the competition.

AnyActor: At least two or more actors are involved in the competition. The competition can take place between two or more people. This actor will have a name, unique ID, role, and category. It has operations such as Compete(), struggle(), playRole(), interact(), request(), explore(), and receive().

AnyRule: Competition takes place between two parties or actors or both, based on written rules, agreeable, or nature rules.

AnyDomain: Competition lies in every sphere of life and it varies from getting admission to a ranked university to becoming a part of any domain. It must be regulated by one or more rules to have discipline and make it fair.

AnyGain: There are different gains that depend on the nature of the competition. This gain maybe cash, a car, or an international tour offered to the winning party. Two or more parties should be involved in a competition and it should be fair. Results should be declared by the judges by giving scores or points.

AnyEntity: Includes anything for which the competition is applied. For instance, it can be applied to two or more universities, its different departments or even within two or more countries. AnyEntity has a type, name, unique ID, classification, or category, etc.

AnyEvent: Competition itself occurs and has an event. AnyEvent is an indicator of when the competition occurs or the date when the competition takes place.

6.2.1.3.2 Nonfunctional Requirements

Competitiveness: A term that is used widely in the domain of international economy, competitiveness offers many meanings in connection with this chapter. It is the ability or performing capacity of an entity to face, challenge, and contest with the opposing party. An entity should have a strong desire to succeed or compete with an adversary or opponent. Without this ability, an entity may not succeed or reach the desired goalpost.

Adaptation: Any entity should have the ability to adapt to any given situation to compete with an adversary. Just as animals and plants adapt themselves to compete for available air, food, and space, any human should develop this trait to successfully compete with opponents. Any entity should be able to use the existing condition and environment to successfully compete to emerge as a winner.

Evolution: Any entity should develop an ability to evolve over time to face competition of any sort. Evolution involves an ability to change to emerging situations, developing capacity to change strategies and techniques midway to defeat an adversary, and enhancing capability to evolve newer ideas and opinions to face a stronger competitor.

Creativity: Competition makes your mind think out of the box. Thus gives rise to creative ideas. Creativity is a phenomenon when something new and valuable is formed. Creativity also means developing a mindset that keeps thinking all the time to generate new ideas and plans to devise much-improved strategies to defeat an adversary.

Recreation: Competition should be enjoyable and entertaining. In fact, this is the essence of fair competition. Good competition always involves bonhomie between two competitors. However, bad competition gives rise to acrimony, anger, disappointment, frustration, hate, and vitriol; two Great Wars are the examples for bad competition. The main goal of competition should be to generate enough recreational opportunities.

6.2.1.4 Challenges and Constraints

6.2.1.4.1 Challenges

Here are a few challenges as shown in Table 6.1.

6.2.1.4.2 Constraints

1. AnyMechanism should be based on one or more rules.
2. One or more parties should utilize the competition to make it useful.
3. AnyDomain should be suitable for one or more mechanisms.
4. AnyActor should specify one or more rules.
5. At least one rule should be related to AnyDomain.
6. There should be atleast two or more AnyParty that are involved in competition.
7. Atleast one AnyRule should exist for examining the competition.
8. AnyParty should comply with at least one AnyConstraint associated with the competition rules and regulations.

TABLE 6.1
Challenges

Challenge ID	0001
Challenge title	Multiple AnyRule() may lead to confusion
Context	Rules should be predefined to avoid confusion
Description	In competition, say soccer game, there are AnyRule() such as scoring mechanisms that are not predefined by (game organization) AnyParty(). Hence, players of the team (AnyParty()) remain confused as what to use as AnyMechanism() to win the game. It leads to confusion in the AnyParty() [teams] involved in the game
Solution	Rules and scoring mechanisms should be properly defined and communicated to AnyParty() to avoid chaos.
Challenge ID	0002
Challenge title	(Success) criteria AnyGain() of the game may not be communicated to (players) AnyParty()
Context	AnyGain() and AnyMechanism() to obtain it should be clearly stated by AnyParty()
Description	Players follow some strategies like which player will stand behind whom to win the competition. AnyRule() like the one mentioned may not be appropriate to the (judges) AnyParty. It leads to violation of game Anyrule
	AnyRule() to the competition should be well defined and communicated to AnyParty()

9. The main constrain in the competition pattern is to identify a connection that is appropriate among all the aspects, AnyMechanism, AnyDomain, AnyParty, AnyEntity, AnyGain, and also according to the binding rules. The point of consideration is to choose a mechanism suitable to the particular domain to get the gain by using entities related to that domain.

6.2.1.5 Solution

The solution shown here utilizes stability model to explain the concept of competition. Figure 6.1 depicts the class diagram for competition pattern.

6.2.1.5.1 Class Diagram

6.2.1.5.1.1 Classes

Competition: This represents the competition itself. It is an EBM.

6.2.1.5.1.2 Patterns

AnyParty: This represents any individual, a country, an organization, or group or a political party, who is competing and utilizes the different outputs of the competition.

AnyActor: This represents any individual or creatures or both, which are involved in and utilizes the different outputs of the competition.

AnyMechanism: This represents the different methods or ways used to help to come with the competition.

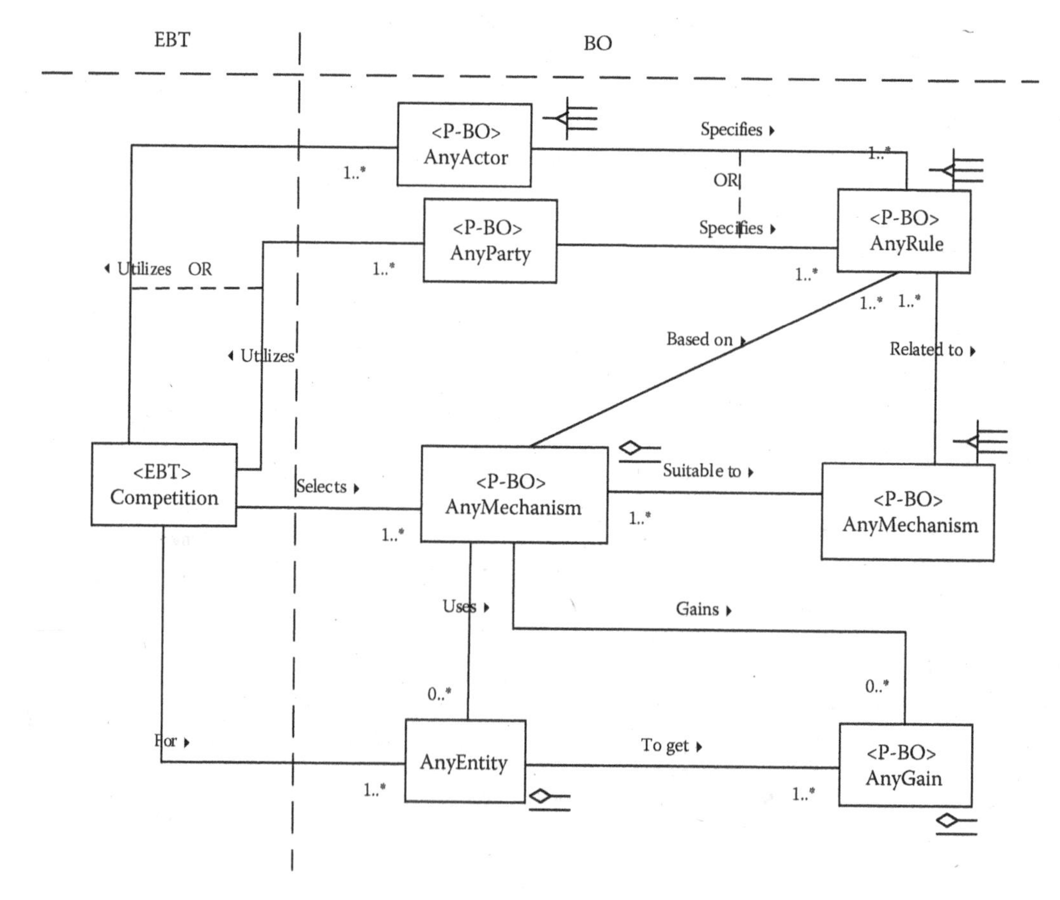

FIGURE 6.1 Class diagram for competition stable analysis pattern. (From the archive of Dr. Fayad, 2007.)

AnyEntity: This represents anything that is needed for the competition or represents what is the competition is all about.
AnyRule: This represents set of rules that control the competition.
AnyDomain: This class represents the domain of the competition.
AnyGain: This *class* represents any kind of success or benefits achieved during competition.
AnyEvent: This represents the time or date or both of the competition.

6.2.1.5.1.3 Class Diagram Description

1. Competition is the EBT of this pattern and a competitive form of the view is provided to AnyParty (BO).
2. Competition (EBT) is done on AnyEntity (BO) to represent the content provided to AnyParty (BO).
3. AnyParty (BO) has certain AnyCriteria (BO), which is necessary for competition (EBT).
4. AnyDomain (BO) relates to AnyRule (BO), which is provided by AnyMechanism (BO) suitable to AnyDomain (BO).
5. Competition (EBT) selects AnyMechanism (BO) to gain AnyGain (BO).
6. AnyParty (BO) specifies AnyRule (BO), which is always related to AnyDomain (BO).
7. AnyParty(BO) takes AnyRisk (BO), which is always related to AnyDomain (BO) and AnyLongTermEffects (BO).
8. AnyLongTermEffects (BO) is the final effect of competition (EBT) as shown in Table 6.2.

6.2.1.6 Consequences

The competition analysis pattern is generic enough to serve as building block for applications in a diverse domain.

The good thing with competition analysis pattern is that the pattern has been derived with stability in mind [1–5]. It has captured the enduring knowledge of business and its capabilities, and will stand the test of time. But, the bad thing with it is that it might result in incorrect or inaccurate results, when the analysis for competition is not done in a proper manner.

6.2.1.6.1 Stability

The competition pattern is a very stable pattern. It can be reused in many different scenarios spread across many different fields.

6.2.1.6.2 Reusability

The model is reusable for the applications of different domains. The model contains the classes suitable for any of the competition type.

6.2.1.7 Applicability

In this section, two examples to illustrate the use of competition analysis pattern are depicted, using use case description and behavior model like sequence diagram.

6.2.1.7.1 Application #1: Competition in Soccer Game

The competition among two soccer teams is to win a game of soccer. There are two teams which follow the soccer rules to play the game of soccer. These rules are being monitored and defined by referee. The soccer rules are defined according to the field games. The game has points related to which are gained by a soccer team to obtain a final goal of victory.

6.2.1.7.1.1 Use Case

Use Case Id: UC-01.01
Use Case Name: Play Soccer as shown in Table 6.3.

TABLE 6.2
CRC Cards

Competition (Competition)EBT

Responsibility	Collaboration: Client	Collaboration: Server
To occur between two parties to achieve gain	AnyParty AnyActor AnyMechanism AnyEntity	contest() gain() desire() determineEffects() select() for()

competitionResult, competitionDomain, competitionLevel

AnyMechanism (AnyMechanism)BO

Responsibility	Collaboration: Client	Collaboration: Server
To symbolize the functionality	Competition AnyEntity AnyGain AnyDomain AnyRule	express() uses() achieve() suitable() basedOn()

mechanismName, mechanismDuration, mechanismDependency

AnyEntity (AnyEntity)BO

Responsibility	Collaboration: Client	Collaboration: Server
To signify the facts for the mechanism	Competition AnyMechanism AnyGain	supports() requires() utilizedBy() about()

entityName, entityForm, entityExistence

AnyGain (AnyGain)BO

Responsibility	Collaboration: Client	Collaboration: Server
Represent the achievement gained in the competition	AnyEntity AnyMechanism	satisfies() gained() classifies()

Quantity, relevance, reason

AnyDomain (AnyDomain)BO

Responsibility	Collaboration: Client	Collaboration: Server
Denotes the collection of the related factors.	AnyMechanism AnyRule	name() categorize() includeRules() limit()

domainName, domainDescription, domainSize

(Continued)

TABLE 6.2 (*Continued*)
CRC Cards

AnyRule (AnyRule)BO

Responsibility	Collaboration: Client	Collaboration: Server
Restrict the usage with specified conditions	AnyDomain AnyMechanism AnyActor AnyParty	relate() operate() restrict() emphasis()

ruleAuthor, ruleAudience, ruleEffectiveness

AnyActor (AnyActor) BO

Responsibility	Collaboration: Client	Collaboration: Server
Monitor the mechanism used to find out the result	Competition AnyRule	utilize() specifies() used()

actorName, actorLifeExpectancy, actorContactInfo

AnyParty (AnyParty) BO

Responsibility	Collaboration: Client	Collaboration: Server
Take part in competition	Competition AnyRule	takePart() follows() seeks() competes() selects() strategies() achieves() does() agrees() complyWith()

partyName, partyConatactInfo, partyDesignation, occupation, idNumber

AnyEvent (AnyEvent) BO

Responsibility	Collaboration: Client	Collaboration: Server
An occurrence of instance	AnyType() AnyMedia()	storedOn() happened() impacts() controls()

type, name, unique ID, classification or category

6.2.1.7.1.2 Use Case Description

1. **Competition** selects **AnyMechanism**(SoccerGame) to be held among any two teams.
   a. Does competition use the correct mechanism?
   b. Does competition consider all the process dependency?

TABLE 6.3
Play Soccer Use Case

Actors	Roles
AnyParty	Referee
AnyParty	SoccerTeam

Class Name	Type	Attributes	Operations
Competition	EBT	competitionResult competitionDomain competitionLevel	select()
AnyMechanism	BO	mechanismName mechanismDuration mechanismDependency	suitable()
SoccerGame	IO	noOfTeams gameVenue tournamentName	held() containFactor()
AnyEntity	BO	entityName entityForm entityExistence	actAsBase()
Point	IO	range value weightage	findWinner()
AnyGain	BO	quantity relevance reason	gained()
Victory	IO	looser winner pointerDifference	differentiate()
AnyDomain	BO	domainName domainDescription domainSize	categorize() includeRules()
FieldGame	IO	fieldType fieldSize noOfGames	fallUnder()
AnyRule	BO	ruleAuthor ruleAudience ruleEffectiveness	utilize()
SoccerRule	IO	lastUpdated ruleSeverity noOfRules	participateIn()
AnyParty	BO	partyName partyConatactInfo partyDesignation	followRule() specify() monitor()
Referee	IO	memberOf experience accuracy	supervise()
SoccerTeam	IO	noOfPlayers originOfTeam uniformColor	UnderstandRule()

2. **AnyMechanism**(SoccerGame) contains the measuring factor, which is **AnyEntity**(Points) to find the winner.
   a. Are data verified for the authenticity by AnyMechanism?
   b. Is AnyMechanism capable of getting the right result?
3. **AnyEntity**(Point) act as a base of **Competition,** which is in turn for **AnyEntity**(Point).
4. **AnyEntity**(Points) is required to get **AnyGain**(Victory) gained to differentiate among the gainer in **AnyMechanism**(SoccerGame).
   a. Does AnyEntity contribute properly for the process?
   b. Is AnyEntity apt for the need?
   c. Is AnyGain useful for the party?
   d. Does AnyGain describe the difference well?
5. **AnyMechanism**(SoccerGame) must be suitable to **AnyDomain**(FieldGame) to fall under its category.
6. **AnyDomain**(FieldGame) includes set of rules **AnyRule**(SoccerRule) for the soccer game.
   a. Does AnyDomain cover the entire requirement?
   b. Does AnyDomain specify the boundaries very well?
7. The rule **AnyRule**(SoccerRule) is being monitored by **AnyParty**(Referee) who also specifies and supervises those rules.
   a. Is AnyRule complete by itself?
   b. Is AnyRule suitable for any scenario?
8. **AnyParty**(SoccerTeam) needs to understand and follow **AnyRule**(SoccerRule) to participate in and utilize the **Competition.**
   a. Is AnyParty suitable for following the rule?
   b. Does AnyParty completely involve in the Process?

6.2.1.7.1.3 Class Diagram The models for this application are shown below in Figures 6.2 and 6.3.

6.2.1.7.1.4 Sequence Diagram Description of the sequence diagram (soccer)

1. Competition selects soccer game.
2. Soccer games are played for points.
3. Point act as base of competition.
4. Competition is for points.
5. Achieving valuable points lead to victory.
6. Victory is obtained by winning the soccer game.
7. Soccer game is suitable to field games.
8. Field games include rules and regulations.
9. Soccer rules are monitored by a referee.
10. Referee specifies soccer rules, which are followed by a soccer team.
11. Soccer team utilizes competition.

6.2.1.7.2 Application #2: Competition in Acquiring Small Company

This application is regarding the competition among several big steel companies to take over a small steel company. The big companies need to follow security and exchange commission rules to buy an outside company. The market gain that they get is the increased market cap, which is straight way related to the shares of the company. The domain steel market includes all the things like shares, rules, etc.

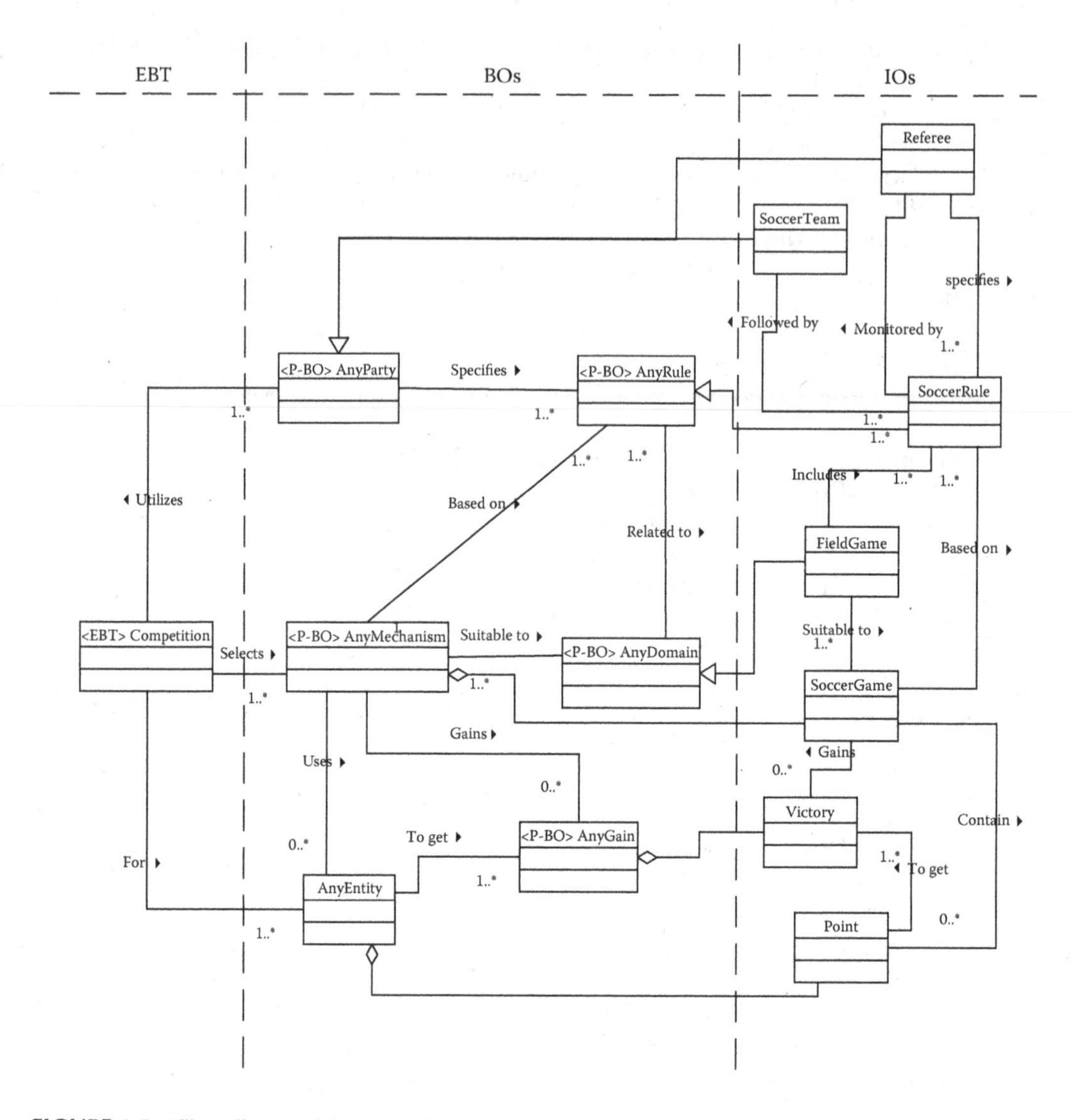

FIGURE 6.2 Class diagram for competition pattern in a soccer game.

6.2.1.7.2.1 Use Case

Use Case Id: UC-01.02

Use Case Name: Compete for acquiring small company as shown in Table 6.4.

6.2.1.7.2.2 Use Case Description

1. **Competition** selects **AnyMechanism**(Acquisition) to acquire smaller company.
   a. Does competition use the correct mechanism?
   b. Does competition consider all the process dependency?
2. **AnyMechanism**(Acquisition) uses **AnyEntity**(Share), which acts as the base for the process.
   a. Are data verified for authenticity by AnyMechanism?
   b. Does AnyMechanism is capable of getting the right result?
3. **AnyEntity**(Share) acts as base of **Competition,** which in turn acts for **AnyEntity**(share).

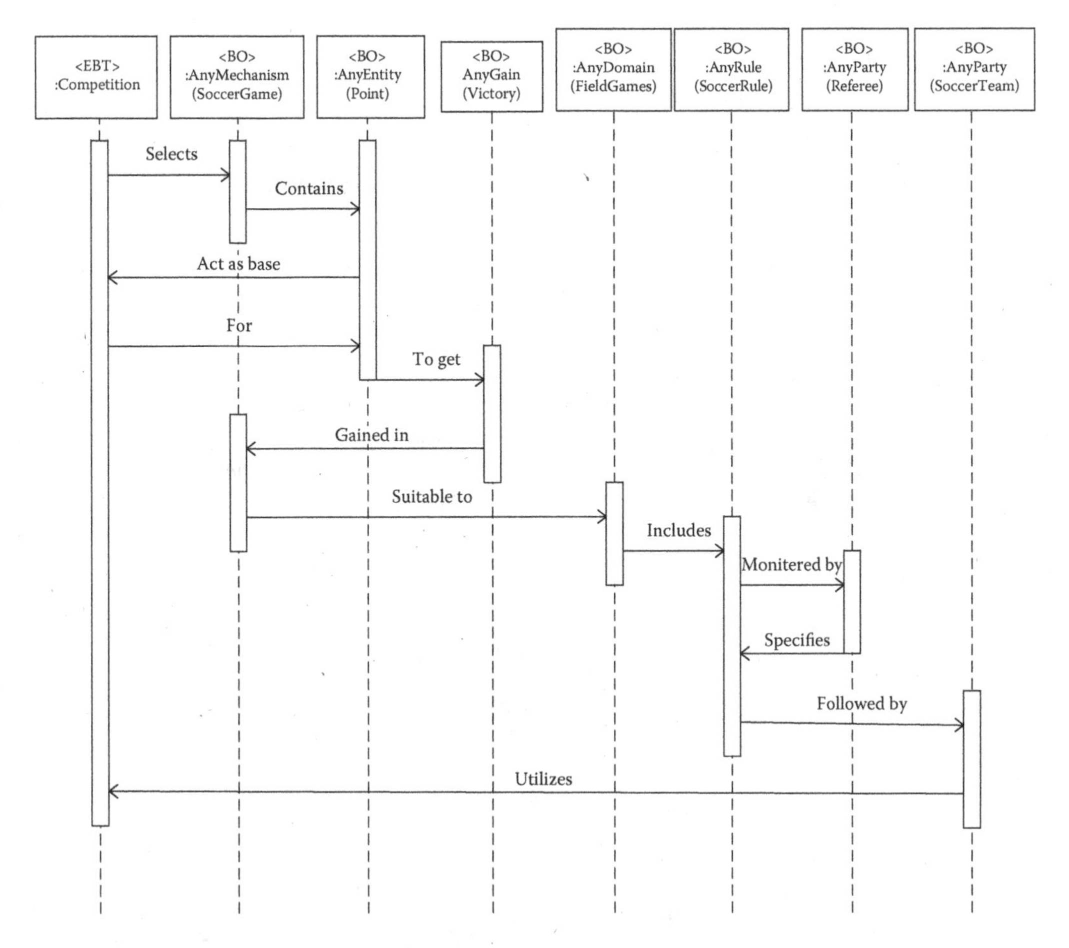

FIGURE 6.3 Sequence diagram for competition pattern in a soccer game.

4. **AnyEntity** (Share) is required to get **AnyGain**(MarketCap), which defines the status gained through **AnyMechanism**(Acquisition).
   a. Does AnyEntity contribute properly for the process?
   b. Is AnyEntity apt for the need?
   c. Is AnyGain useful for the party?
   d. Is AnyGain is differentiator among the gainer?
5. **AnyMechanism**(Acquisition) must be suitable to **AnyDomain**(SteelMarket), which enables the smooth processing.
6. **AnyDomain**(SteelMarket) includes set of rules **AnyRule**(Sec&ExCommRules), which explains steps for the company acquisition.
   a. Does the AnyDomain cover the entire requirement?
   b. Does the AnyDomain specify the boundaries very well?
7. The rule **AnyRule**(Sec&ExCommRules) is being monitored by **AnyParty**(Sec&ExCommOfficer), who also specifies and confirms those rules.
   a. Is AnyRule complete by itself?
   b. Is AnyRule suitable for any scenario?
8. **AnyParty**(SteelCompany) needs to follow **AnyRule**(Sec&ExCommRules) to take part in and utilize the **Competition**.
   a. Is AnyParty suitable for following the rule?
   b. Is AnyParty completely involved in the process?

TABLE 6.4
Compete for Acquiring Small Company Use Case

Actors	Roles
AnyParty(Human)	Sec&ExCommOfficer
AnyParty(Organization)	SteelCompany

Class Name	Type	Attributes	Operations
Competition	EBT	competitionResult competitionDomain competitionLevel	select()
AnyMechanism	BO	mechanismName mechanismDuration mechanismDependency	suitable()
Acquisition	IO	dateOfOccurance acquiringCompany acquiredCompany	happenAmongst() acquire()
AnyEntity	BO	entityName entityForm entityExistence	actsAsBase()
Share	IO	marketValue typeOfShare noOfSplitsInLastYear	requirement()
AnyGain	BO	quantity relevance reason	gained()
MarketCap	IO	category lastChanged overLastQuarter	defineStatus()
AnyDomain	BO	domainName domainDescription domainSize	categorize() includeRules()
SteelMarket	IO	noOfCompanies leadingCompany annualTurnover	enableProcess()
AnyRule	BO	ruleAuthor ruleAudience ruleEffectiveness	utilize()
Sec&ExCommRule	IO	revisionNumber numberOfRules yearOfEstablishment	explainSteps()
AnyParty	BO	partyName partyConatactInfo partyDesignation	followRule() specify() monitor()
Sec&ExCommOfficer	IO	officeDetails age NoOfReportees	confirms()
SteelCompany	IO	noOfEmployees nameOfFounder yearFounded	takePart()

6.2.1.7.2.3 Class Diagram The models for this application are shown below in Figures 6.4 and 6.5.

6.2.1.7.2.4 Sequence Diagram Description of the sequence diagram (Acquisition)

1. Competition selects acquisition.
2. Acquisition uses share for evaluation.
3. Share acts as base of competition.
4. Competition is for share.
5. Share leads to capturing of market share.
6. Market share is gained by acquisition.
7. Acquisition is suitable to steel company.
8. Steel market includes Security and Exchange Commission rules and regulations.
9. Security and Exchange Commission rules and regulations are monitored by the Security and Exchange Commission officer.

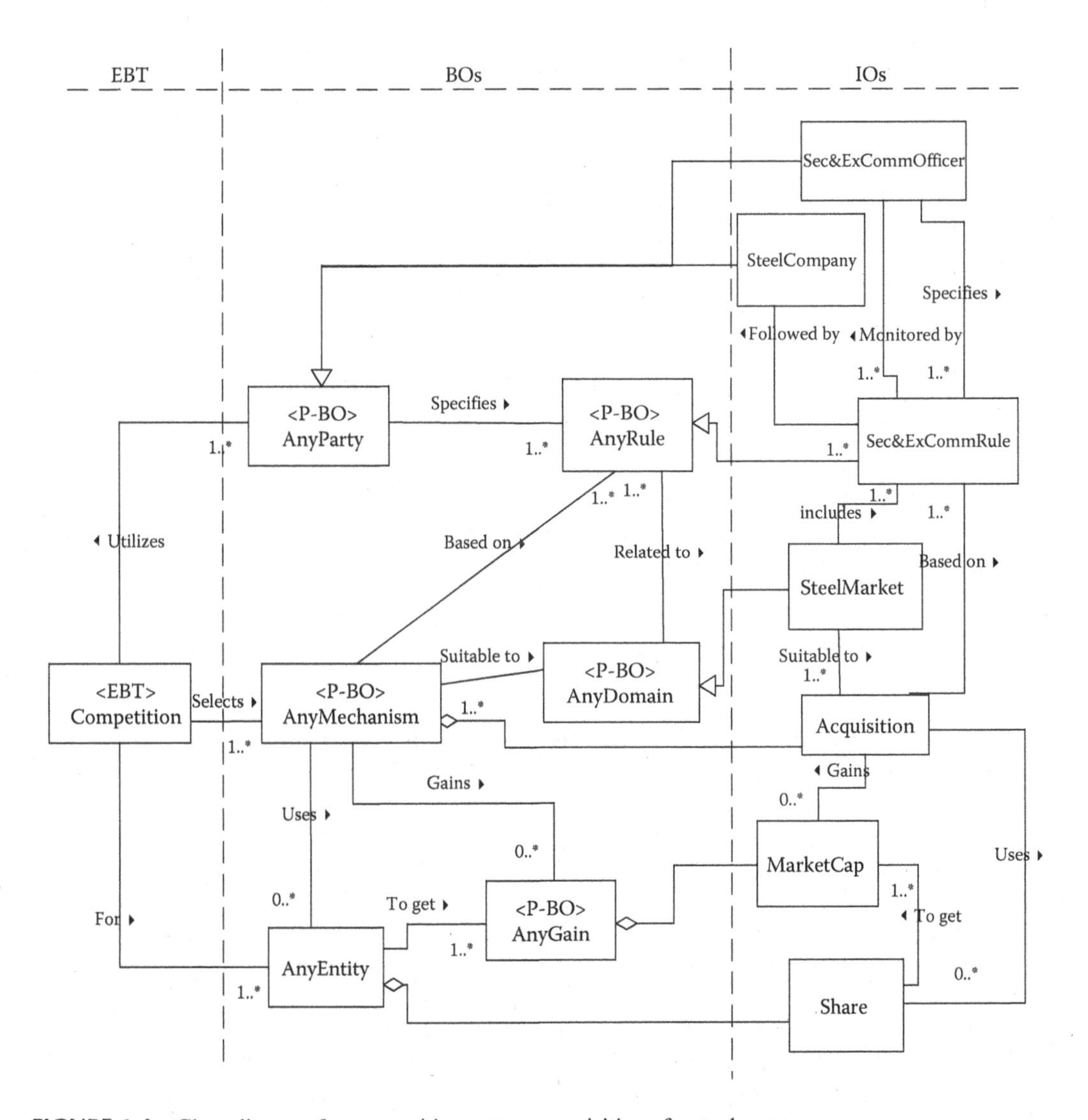

FIGURE 6.4 Class diagram for competition pattern acquisition of a steel company.

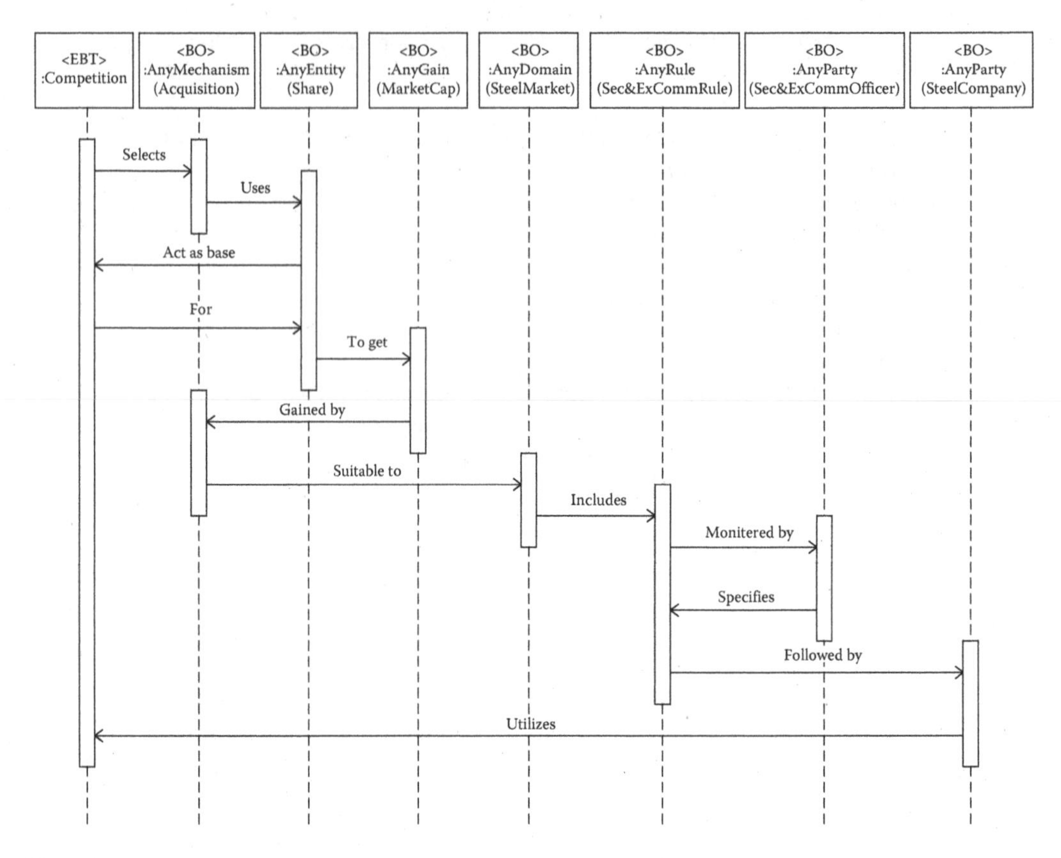

FIGURE 6.5 Sequence diagram for competition pattern for acquisition of small steel company.

10. Security and Exchange Commission officer specifies Security and Exchange Commission rules and regulations.
11. Security and Exchange Commission rules and regulations are followed by the steel company.
12. Steel company utilizes competition.

6.2.1.8 Related Pattern and Measurability

6.2.1.8.1 Related Pattern

The above class diagram explains the traditional model for competition pattern. This is also related to one of our applications in which large steel companies compete to acquire small steel companies. In the traditional model, it can be seen that this includes the specific classes related to that pattern only. ChiefExecutiveOfficer organizes the company, which gains the MarketCap. The ChiefExecutiveOfficer also initiates the process merge based on some rules defined by Security and Exchange Commission. These rules are a part of Security and Exchange Commission agency. The company converses with the agency to go ahead with the merger procedure. The market cap of the company enhances because of the shares gained by the ChiefExecutiveOfficer as shown in Table 6.5.

Looking at the above traditional patterns, we can easily figure out the process of acquisition, but not the process of competition, which is the main theme of the pattern. The pattern above can be only used for the specific application as all the classes defined in there are related to the application only and not the generalized classes. But, in the stability model (Stability Model [SSM] is shortened to Stability Model throughout this book), all the business objects are stabilized and can be applied

TABLE 6.5
Traditional Model versus Stable Model of Competition

Criteria	Weight	Traditional Model	Traditional Weight	Stability Model	Stability Weight
Simplicity	20	Traditional model does not have separate EBTs, BOs or IOs. So it becomes complex in terms of simplicity.	8	Comparing to a traditional model, stable model contains well-defined EBTs, BOs, and IOs that makes it simple and more understandable.	18
Understanding	25	Traditional models are complex, when compared to stability models, which makes them less understandable.	8	Stability model focuses on EBTs, BOs that are easy to understand	22
Notational	15	Traditional model uses inheritance, association, aggregation, and composition for representing the model	5	Stability model uses EBTs, BOs, IOs along with inheritance, association, aggregation, and composition for representing the model	14
Extensibility	20	Only inherited classes can be extended in traditional model	0	All the EBTs and BOs can be extended in stability model	18
Systematic	20	Traditional models are lesser systematic in terms of separation and management of tangible and intangible classes	0	Stability models are most systematic in terms of separation and management of tangible and intangible classes	18
Total			21		90

to any of the application related to that EBT. There may be several IOs, which can be connected to those BOs to make that application fit into that stable model. Thus, stable model includes all the application related to that domain but the traditional model is just specific.

6.2.1.8.1.1 Traditional Model (Meta Model) with the Stable Model (Pattern)

1. Traditional model does not highlight goals of the system, whereas stability model recognizes important goals of the system.
2. Stability model needs us to find set of EBTs, BOs, and IOs.
3. The system built by using stability model is easily adaptable to reuse.
4. Model is built upon what is it for, why (purpose), and how they are stable.
5. The size of the problem does not matter much because even if it increases the solution in the model, it remains same and there is no need to add more EBTs, BOs, and IOs.

6.2.1.8.2 Measurability

6.2.1.8.2.1 Quantitative Measurability Cyclomatic Complexity: Equation for cyclomatic complexity is

$$M = E - N + 2P$$

where

M = cyclomatic complexity
E = the number of edges of the graph
N = the number of nodes of the graph
P = the number of connected components

Description: Cyclomatic complexity is software metric (measurement). It is used to indicate the complexity of a program. It is a quantitative measure of the complexity of programming instructions. Here with respect to modeling problem, it is used to measure the complexity of specific model. We can compare traditional model and stability model on complexity bases.

1. *Traditional model*:
   E = 8
   N = 7
   P = 1
   Cyclomatic complexity,
   M = E − N + 2P
   = 8 − 7 + (2*1)
   = 3
2. *Stability model*:
   E = 7
   N = 7
   P = 1
   Cyclomatic complexity,
   M = E − N + 2P
   = 7 − 7 + (2*1)
   = 2

Impact: As per the calculation, the cyclomatic complexity of traditional model is higher when compared to the stability model. It believes that traditional models are more complex when compared to stability models. Complex nature of traditional models usually influences stability, understandability, extensibility, and many more factors. Based on the magic index card generator, traditional model is not too effective, because we cannot add additional IOs to the system. Its structure is also less effective to understand.

6.2.1.8.2.2 Qualitative Measurability *Reusability*: One of the main goals of any project is that the program should be reusable. Companies try to make strategy to increase reusability factor in every program to increase productivity and profit. This is done with reuse metrics and models. Reuse models and metrics can be categorized into six types. They are reuse cost-benefits models, maturity assessment, amount of reuse, failure modes, and reusability and reuse library metrics.

In traditional models, only inheritance super class of inheritance relations are reusable. In stability model, all the EBTs and BOs are reusable.

Reusable classes in traditional model = 2
Reusable classes in stability model = 15

This calculation shows that stability model is much more reusable than traditional model.

Stability: A system is called stable, if it can run across different environments, which is only possible if the number of tangible classes are minimum, or else they should be easily removable. Otherwise, it will create problem, when any changes are to be made anytime.

So, we can decide the stability of which model is more after comparing the number of tangible classes in each.

Number of tangible classes in traditional model = 8
Number of tangible classes in stability model = 7

6.3 SUMMARY

Traditional models, as they are used still now, are beset with a higher degree of complexity with regard to their usability factor as shown in Figure 6.6. The usability factor, per se, is not limited to their deployment in practice, but it also relates to several other factors like stability, understandability, and extensibility. On the contrary, stability models represent a new generation system of allotting an array of benefits to the resultant model like stability, understandability, and extensibility apart from endless and unlimited usability without investing much time, effort and money. Another advantage of using the model of stability is its varied applications to a wide array of situations that might come across while designing meaningful patterns as shown in Figure 6.1.

6.4 OPEN AND RESEARCH ISSUES

1. *e-Competition:* Utilize stability model or knowledge map methodology as a way for developing e-Competition Engine. Building this engine by using traditional development approaches is not an easy exercise, specifically when several factors can undermine their quality success, such as cost, time, and lack of systematic approaches.

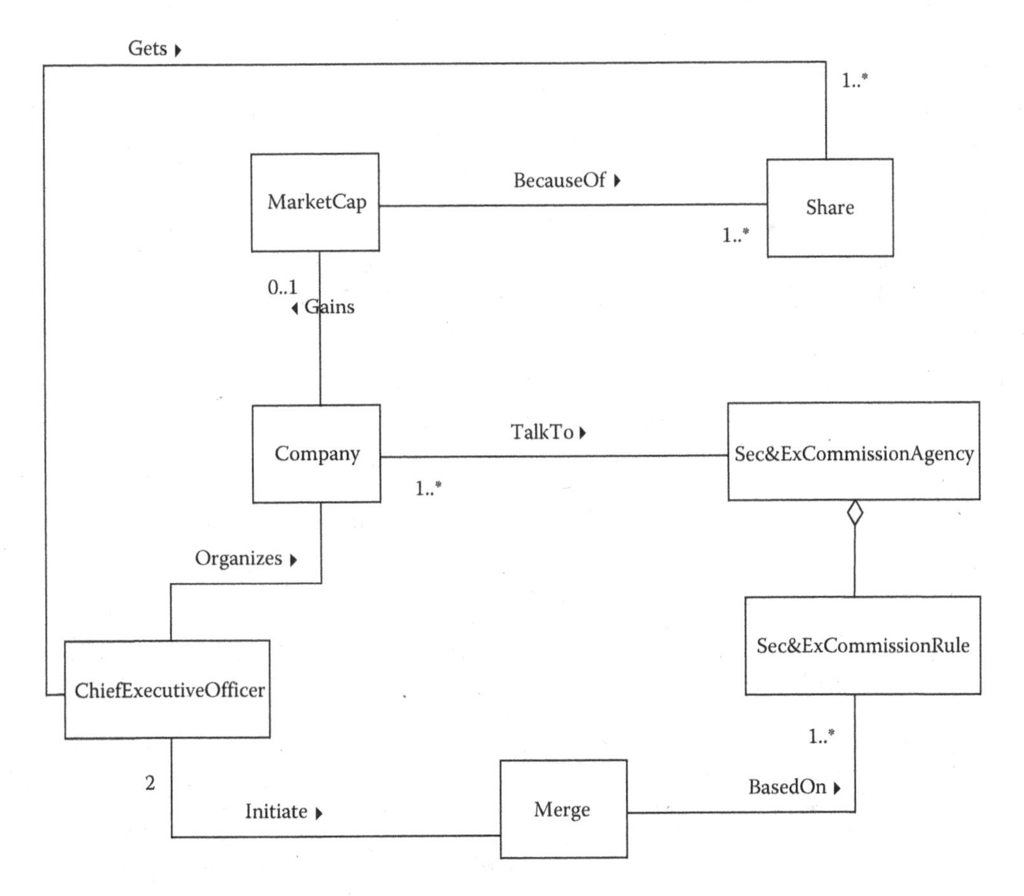

FIGURE 6.6 Traditional model for competition.

2. *Competition unified software engine (C-USE):* Utilize stability model or knowledge map methodology as a way for developing C-USE. The engine mainly focuses on several patterns: competition, analysis, adaptability, extensibility, customizability, etc. The proposed solution attempts to extract the commonality from all the domains and represent it in such a way that it is applicable to a wide range of domains without trivializing or generalizing the concepts. The engine is a stable structural pattern and it provides a generic engine that can be applied and/or extensible to any application, such as economic competition, biological competition, forecasting competition, political competition, and sports competition by plugging application-specific features. The C-USE can be applied to unlimited context and reusability.

REVIEW QUESTIONS

1. What are different types of competition that are discussed in this chapter?
2. What makes competition an EBT?
3. What could be IOs for AnyGain? List any two.
4. Explain competition in the context of AnyRule, AnyMechanism, and AnyDomain.
5. Rewrite each of the scenarios with EBT and BOs in mind. Make sure that each BO is attached to one or more IOs, except AnyParty that is attached to two or more roles.
6. List and describe all the terms that are similar to competition, if any.
7. List and describe two scenarios of competition of each of the following areas:
    1. Culture
    2. Economy
    3. Law
    4. Social life and society
5. List and explain two scenarios of competition that are not covered in this chapter.
6. Define internal requirements. List and explain the internal requirements for competition.
7. Define external requirements. List and explain the external requirements for competition.
8. Examine attributes and operations of competition. Are there any missing attributes and operations? Discuss them.
9. Examine the functional requirements of dignity pattern—Are there any missing requirements? Discuss them.
10. Examine the nonfunctional requirements of competition pattern—Are there any missing requirements? Discuss them.
11. What is the difference between inheritance and delegation? Why is delegation considered better than inheritance? Explain with the help of an example.
12. What is an interface? How are interfaces implemented for stability model?
13. Why is it easy to test stability model as compared to traditional model?
14. What are business rules? List some business rules for competition pattern.
15. What are the key elements of business rules?
16. What is meant by business integration? Why is it important?
17. "Developing any application using stability approach simplifies things." Explain and justify.
18. "It is much easier to integrate the dignity pattern in any business model." Explain and justify
19. Explain some of the known usages of competition pattern in day-to-day life. List a few that are not covered in the chapter.
20. What is a domain model?
21. What are the properties of domain models?
22. T/F: developing conceptual models requires both domain knowledge and modeling skills.
23. T/F: Analysis patterns are conceptual models that can be used to model and share domain knowledge.

24. T/F: Stable analysis patterns (SAPs) separate the core concepts of the domain from business-specific concepts.
25. T/F: Traditional analysis patterns do not separate the core concepts of the domain from business-specific concepts.
26. T/F: Traditional analysis patterns have a very limited reuse.
27. T/F: Stable analysis patterns model the knowledge of the problem domain.
28. T/F: Stable analysis patterns aid the understanding of the problem rather than showing how to design a solution.
29. Traditional analysis patterns have limited reuse. Why?
30. What is the need for a stable analysis pattern?
31. What EBTs can be brought under the umbrella of stable analysis pattern?
32. How does competition qualify as a subject for a stable analysis pattern?
33. How do we start with the identification of competition and its related terms?
34. What are BOs and what are IOs?
35. How are the above terms associated with our EBT of competition? List possible application domains for the SAP competition.
36. How reusable is the competition stable analysis pattern?
37. How sustainable is the competition stable analysis pattern?
38. What are the most practical and prominent examples of the competition stable analysis pattern?
39. What are the challenges in such a type of analysis?
40. Are there any constraints that need to be handled by the competition pattern?
41. Do these constraints reduce the number of applications for the competition pattern?
42. What are the constants involved in the design pattern, if any?
43. Is our analysis scalable according to application?
44. What are the limits of scalability that we can achieve with this model? Is it infinitely scalable?
48. Is our model technology-independent?
49. Are there any existing anomalies to our design pattern?
50. Identify the existing anomalies (if any) and find the reason for their being anomalous?
51. Are certain modifications required in our stability model so that even the anomalies can fit in?
52. What are such modifications and their effects?
53. Can you think of any example application for the competition pattern that is seemingly outside the scope of the pattern?
54. The competition pattern can be applied to modern video game development and design. (T/F)
55. The competition pattern is not suitable for sport for multi-team competitions. (T/F)
56. The competition pattern is not suitable for describing the card game solitaire. (T/F)

EXERCISES

1. Choose a sport that you are familiar with (baseball, basketball, rugby, cricket, etc.) and apply the competition pattern to your choice. List the IOs below. Take the IOs from the previous question and create an application class diagram.
2. Draw a class diagram based on the competition SAP to show the application of the following application areas of competitions:
   a. Economic competition
   b. Forecasting completion
   c. Aerobatic competition
   d. Student competition

PROJECTS

1. Consider a Robot for an application of competition SAP and design the following for each application:
   a. Draw a class diagram based on the competition pattern to show the application of Robot competition.
   b. Document a detailed and significant use case as shown in Case Study 1.
   c. Create a sequence diagram of the created use case of b.
2. Consider the card game solitaire for an application of competition SAP to the following for each application:
   a. Draw a class diagram based on the competition pattern to show the application of card game solitaire.
   b. Document a detailed and significant use case as shown in Case Study 1.
   c. Create a sequence diagram of the created use case of b.
3. Consider athletic competition for applications of competition SAP and design the following for each application:
   a. Draw a class diagram based on the competition pattern to show the application of athletic competition.
   b. Document a detailed and significant use case as shown in Case Study 1.
   c. Create a sequence diagram of the created use case of b.
4. Consider a dance competition for applications of competition SAP and create the following for each application:
   a. Draw a class diagram based on the competition pattern to show the application of dance competition.
   b. Document a detailed and significant use case as shown in Case Study 1.
   c. Create a sequence diagram of the created use case of b.
5. Consider academic competition for applications of competition SAP and design the following for each application:
   a. Draw a class diagram based on the competition pattern to show the application of academic competition.
   b. Document a detailed and significant use case as shown in Case Study 1.
   c. Create a sequence diagram of the created use case of b.
6. Consider an eating competition for applications of competition SAP and design the following for each application:
   a. Draw a class diagram based on the competition pattern to show the application of eating competition.
   b. Document a detailed and significant use case as shown in Case Study 1.
   c. Create a sequence diagram of the created use case of b.
7. Consider the described competition in weaponry below as an application of competition SAP.
   This application defines the competition in arms and weapon technology that usually occurs among different developed countries. The particular incident relates to the post-World War II competition in arms race between the Soviet Union and the United States. This competition became so severe during the cold war that any movement by these two countries or any kind of weapon deployment or testing by a particular country was met with immediate reaction by the other. The war between Capitalism and Socialism had a direct consequence on the arms development and deployment.
   a. Draw a class diagram based on the competition pattern to show the application of competition in weaponry.
   b. Document a detailed and significant use case as shown in Case Study 1
   c. Create a sequence diagram of the created use case of b.

REFERENCES

1. M.E. Fayad, and A. Altman, An introduction to software stability, *Communications of the ACM*, 44, 2001, 95–98.
2. M.E. Fayad, Accomplishing software stability, *Communications of the ACM*, 45, 2002, 109–112.
3. M.E. Fayad, How to deal with software stability, *Communications of the ACM*, 45, 2002, 111–115.
4. M.E. Fayad, and S. Wu, Merging multiple conventional models in one stable model, *Communications of the ACM*, 45Z, 2002, 95–98.
5. M.E. Fayad, H.A. Sanchez, S.G K. Hegde, A. Basia, and A. Vakil, *Software Patterns, Knowledge Maps, and Domain Analysis*, Boca Raton, FL: Auerbach Publications, 2014.
6. A. Mahdy, and M.E. Fayad, A Software Stability Model Pattern. *Proceedings of the 9th Conference on Pattern Language of Programs (PLoP02)*, Monticello, IL. 2002.

7 Corruption Stable Analysis Pattern

There is no compromise when it comes to corruption. You have to fight it.

A. K. Antony

Corruption pattern models the pattern to describe the ill effects of the corruption to the economic and social growth of the society. The goal of the corruption pattern is to define a stability model [1,2]. This sequence pattern can be extended into any of the existing or new corruption applications. We have discussed two case studies in this pattern to explain how to achieve the goal by applying different and appropriate BOs, which are associated to EBT, although IOs vary from one domain to another. The first case study is about the corruption involved in public and the second case study involved in business.

7.1 INTRODUCTION

The existence of corruption varies from simple data or file corruption to very complex, such as economic and political corruption. It is common in many areas of our life, like corruption in public and private offices and corruption in governmental agencies. Corruption can be found in many cases on different levels and ranges from minor local ones to very serious ones and in tightly coupled hardware networks. Here are some samples of the existence of corruption in different walks of life and different domains.

Corruption is not a new phenomenon, it is age old. Corruption exists in developing countries, industrialized countries, and less-developed countries too. Some of the common corruptions that take place in society are business corruption, public corruption, political corruption, and computer data corruption out of which some are explained briefly.

Political corruption: Misuse of the government power by politicians to increase their wealth or power. It includes a wide variety of corruption crimes committed by political officials before, during, and after leaving the office.

Data corruption: Deterioration of computer data as a result of some foreign agents like a virus or malware.

Business corruption: The presence of business corruption in a market provokes firms to make choices between legal business approaches and illegal bribery.

Public corruption: The misuse of public office for private gain.

7.2 CORRUPTION STABLE ANALYSIS PATTERN

The concept *Corruption* describes any interdependent system not performing its duties properly. Corruption is disseminated all over the world, causing a big toll on the economic growth. Corruption can occur either in a database system that cause the data corruption or in an organization resulting in political corruption. Corruption can be viewed in different ways, such as political corruption, economic corruption, and social corruption; it necessitates the requirement for corruption analysis pattern that can be used in all scenarios.

7.2.1 Context

Any corruption deals with bribing humans for performing a task in an organization. The person involved in bribery or corruption may either belong to a government organization or a private organization. This leads to a political corruption that affects the entire organization and the members involved with the organization.

Corruption takes place due to the mismanagement and misappropriation of a person involved in an organization. This leads to a system corruption that has a great toll on the economic aspect of the organization. Corruption also involves data corrupted in the process of transmission and retrieval of data between systems. Data corruption impacts the entire system, thereby leading to any system and economic impact on the organization.

Corruption usually crops up when hackers try to corrupt the information archived in any system and thereby causing a great deal of impact on the system.

Corruption can be of different types. Let us see some of the examples.

7.2.1.1 Executive Branch Political Corruption

It was discovered that officials (AnyParty) in Phoenix, VA hospital lied about how long the wait times were there for veterans to see a doctor. An investigation team (AnyParty) was formed by the veteran's affairs office of the general inspector. It was found that many veterans were not given admission papers (AnyEntity) at all in the hospital, which made their health condition even more worse (AnyLevel) causing death (AnyDamage) of three veterans. Veterans had problems ranging (AnyLevel) from eye infections, body aches, ulcers, to kidney problems (AnyType), which required immediate medication or even hospitalization. The inspection (AnyMechanism) was conducted by directly talking to the relatives (AnyEvidence) and taking from them in written form (AnyEvidence) about what happened in the hospital. The damage caused was serious and of alarming intensity. The cause behind ill treating all those veterans was that there was scantier availability (AnySituation) of doctors and nurses in the hospital, because of the thanksgiving vacations (AnyReason).

7.2.1.2 Public Corruption

In San Carlos, CA, on December 18, 2010, Nancy was sentenced (AnyMechanism) to 22 months in prison and was ordered to pay $1,720,000 in restitution. This was published in newspapers (AnyEvidence) and viewed (AnyEvidence) by thousands of US citizens. Nancy previously pleaded (AnyEvent) guilty to wire fraud and filing a false tax return. According to court documents (AnyEvidence), Nancy served as the village of Burnham's elected clerk (AnyLevel) from 1985 until she resigned in 2000. As clerk, Nancy was responsible for managing Burnham's finances and depositing cash and checks collected by the clerk's office into the village's bank accounts.

Between 1988 and 1999, Nancy took cash that the village received as payment for fees and fines from the public and used most of the cash to gamble at casinos (AnyReason). Nancy took cash from both the villages' (AnyParty) cash register and the collection of money received as tow bonds. She also filed a false federal income tax return for the years 2005–2010, knowing that her total income was substantially greater than what she reported, because she failed (AnySituation) to report the cash she misappropriated (AnyDamage) from the village as income.

7.2.2 Problem

The corruption pattern represents the distortion caused by corruption. The corruption can be either a political corruption, economic corruption, or a social corruption.

Political corruption distorts the whole nation and has a social, political, and economic impact on the people of the corrupted nation.

Data corruption resulting due to the mismanagement of a system, discloses both the system impact and the economic impact posed on the involved organization. Social corruption distorts the reputation of an organization.

The corruption pattern represents how corruption influences the economic growth of the society in different areas.

Public corruption: Government officials for their personal gain, demand or receive anything to perform their duties. This effects the economic growth of the country.

Business corruption: Business corruption has a corrosive impact on market and economic growth.

7.2.2.1 Functional Requirements

Corruption (EBT): Corruption is the abuse of public resources to enrich or give unfair advantage to individuals, their family, or their friends.

AnyParty: It represents four types of legal users: human, organization, country, and political party. Any of these parties are involved in the corruption. AnyParty is the legal user of the system. He/she has a unique ID, a name, and an interest. AnyParty fulfills his/her desire to commit corruption through AnyMechanism. AnyParty gets seduced, can lose self-control, can commit corruption, and can resist corruption.

AnyMechanism: In philosophical, theological, or moral discussion on corruption, *it* is the abuse of bestowed power or position to acquire a personal benefit. Corruption may include many activities including bribery and embezzlement. Government, or "political", corruption occurs, when an office-holder or other governmental employee acts in an official capacity for personal gain. Corruption can include giving or accepting of bribes or inappropriate gifts, double dealing, under-the-table transactions, manipulating elections, extortion and blackmailing, diverting funds, laundering money, and defrauding investors.

AnyMechanism is a way to commit corruption. AnyParty/AnyActor uses AnyMechanism to fulfill his or her desire of getting rich by indulging in corrupt practices. AnyMechanism can be of many types. It can have attributes and should be within the context of the system. AnyMechanism is operated, build, and executed by AnyActor/AnyParty.

AnyLevel: Corruption can occur on different levels. There is corruption that occurs as small favors between a small number of people (petty corruption), corruption that affects the government on a large scale (grand corruption), and corruption that is so prevalent that it is part of the everyday structure of society, including corruption as one of the symptoms of organized crime (systemic corruption).

AnyLevel is the level returned by the technique used to carry out corruption and it also represents the level that impacts the factors used to affect AnyParty/AnyActor. AnyLevel has a unique ID, a description, and ingredients. It can be found, revealed, quantified, and measured. It gives the proportion of the committed case of corruption by AnyParty.

AnyReason: Corruption can occur for many reasons. In politics, big time corruption is witnessed through the funding money collected without any accounts; in reality, political funding could be a future form of lobbying done by big time corporations to get future corporate benefits. In governmental departments, corruption can occur by officials to get rich quickly by illegally asking money from people who are looking forward to get some important work done by these officials. In the domain of software industry, data corruption may occur due to reasons like cyber attack by hackers who may exploit deficiencies in system loopholes and security leakages.

AnyReason is the reason by which corruption occurs. It also represents the reason for which the corruption is created and thus impacts AnyMechanism. AnyReason can have a unique ID, description, type, and attributes. It can be implemented, controlled, reduced, and increased.

AnyDamage: Damage due to corruption could be very insignificant to humungous depending on the degree of damage. Political and public corruption can result in misappropriation that may drain a huge dent in the government treasury. This will not only destroy the economy of a state or country, it might even make the entire society crippled and immoral. In software industry, damages due to data corruption because of security breaches may cause huge losses to businesses and their customers. One best example is the banking sector.

AnyDamage is the degree of damage caused due to corruption. It can have a unique id, description, type, and attributes. It can also be implemented, controlled, managed, and could be either reduced or increased.

AnyEvidence: It is the proof and evidence that confirms the occurrence of AnyCorruption. AnyEvidence eventually leads to AnyEvent when the guilty confesses to the crime of AnyCorruption. For example, in the case of embezzlement by Nancy, evidences published by newspapers and read by thousands of readers (AnyEven) resulted in her confession.

AnySituation: The cause behind ill treating all those veterans was that there was scantier availability (AnySituation) of doctors and nurses in the hospital. It is the AnyReason that is attributed to defend corruption. AnySituation usually leads to AnyMechanism to commit to AnyCorruption. It has a special id, a reason, an effect, and a type. It defines reasons and is used by AnyMechanism.

AnyEntity: It represents anything toward which AnyParty/AnyActor is denied the freedom of rights and which impacts AnyLevel that usually ends up in causing serious damage to AnyActor/AnyParty. AnyEntity has a unique ID, a name, an effect, and a type. It defines characteristics and is used by AnyLevel.

AnyEvent: It is the event when an instance of corruption is confirmed as in the case of Nancy or in any other cases, where the guilty commits the crime of corruption. AnyEvent has a unique ID, an occurrence date, time and address, and a name. It occurs at a specific time, is organized for a specific reason, and it effects AnyActor/AnyParty by increasing or decreasing corruption.

7.2.2.2 Nonfunctional Requirements

1. *Morality:* Corrupt practices are immoral and illegal. In fact, there is morality associated either with instances of corruption or with people who commit corruption. Corruption usually involves embezzlement, swindling, cheating, and misappropriation, mismanagement, stealing, or duping gullible people of their wealth.
2. *Confrontation ability:* Confrontational ability of a person or an institution to reduce or eradicate corruption is very limited as corruption always takes an invisible form while people who commit crimes of corruption hide behind the cloak of legality. It requires inhuman efforts to confront the ugly tentacles of corruption.
3. *Prevention:* Preventing corrupt practices is very difficult and it requires solid proof and evidences to bring corrupt institutions (like Enron) and corrupt persons (like bank fraud individuals). However, corruption should be preventable by presenting solid evidences and public support.
4. *Controllable:* Corruption should be controllable. Otherwise, it can create havoc in the entire institutional and societal system. For example, a failure to foresee and prevent corrupt corporate practices of Enron officials eventually resulted in huge losses to shareholders and a deep dent to corporate ethics and morality.
5. *Measurable:* One should be able to measure the degree or extent of corruption either in monetary value or in losses incurred to the society and institutions. To prove instances of corruption, the court of law should specifically state the amount of losses that occurred because of corruption. A failure to measure corruption level may not prove the crime in legal systems.
6. *Improvement:* A deep commitment should be made to improve the nation and society by controlling corruption. An improvement is possible only with the support of the people, legal system, media, and governmental intervention.

7. *Learning:* Corruption should provide a future platform to learn the debilitating effects of corruption on the nation, its people, and society at large. By learning more about the aftermath of corruption, it will be easier to control and manage corruption in the future and it will also help authorities to put in place newer mechanisms of preventing corruption.
8. *Impact:* The impact or influence of corruption could be potentially huge or insignificancy low depending on the situation. Societal corruption by ordinary people may be restricted to smaller domains of daily life, while institutional and organizational corruption could be disastrous for the country and corporations as the money involved could be humongous.

7.2.3 Challenges and Constraints

Challenges as shown in Table 7.1.

7.2.3.1 Constraints

1. Corruption is caused by one or more party mismanaging any organization.
2. Any corruption has one or more impact on the organization involved.
3. Any situation involving any corruption consists of at least one or more evidence.
4. AnyType has to be defined for any corruption to distinguish with other corruptions to know the level of that corruption.
5. AnyDamage has to be examined properly to know the effects of the corruption.

7.2.4 Solution

The solution for *AnyCorruption* pattern is demonstrated by using a class diagram and CRC cards that includes the role of each classes and its operation as shown in Figure 7.1 and Table 7.2.

7.2.5 Applicability with Illustrated Examples

CASE STUDY 1 Public Corruption

Involves a breach of public trust and/or abuse of position by federal, state, or local officials and their private sector accomplices.

Use Case as shown in Table 7.3.

TABLE 7.1
Challenges

Challenge ID	001
Challenge Title	Situation led to corruption; if unclear, leads to confusion.
Context	Situation, under which corruption occurred, should be clear to AnyParty().
Description	Political leaders AnyParty() perform corruption to gain votes. This leads to unpopularity of the party.
Solution	AnySituation() that led to corruption must be communicated to AnyParty().
Challenge ID	002
Challenge Title	Event that leads to corruption causes damage to a reputation of the party.
Context	Events leading to corruption should not be undertaken.
Description	Donating money to needy people may lead to misuse of money by the people. This damages the reputation of political party.
Solution	Organize events that will not damage reputation of AnyParty.

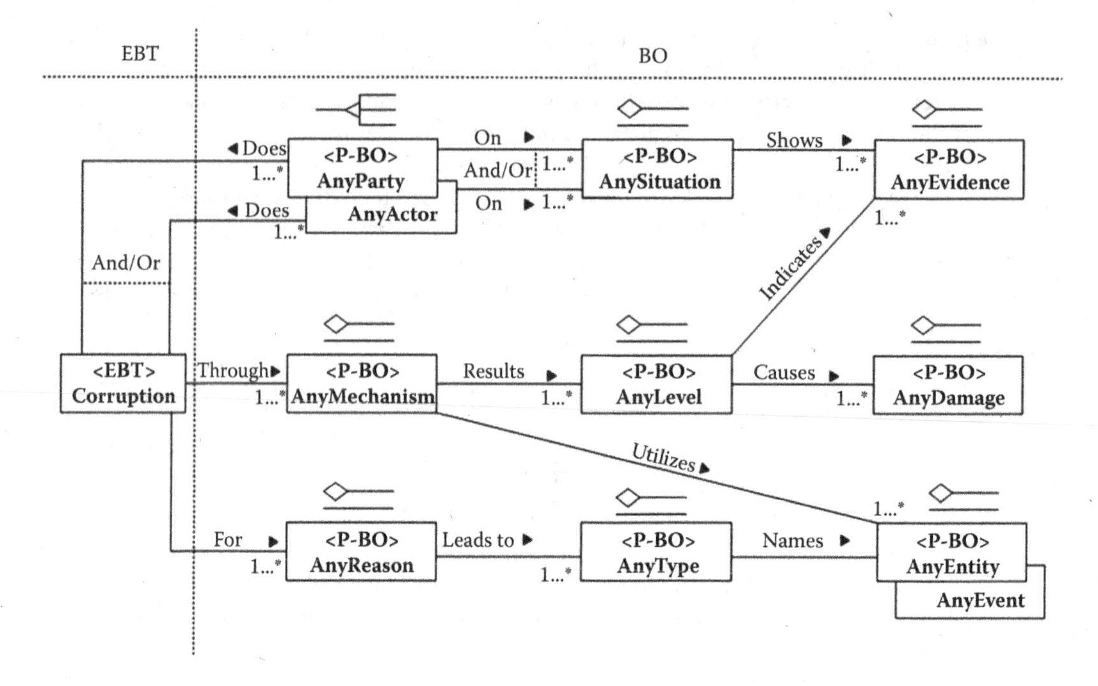

FIGURE 7.1 Corruption stable analysis pattern class diagram.

Use Case Description

1. Corruption can be performed through any mechanism and will be of different kinds, which affect the society.
   Can corruption affect the society?
   Are there different kinds of corruption?
   Does corruption use any mechanism?
2. Anyparty(AnyParty) will practice corruption and specify the amount for the work done by them.
   Does AnyParty practice corruption severely?
   Does AnyParty perform their work?
   Can AnyParty accept amount?
3. AnyType (AnyType) possess the AnyParty(GovernmentOfficial), determines AnyEntity (AnyEntity), and carries corruption.
   Does AnyType have many corruptions?
   Does AnyType possess the government officials strongly?
   Does AnyType determine AnyEntity properly?
4. AnyMechanism (AnyMechanism) is implemented by corruption and will have AnyDamage(AnyDamage) and decline the AnyDamage(EconomicGrowth).
   Does AnyMechanism implement corruption?
   Does AnyMechanism have damages?
   Does AnyMechanism decline the economic growth?
5. AnyEntity(AnyEntity) will bring out some amount of money to deliver to AnyParty (GovernmentOfficial).
   Does AnyEntity have money?
   Does AnyEntity provide money to AnyParty?
   Does AnyEntity deliver money properly?

TABLE 7.2
CRC Cards

Corruption (Corruption) EBT

Responsibility	Collaboration	
	Client	**Server**
Disrupts the functionality of any organization	AnyParty AnyActor AnyMechanism AnyReason	disrupt(), inform(), providePerpetrator uncoverDisruptionType()

Attributes: engagedOrganization, perpetratorInformation, corruptionType

AnyParty(AnyParty) BO

Responsibility	Collaboration	
	Client	**Server**
Indulges in any corruption	Corruption AnySituation	indulge(), work() mismanage() investigate()

Attributes: name, designation, contactInformation, systemWorkingFor situationInvolvingCorruption

AnySituation(AnySituation) <BO>

Responsibility	Collaboration	
	Client	**Server**
Hold any evidence of corruption	AnyEvidence AnyParty	includeEvidence() involve()

Attributes: evidenceIncluded, personInvolved

AnyEvidence(AnyEvidence) <BO>

Responsibility	Collaboration	
	Client	**Server**
Reveal corruption	AnySituation AnyLevel	linkSituation() reveal() holdDistortedInformation()

Attributes: linkedSituation, distortedInformation

6. AnyDamage(AnyDamage) will show result, and cause impact for the development of a AnyParty(Society).
   Does Any Damage have result?
   Does Any Damage cause any impacts?
   Does Any Damage affect the society?
7. AnyParty(Society) will offer AnyMechanism(Bribery) and cause Any Damage(EconomicGrowth) on which it depends.
8. AnyParty(GovernmentOffical) will affect AnyMechanism(Bribery) to determine AnyType(PublicCorruption) to execute their work.
9. AnyType(PublicCorruption) involved by talking AnyMechanism (Bribery), which influences and reduces AnyDamage(Economic Growth).

TABLE 7.3
Perform Corruption Use Case

Actors		Roles	
AnyParty		Society, GovernmentOffical	
Class	**Type**	**Attributes**	**Operations**
Corruption	EBT	Level, type, location, sector, name	perform() have() affect()
AnyParty	BO	Name, address, url contactNumber idNumber	practise() specify() done()
AnyType	BO	Rate, type, characteristic, kind, level	possess() determines() carry()
AnyMechanism	BO	mechanismName mechanismType, mechanismDate, process, result	implement() have() decline()
AnyEntity	BO	Name, description, kind function property	bringAbout() amount() deliver()
AnyDamage	BO	Level, location, situation, result, source	showResult() simulate() develop()
Society	IO	Region, development, educated, status, history	offer() cause() depend()
GovernmentOffical	IO	Destination typeOfReason idNumber phoneNumber departmentName	accepts() determine() execute()
PublicCorruption	IO	Type, region, consequence, reason, department	involves() influence() reduce()
Profit	IO	How, region, departmentName, reason, transaction	present() satisfy() received()
EconomicGrowth	IO	Status, development, rate, region, type	depend() result() simulate()
Bribery	IO	Kind, departmentName, situation, location, personName	result() given() issue()

10. AnyMechanism (Bribery) will result in AnyDamage(AnyDamage), given by AnyParty(Society) to consider their work.
11. AnyEntity(Profit) will present the amount and satisfy to AnyParty(GovernmentOfficial) and receive from AnyParty(Society).
12. AnyDamage (EconomicGrowth) depends on AnyType (Public Corruption) will cause AnyDamage(AnyDamage) to simulate Anyparty(Society).

Class Diagram

1. Corruption will be practiced by AnyParty(AnyParty).
2. AnyParty(AnyParty) is a AnyParty(GovernmentOfficials).
3. AnyParty(GovernmentOfficials) for their AnyEntity(Profit).
4. AnyEntity(profit) can be AnyEntity(AnyEntity).
5. AnyEntity(AnyEntity) determined by AnyType(AnyType).
6. AnyType(AnyType) will be AnyType(PublicCorruption).
7. AnyType(PublicCorruption) involves AnyMechanism(Bribery).
8. AnyMechanism(bribery) affects AnyDamage(EconomicGrowth).
9. AnyDamage(EconomicGrowth) was AnyDamage(AnyDamage).
10. AnyDamage(AnyDamage) results by AnyMechanism (AnyMechanism).
11. AnyMechanism(AnyMechanism) can be done through corruption as shown in Figure 7.2.

CASE STUDY 2 Corruption in Court Rooms

Corruption is a perennial disease destroying our society, culture, and traditions. More and more crimes of various kinds take place, because people committing these crimes know that there is always a way out of the law. This ability of criminals to find loop holes from within the law has increased due to the spread of corruption into the sanctity of court rooms. There are various ways to corrupt a court room from buying the judges, the jury, and the prosecutor by intimidating them. Though intimidation leads to corruption, it is not a direct form of it and has not been considered for discussion in this section. This section primarily deals with corruption in court rooms due to bribery. *Use Case* as shown in Table 7.4.

Use Case Description

1. AnyParty(Criminal) spreads corruption by AnyType(Bribe).
   Does the criminal bribe other parties?
2. AnyParty(Criminal) spreads AnyType of Corruption through AnyMechanism.
   Does the criminal make use of AnyMechanism to spread corruption?
3. AnyType(Bribe) is paid to AnyParty(Court officials) and AnyParty(Jury) by.
   AnyParty(Criminals) at AnyLocation.
   Is the bribe paid at the specified location?
4. AnyParty(Jury) rules leading to AnyConsequence(Criminal Acquital).
   Does the jury vote to acquit the criminal.
5. AnyParty(court officials) carryout the ruling to acquit the AnyParty(criminal).
   Is the criminal allowed to go free?

Class Diagram as shown in Figure 7.3.

7.2.6 Business Rules

1. Corruption is practiced by some AnyParty.
2. AnyMechanism is used to perform the corruption.
3. Each AnyMechanism results in some AnyDamage.
4. Each AnyDamage can have only one AnyLevel.
5. Each corruption has some AnyType.
6. One or more AnyType determines one AnyEntity.
7. Each AnyDamage is related to one or more AnyDamage.
8. Each AnyMechanism utilizes one or more AnyEntity to cause corruption.

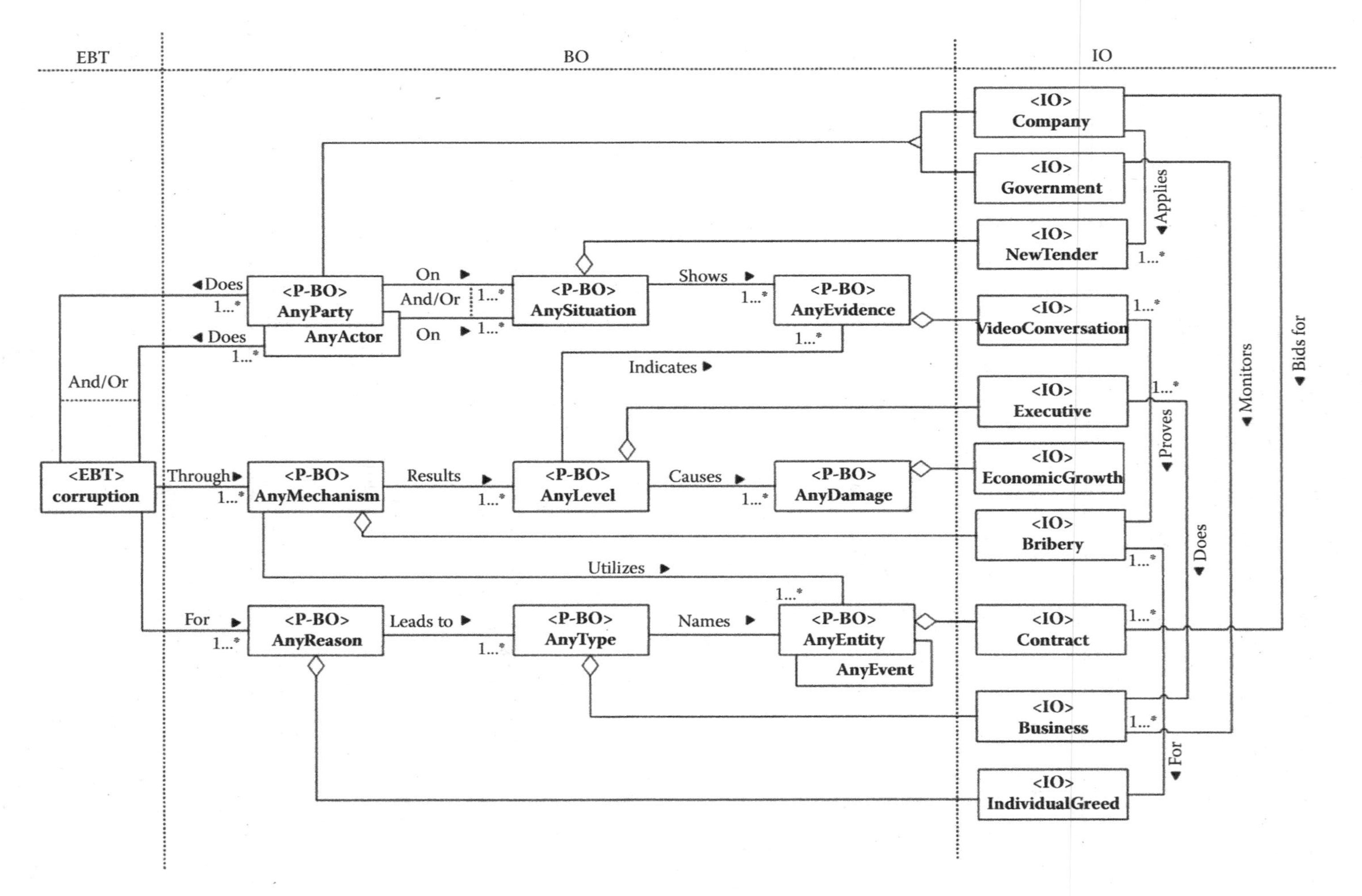

FIGURE 7.2 Public corruption application.

TABLE 7.4
Corrupt in Courts Use Case

Actors	Roles
AnyParty	Court Official Jury Criminal

Class	Type	Attributes	Operations
Corruption	EBT	level type extent name result	Pay() Accept() Performillegal()
AnyParty	BO	Name address greed	practise() decide() accept() bribe() acquit()
AnyType	BO	Rate type characteristic kind level	possess() determines() carry()
AnyMechanism	BO	mechanismName mechanismType mechanismDate process result	implement() have() decline()
AnyLocation	BO	Address location	Locate() Exist()
AnyConsequence	BO	Result	Criminalacquit() depend()
CourtOfficals	IO	Name Designation Power rate	accept() decide() execute()
Jury	IO	Name Address rate	involve() accept() decide()
Criminal	IO	Name Crime Wealth	Bribe() Escape()
Criminal Acquittal	IO	Crime Value result	depend() result() escape()
Bribery	IO	Name department situation location amount	Give() Take() Demand() Decide()

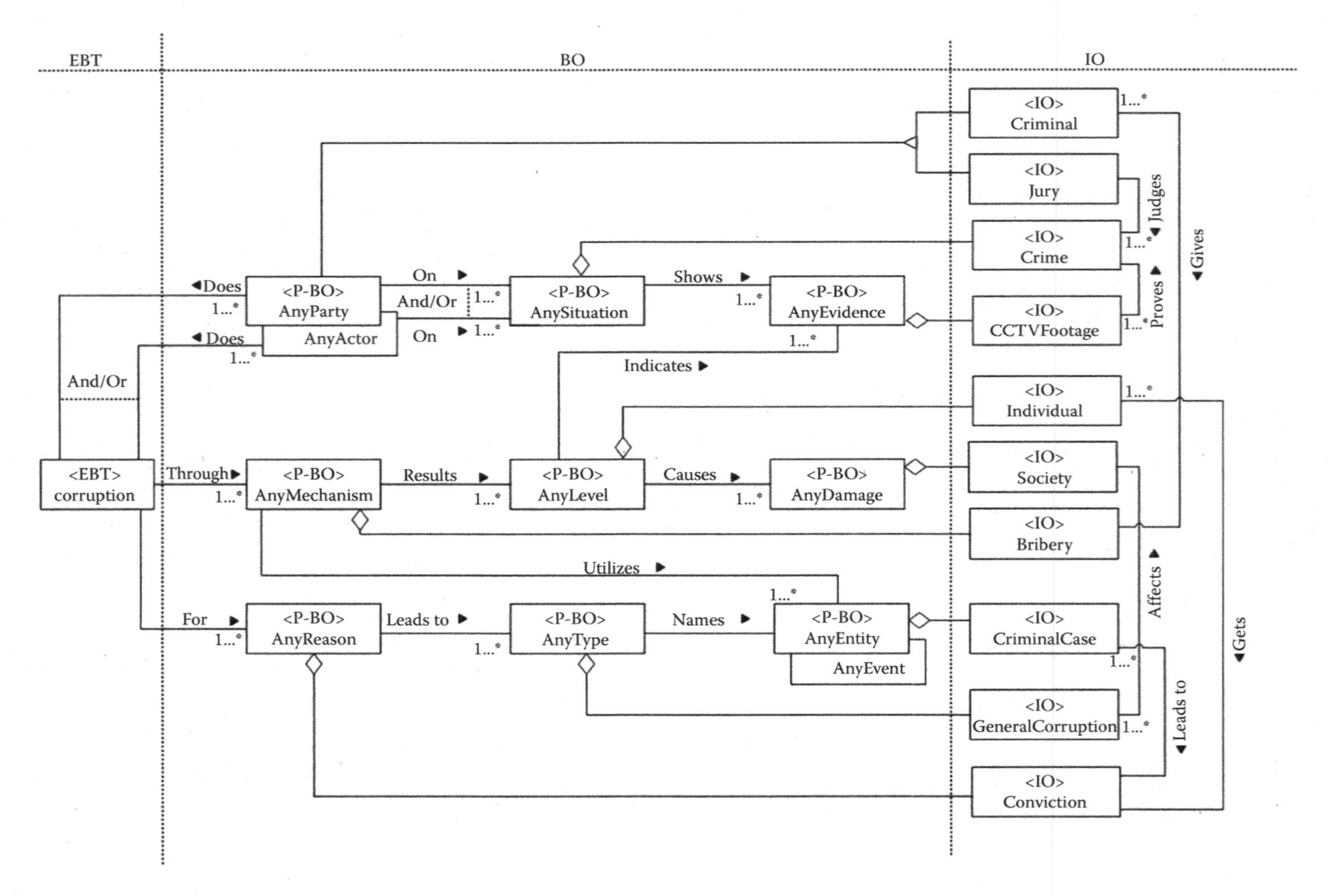

FIGURE 7.3 Class diagram of corruption in courts.

9. Each AnyMechanism results in one or more AnyDamage.
10. AnyMechanism results in one or more AnyLevel of AnyDamage.
11. Some AnyParty achieves one or more AnyLevel of AnyDamage.
12. AnyLevel measures the effect of the AnyDamage caused by the corruption.
13. If there are systems to monitor AnyMechanism and AnyEntity, corruption can be reduced.
14. AnyParty takes part in corruption as the AnyEntity provides a material gain to AnyParty.
15. Corruption can also happen, when AnyMechanism is not working properly and hence, causing the corruption. This may not be the true behavior of the AnyMechanism, but other factors are forcing it to cause corruption.

7.2.7 Known Usage

- Data corruption in a database system
- Political corruption in a government
- Social corruption due to the greediness of the people
- Business corruption takes place from small to big business companies
- Public corruption will be in any organization

1. *Enron accounting scandal*: Enron is a classic case of corruption. They perpetuated fraudulent accounting practices and misled the investors. They created offshore units, which could be used for planning and avoidance of taxes. This was used for hiding their losses and making Enron look more profitable. It created an illusion of billions of profits for the company, and it was done just to drive stock price higher up. At this point, the executives started working on the insider information and started trading Enron stock. The investors did not know anything about the account practices. So, the executives gained millions of dollars, while the ordinary investors lost all their money.
2. *Linguistic corruption*: This is a term used for changes in a language by cumulative errors in writing or the way some words are used or spelled. This is very rampant in this Internet age, where the words get shortened to reduce the time it takes for typing the word. This is taking away the purity of the language. But, the argument against this could be that all languages are evolving and has to be changed with the times based on the people who practice the language.

7.3 SUMMARY

By using the stability model we have developed a corruption pattern, which includes use cases, CRC cards, class diagram, and sequence diagram. While developing the model with many numbers of iterations, we have figured out relevant BOs for EBT, as well as IOs. We have also built the pattern in general, so that it fits into the other domain context easily.

7.4 OPEN AND RESEARCH ISSUES

The chapter has discussed the application of the corruption analysis pattern to data and interactions between people. However, can the pattern be applied to more than these two examples?

1. Application of the corruption pattern to the analysis of illness in living organisms. Diseases such as cancer can be thought of an entity that invades an organism and corrupts the normal bodily functions of the host organism. Can the corruption analysis pattern be applied to measure the affliction of cancer and similar diseases?
2. The term corruption itself can be applied to inanimate objects and systems. For example, a clean drinking water supply can be corrupted due to unnatural pollutants or other natural disasters. How well can the corruption pattern be applied to track the quality of drinking water, or food sources?

3. Corruption-unified software engine (Corruption-USE): Utilize stability model [1,2] or knowledge map [3] methodology as a way for developing *Corruption*-USE [4]. The engine mainly focuses on several patterns: *Corruption*, Analysis, *Recording, Legality*, etc. The proposed solution attempts to extract the commonality from all the domains and represents it in such a way that it is applicable to a wide range of domains without trivializing or generalizing the concepts. The engine is a stable structural pattern and it provides a generic engine that can be applied and/or extensible to any application *of Corruption. The Corruption*-USE can be applied to unlimited context and reusability.

REVIEW QUESTIONS

1. (T/F) The Corruption design pattern is domain-specific.
2. (T/F) The corruption design pattern can only be applied to human activities.
3. (T/F) The corruption design pattern cannot be applied to inanimate entities.
4. (T/F) The corruption design pattern describes a system.
5. (T/F) The corruption design pattern can be applied and extended to any domain.
6. List two major applications of the corruption analysis pattern.
7. Why do the BOs of the pattern have "Any" as a prefix?
8. List three challenges in formulating the corruption pattern.
9. List four constraints during the formulation of the corruption analysis pattern.
10. Draw the design pattern solution for the corruption analysis pattern.
11. List the participants (EBTs and BOs) in the corruption analysis pattern.
12. Illustrate a detailed model with a diagram for the corruption analysis pattern.
13. Draw the CRC cards for two of the BOs in the corruption analysis pattern.
14. List four main benefits and usage notes for the corruption analysis pattern.
15. List the use cases in classifying politician data applying the corruption analysis pattern.
16. List some of the patterns related to the corruption analysis pattern.
17. List two design issues for the corruption EBT during the process of linking from the analysis phase to the design phase.
18. List two important implementation issues for the AnyMechanism BO, during the process of linking from the design phase to the implementation phase.
19. List two scenarios that would not fit within the context of the corruption analysis pattern.
20. List three business issues in the context of the corruption analysis pattern.
21. Briefly explain how software stability concepts have been incorporated in the corruption analysis pattern.
22. List a couple of research issues relevant to the corruption analysis pattern.
23. How will the corruption analysis pattern be stable over time?
24. Explain briefly how the corruption analysis pattern provides a high level of extensibility.
25. List the synonyms of "Corruption." Can these terms be used interchangeably with respect to the corruption pattern?
26. What are the requirements for AnyCorruption? Describe each of them?
27. List differences between the corruption analysis pattern described here and the traditional methods.
28. List some design and implementation issues faced, when implementing the corruption pattern. Explain each issue.
29. What do you think are the implementation issues for the AnyEvidence BO, when used in the corruption analysis pattern?
30. What do you think are the implementation issues for the AnyReason BO, when used with the corruption pattern?

31. List some of the testing patterns that can be applied for testing the corruption analysis pattern.
32. List three test cases to test the class members of the corruption pattern.

EXERCISES

1. Explain with an example the applicability of the corruption analysis pattern in classifying politicians participating in national elections. Draw the class diagram include all the EBTs and BOs.
2. Illustrate with a class diagram the applicability of the corruption analysis pattern to a major league sports association club member. The pattern will track club members' association with illegal betting and match fixing.
3. Think of any scenarios that have not been described in this chapter. List them with a short description.
4. Try to create a use case and interaction diagram for each of the scenarios you thought of in the above question.

PROJECT

1. Create a political website that tracks the associations of politicians and the private groups that donate campaign money. This website will also trace the voting patterns of the politician and relate them to the respective donations and perks they received during their campaign. With this information create a system for scoring politicians and their level of corruption based on the collected data. Use the US political system and its PACs as examples for formulating the website. Create class diagrams, sequence diagrams, and CRC cards to describe the problem.
2. http://en.wikipedia.org/wiki/Political_action_committee
3. Design a Civil Group Corruption Watch website that provides updated information about political and departmental corruption in offices. The main aim of this group is to track electoral politics and illegal activities in governmental departmental and take action against erring officials and politicians who indulge in corrupt practices. With the information gathered, create a rating system to track and tag erring individuals based on the published primary data. Design class diagrams, sequence diagrams, and CRC cards to highlight any problems.

REFERENCES

1. M. E. Fayad and A. Altman, An introduction to software stability, *Communications of the ACM*, 44(9), 2001, 95–98.
2. M. E. Fayad, Accomplishing software stability, *Communications of the ACM*, 45(1), 2002, 95–98.
3. M. E. Fayad, H. A. Sanchez, S. G. K. Hegde, A. Basia, and A. Vakil, *Software Patterns, Knowledge Maps, and Domain Analysis*, Boca Raton, FL: Auerbach Publications, 2014.
4. M. E. Fayad. Unified Software Engines (USEs). In Progress.

BIBILOGRAPHY

http://info.worldbank.org/etools/docs/library/35970/mod03.pdf
http://www.raids.org/gen00343.htm
http://www.fbi.gov/hq/cid/pubcorrupt/pubcorrupt.htm
http://209.85.173.104/search?q=cache:6-DQUI4mPYMJ:unpan1.un.org/intradoc/groups/public/documents/APCITY/UNPAN019105.pdf+administrative+corruption&hl=en&ct=clnk&cd=1&gl=us&client=firefox-a

8 Dignity Stable Analysis Pattern

A dignity can be assessed for anything that cannot be replaced.

M. E. Fayad

Dignity is the state of being worthy of honor and respect. It is a term that defines the respectability and worthiness factor. Generally speaking, dignity has many shades and hues. It applies to all spheres of life, some examples being in religion, humans, race, caste, country, academics, sports, organizations, and even also in animals. The concept of dignity mostly identifies with humans. Human nature has a tendency of being dignified. Every human wants to be dignified regardless of age, sex, health status, financial status, social or ethnic origin, political ideas, or criminal history. Every country wants to be superior and dignified [1].

Dignity is a concept that can be assessed to some extent. In other words, dignity can be assessed for anything that cannot be assessed (as quoted by M.E. Fayad). Humans try to fit in this dignified image by means of some metrics for assessment such as financial position, political hold, age, and experience. All these parameters provide a state that gives humans a way to assess and judge themselves. Dignity implies that every human being should be acknowledged as a valuable member in human society. This principle applies to animals too. In other words, even animals, from highest to the lowest order, have a sense of dignity. *Dignity can be assessed for anything that cannot be replaced.* They always strive to be superior and perfect in their community/group. This right is best exemplified by Immanuel Kant's celebrated quote—"Everything has either a *price* or a *dignity*" [2,3]. What has a price can be replaced by something else as its *equivalent*; what on the other hand is above all price and therefore admits of no equivalent has a dignity." In spite of their lower status compared to humans, animals try to establish their own identity and dignity in a natural world that is full of competition and survival instinct. From the basic human point of view, dignity is the state of integrity and virtue. It is an admirable quality that everyone strives to achieve. The main objective of this chapter is to create a generic model of dignity, which can be applied to any area. The best way to present this model is stability model.

8.1 INTRODUCTION

Dignity is state of virtue and self-esteem. It is an estimation of one's respect and merit in society. As dignity changes, it can take different meanings under different contexts, but basic concept remains similar. Dignity for a country can be in the form of superiority, unity, power to protect the nation and its people, but the term dignity can take a very different meaning when it is referred in terms of dignity of a woman. Dignity for a woman, is her self-respect.

An example of dignity in human life is the dignity that is associated with academic positions. Doctors and surgeons are considered as persons next to God, as they are the ones who save life, when someone is in the midst of a critical health status. This provides them dignity in society. They are revered and worshipped in society because of their position, capability, and academic excellence. Same thing applies to scientists, engineers, lawyers, physicians, and astronauts.

This does not imply that people, who are normal human beings doing the simplest job and earning a meager amount of money, are not dignified. Dignity, in this context, is identical to everyone. The only dynamic that is considered here is the measurability, that is, the assessment associated with that dignity. This is the sole reason why dignity gives an individual a certain place in society, thereby marking the very existence of that individual.

One another example of dignity is marriage. When two people come together in an institution of marriage, they demonstrate dignity to each other and they also shower respect toward each other. This is the only and ultimate way to survive the marriage. If the dignity is lost for any individual, the marriage falls apart and harmony/synchronization is lost forever. Therefore, it is very important to assign dignity to an individual and maintain it. In other words, dignity can be termed as the right of an individual to survive and achieve certain degree of esteem in life.

8.2 PATTERN DOCUMENTATION

Dignity is the term used to show the state of respect and worthiness. Dignity is an enduring quality of humans and animals. It applies to other entities also. Other names for dignity are respect, esteem, nobility, honor, self-respect, and prestige.

The basic idea behind choosing the term dignity is to give this pattern a general concept. Generality is the driving force for choosing the term dignity, as this term applies to all fields with its different types, and it takes different values, yet leads to the same meaning. This generality will also lead to a stable analysis pattern for dignity by using it as EBT and it will help in categorizing BOs. Hence, it applies to all other names of dignity and application using these names. More generic term will always lead to greater applicability.

8.2.1 Known As

The most common form of dignity is respect that can be used interchangeably sometimes as it leads to the same meaning. Respect is a personal quality in man, which shows his or her esteem or worth in a society. The only difference between dignity and respect is that there can be a manifestation of respect, whereas dignity is something that is an intrinsic part. Dignity can be achieved through any rules, which follow some basic criteria to get that quality. It is achieved through some actions or deeds. Actions greatly affect the way we perceive dignity of a person. Respect is also the equivalent. Some rule/standard has to be there in order to be respectful in a society. Deeds greatly influence the degree of respectability.

Dignity can be possessed by any party and can take any state by using some of the common metrics. Another common word that is similar in context to dignity is esteem. Esteem is similar to dignity, but it specifically emphasizes on estimation and valuation of respect or worth. It is associated with some level or degree. But more or less, we can combine the meanings of dignity and esteem.

Worthiness is the word that has identical meaning as dignity, but is used rarely as a substitute for it. The reason for this is when we consider worth, some values get associated thereby limiting its generality and the main concept of dignity is entirely lost. More or less, state of worthiness is somewhat similar to dignity, but has some differences. Worthiness possesses some merit. When we consider the worth of a person, it may imply his or her financial status, but when we consider it in terms of dignity, a wide array of terms comes into picture, which may include person's power, position, academic excellence, age, and experience. From this example, we can in all likelihood, portray the fact that dignity is the proper generic term.

The word that is mostly used as substitute for dignity is pride. However, it is not the proper way of referring to the dignity of a person, as it relates to something that a person can be proud of. This does not necessarily imply dignity. Thus, learning from points listed above, it can be well-observed that dignity is the most comprehensive term for describing state of respect for any party or any precious entity.

8.2.2 Context

Dignity is the quality that all strives to achieve and possess. It is the state of respect that everyone aspires to get in a society. One of the proverbial saying goes like this,

> Dignity is not in what we do, but what we understand.

The same applies to everyone in all domains. Below are some of the contexts that apply to dignity pattern.

- *Dignity of country:* When dignity of a country is referred, it states the power of the country, superiority, ability to protect its people, provide its citizens with important basic amenities in life such as health care, new developments, and research in various fields, its military power, its hold and regard in whole.When war occurs between two countries, the parties involved consider it a matter of dignity to win. Here, the requirement for dignity is a victory. Providing human rights to its citizens also dignifies the country. It shows the capabilities and concern for its citizens. Many organizations such as WHO work toward this cause and help a country fight serious outbreaks of diseases. Hence, providing human rights dignifies the country in the sense it shows its efficiency to tackle the problems facing the country.
- *Dignity of person:* Dignity of person implies his or her designation, experience, age, financial status, academic excellence, and political hold. The simple example of this is, while considering marriage proposal, one takes into account first the position/designation/job of the person, then his financial status and so on. In other words, the more dignified the person is, the better the prospects and options for marriage are.Other context in similar terms is of dignity at work. A person (employee) will get a job and attends the office, only when the position offers dignity. He or she does work and displays the expertise and skills that are necessary to contribute something to society. The work also leads to an employee's workplace appraisal. A positive appraisal will motivate the worker to perform well in the office and in a sense his or her stand gets dignified in the workplace thereby making him or her efficient and worthy employee.
- *Dignity in sports:* Various sports events are held every year around the world at different places. Every country encourages sports, sportsmen, and women. They send their best of team to the event to represent their country. The idea behind this is to present before the world the country's best performers, to deliver their best, and bring a crown of glory to their country. To bring good reputation to one's own country is the requirement for dignity here. Best example is the Olympics or Commonwealth Games.
- *Dignity in profession:* Every profession has some degree of dignity associated with it and hence, the respect accorded to each profession. The most dignified profession is of doctors and nurses. They are considered next to God. Doctors are respected everywhere, as they are the ones who save life. Similarly, teaching profession is also highly respected, as a teacher disciplines a child from the very beginning and makes him or her a good human being. Thus, according to an individual's profession, they are given respect in society.
- *Dignity in animals:* In some animals, the male members fight with each other in order to impress the female member of the same species. Whoever wins gets companionship of the female animal. For them, it is a matter of dignity too.

As illustrated here, dignity stable analysis pattern can be applied in many contexts mentioned above. It should also be applicable to all contexts related to dignity. Only then will its purpose of commonality be satisfied.

8.2.3 Problem

To build a stable analysis pattern of dignity, we should first understand the core purpose of the pattern. To understand the essence of the pattern we should first focus on what the actual requirements of the pattern are and how they should be achieved.

Building a stable pattern for dignity is not an easy task. We will need to first project the concept of dignity in a general way, so that it can be applied to any application in this context. To achieve this objective, we will also need to find selected business objects in a model, which in combination gives an idea of generic dignity. Then, we should be able to use the same model pattern and apply it to different dignity context. If the model fits into any context of dignity, then the generality of the whole concept is gained.

The second problem to focus on is the idea that this pattern applies to any party. Hence it should be general enough for every entity, be it hardware, person, or animal. Next is the analysis domain. The model should analyze the domain and context in which dignity is used in a pattern. Analyzing the domain involves identification of BOs and IOs and how the objects communicate with each other using different relation.

The points listed below give some means of reducing red flags in modeling dignity stable pattern.

- Pattern should understand the core part of the concept, which is EBT of the pattern.
- It should focus on actual requirements of the pattern.
- Pattern must be general and flexible enough to be applicable to variable contexts.
- Analysis of the pattern should be in a domain specifically.
- Dignity is based on assessment, so it becomes very important to define assessment accurately and the rules that influence assessment are correctly defined.
- The pattern should be able to evolve with changing needs.
- The pattern should possess complete coverage, thus it should be able to define and provide dignity in any application.
- While using dignity stable analysis pattern, this pattern must show some kind of specificity to the domain. This means that while maintaining enough generality, it should also exhibit specificity, so as to apply the pattern in any specific domain.
- As dignity is used in a wider context and in varying domains, building a generic model without losing functionality is quite challenging. By using stability model, this problem is solved and a generic model is modeled for different domains.
- The pattern should be able to explain the context of dignity when used.
- It should be applicable to AnyParty.

Dignity requirements include the following:

1. *Requirements: Functional Requirements:* Functional requirements may be classified as:
   a. Internal requirements: As the name suggests, these requirements are internal to the goal; it indicates the nontangible things needed to achieve dignity. This requirement is directly related to the current position of the party or actor involved. As the status of the party or actor changes, internal requirements also change. These requirements are not directly visible to others, but are tightly intertwined with the EBT. Some of them for dignity can be
      i. Work: The first and foremost thing through which a party or person is recognized is work. Good deeds always show up. Thus, to attain dignity, a person should always work for the benefit of the community.
      ii. Designation or Position: It refers to the status of a party or actor in society, as it directly influences its work. Depending on the current position, the next level of dignified position is determined and the person is given recognition.
      iii. Responsibilities: With dignity also comes important responsibilities. They form an internal part of dignity and the party or actor has to abide by its responsibilities.
      iv. Community: It also plays an important role in attaining dignity. It basically decides the spectrum of it. It is possible that a person who is recognized and famous in one community may not be recognized by some other community.
   b. External requirements (common requirements for all possible scenarios): These represent all the BOs that are important for gaining dignity. Based on the goal to

achieve, the BOs change their definition and scope of application. The BOs related to dignity, are described below

i. *State:* There are different states of dignity based on many factors that are related to rules and regulations: domain, entity, and/or the parties and actors (the users) of the pattern. Dignity of human is the state of being honored, esteemed, respected, or inherently worthy. This state of dignity keeps on changing with time and also with numerous changes in rules and regulations. So, the first and foremost requirement of this pattern is to remain stable with changing states and the pattern should also reflect the change in state.

ii. *Type:* There are different types of dignity based on many factors that are related to the domain, entity, and/or the parties and actors (the users) of the pattern. Dignity is either essential or accidental. Dignity can be awarded or granted. So, the pattern should be able to represent all the types of dignity and their context.

iii. *Parties/actors:* Dignity can be of a country, party, or a single person. Even animals and plants have their own dignity. So, the pattern should be able to stand for all, be it plant, animal, human, or country.

iv. *Rules:* There are some constraints imposed on dignity in any domain and these rules influence the means for achieving dignity, and also lead to change in state. Thus, the pattern should be able to accommodate all these changes with the change in rules.

v. *Domains:* Dignity can be applied to any domain such as knowledge, environment, etc. Every domain has its own dignity with certain degree or level, and it defines its own rules for achieving dignity. The stable pattern should also represent dignity in this context.

vi. *Entities:* Dignity can be applied even to any entity, for example, a building. If the building is of a hospital, then it has different dignity, but if it is President's office then it shows different level of dignity. The pattern should be able to convey dignity of any entity in correct context.

vii. *Assessment:* It is the means for achieving dignity and depends on rules and context of dignity. Different types and levels of assessment lead to different states of dignity. A good pattern should be able to show any kind of assessment and the state it leads to.

Nonfunctional Requirements:

a. *Fulfillment of rules/rights:* The pattern should define properly the influence of rules on achievement of dignity. Any type of dignity has some rules that are defined for its achievement, so the pattern should exhibit type of rules defined and how they influence the assessment of dignity.

b. *Quality:* In the definition of dignity, it is the quality of one being worthy of esteem and respect, especially humanness, but it could also be, for example, augustness, nobility, majesty, grandeur, glory, superiority, wonderfulness [4–6]. It is also the quality or state of being excellent; the quality of being poised, or formally reserved in appearance and demeanor; a high rank, office, or title.

c. *Appropriateness:* The pattern should be able to present the correct meaning of dignity when applied. It should also determine the correct state, depending on the assessment level and rules that are defined by the application.

d. *Completeness:* The pattern should be complete in the sense it should be able to present all the meanings of dignity and the areas where it can be applied. This means that the pattern is applicable to many areas and it will have different meaning everywhere. Subsequently, the pattern should also possess the quality of inferring correct meaning in different contexts and should define the state, rules, assessment, and type according to the context in which it is applied.

e. *Worthiness:* Acknowledging your self-worth allows you to still feel dignified even under the most extreme or critical situations, especially when people humiliate or offend you. Unfair treatment and violation of rights can torment and cause anguish to you. Having a proper sense of pride and dignity will allow you to continue to walk tall and proud, and most importantly not take insults on a personal level.
f. *Autonomy:* Autonomy refers to the capacity of a rational individual to make an informed decision. Autonomy is often used as the basis for determining moral responsibility for one's actions [7,8]. An autonomous person has dignity conferred upon him or her by others [7,8].
g. *Virtue:* Virtue is the moral excellence and personal virtues are characteristics valued as promoting individual and collective well-being. Each individual has a core of underlying values that contribute to our system of beliefs, ideas, and opinions [9]. Integrity in the application of a value ensures its continuity and this continuity separates a value from beliefs, opinion, and ideas. In this context, having fundamental values is a very important issue in order to observe the true qualities of a person's dignity.
h. *Respect:* Respect is the esteem for a person, a personal quality, ability, or a manifestation of a personal quality or ability. In certain ways, respect manifests itself as a kind of ethic or principle. Possessing ethics and principles is a key quality of a person's true character and dignity.

2. *Properties:* Properties of the goal refer to operations and attributes associated with it. Each of the operations and attributes defined for EBT has some constraints or pitfalls attached to it. Hence, it becomes important to determine all the operations, attributes, and their constraints that can affect the pattern of dignity.
   a. Operations of dignity and the constraints associated with them are as follows:
      i. showsWorthiness():
         A. Worthiness is defined according to whether it is defined for a party or actor and hence, it keeps changing its meaning
         B. It also depends on the application of dignity, meaning the domain in which dignity is defined
      ii. defines():
         A. Definition of dignity depends on the assessment and the type of assessment carried out
         B. It also depends on the number of actors or parties involved
         C. Rules defined for attaining dignity also change its meaning
      iii. gives():
         A. This operation is restricted by the party or actor to whom dignity is given
         B. Type of dignity also restricts this operation
         C. The type of assessment done also affects
      iv. respects():
         A. Depends on party or actor involved
         B. Rules defined
         C. Mechanism used
      v. increasesPopularity():
         A. Domain of dignity
         B. Actor or party involved
         C. Number of entities involved
         D. Type of dignity
      vi. meetRules():
         A. Number of rules defined
         B. Factors involved in attaining dignity
         C. Domain of dignity

vii. typeIndication():
   A. Rules defined
   B. Domain of dignity
   C. Entities involved
viii. determineState:
   A. Number of states involved
   B. Determination of ultimate state
   C. How state changes
   D. Number of entities involved

b. Attributes of dignity and constraints associated with them:
   i. dignityAssessment:
      A. How to determine the appropriate assessment for dignity?
      B. How to distinguish among different types of assessment?
      C. How to determine minimum amount of assessment required to attain dignity?
   ii. possessedBy:
      A. How to determine the correct party or actor for dignity?
      B. What factors influence the dignity possession?
   iii. stateOfDignity:
      A. How to determine different states involved in achieving dignity?
      B. How does state change?
      C. What factors influence the state change?
      D. How to distinguish new state from the old one?
   iv. typeOfDignity:
      A. How to define the various types of dignity?
      B. How Domain affects type?
      C. How to distinguish various types from each other?
   v. value:
      A. What are the factors influencing value of dignity?
      B. How to determine exact value of dignity and in what terms?
   vi. context:
      A. How to separate different contexts of dignity?
      B. Conflicting issues arises when one form of dignity in one context has conflicting form in another context.
   vii. givenBy:
      A. Value of same dignity or position given by different people

3. Constraints
   a. The first and foremost constraint in requirement is, understanding the context of the pattern. The pattern adopts different forms depending on whether it is applied to a person, organization, entity, or some domain.
   b. In the case of requirement, the rules defined for pattern should also be laid down properly.
   c. The changes in state and their transition order should also be defined at this stage.
   d. The final state that will lead to attainment of dignity should also be defined properly.
   e. Any conflicting issues related to definition of dignity in the applied context should also be stated out clearly in the requirement phase.
4. ExampleIn order to provide a broader view of BOs and show how they relate to the goal of dignity, some examples are provided:
   a. AnyParty—Dignity can be associated with anyone, be it humans, plants, animals, organization, or any nonliving things. Everything has its own dignity in its own way, for example, road, hospital.

b. AnyDomain—Different fields have different meaning of dignity, for example, dignity in knowledge is different from that of sports. Dignity in economy/business is measured through wealth.
c. AnyAssessment: This BO represents the means by which dignity can be achieved and it takes shape according to the domain of dignity. For example, in business, business deals are one way through which dignity can achieved. In knowledge, dignity is attained through some honor/award.
d. AnyState: It represents some intermediate state, as well as initial and final state in attaining dignity. The state can be anything like various stages in book writing, experiment, etc.
e. AnyRule: In every field, rules have to be followed in order to climb the steps of success. So is the case with dignity. These rules can be enforced in any manner like rules and regulations that are imposed by government and rules at international level.
f. AnyType: This represents various forms of dignity, like in knowledge a person is dignified by giving titles.
g. AnyEntity: It is anything that helps in achieving dignity at any stage. For example, while discovering anything, experimental results from other scientists act as entity.

8.2.4 Challenges and Constraints

8.2.4.1 Challenges

Here are a few challenges as shown in Table 8.1.

8.2.4.2 Constraints

- There must be at least one party/actor/domain/entity with whom dignity is associated. Party, actor, domain, or entity forms the core concept, as it determines the context in which dignity pattern can be applied.
- Dignity must be able to define one or more assessments, which can ultimately lead to dignity.
- These one or more assessments should relate to any party/actor/domain/entity that relates to dignity.
- There are many rules that affect dignity and the party/actor/domain/entity associated with dignity must define one-to-many rules.
- Assessment should be influenced by at least one rule.
- Assessment must evaluate one-to-many states and these states keep on changing.
- One or more rules must influence the assessment of a party/actor/domain/entity that causes the state to change.
- Dignity is a form of self-respect, esteem, and regard for oneself that is assessed based on a set of predefined rules.
- There must be one or more parties/actors/domains/entities that do the assessment.
- A party/actor/entity/domain must have dignity that is defined by the assessment of oneself.
- A party/actor/entity/domain must be in at least one state depending on the assessment.
- AnyParty/Actor/Domain/Entity must have some level of dignity.
- The operations in the CRC cards should be generic enough to accommodate all possible applications.
- Dignity should have at least one type when applied in any context. There can be any number of types of dignity in a particular context.
- Any type of dignity should have some state to represent dignity. These states change depending on the assessment.
- The attributes and operations of dignity may not be applicable for every case. This is because it has very wide applicability and the pattern so developed should be generic to depict all those applications.

TABLE 8.1
Challenges

Challenge ID	0001
Challenge Title	Increase in number of AnyRule
Scenario	Marriage Proposal
Description	In the scenario where AnyParty is considering a marriage proposal, he/she takes into account the position, designation, job, financial status etc. of the proposing party. These are AnyRule that defines how dignified the proposing party is. If there are 10 such rules, then those can be met and the party can accept the proposal. On the contrary, consider a case when there are thousands of such rules imposed by AnyParty on the proposing party. As the number of rules increases, AnyMechanism of being dignified becomes more and more difficult, and after a particular threshold it will become impossible to meet all those rules.
Solution	AnyParty should set a threshold on number of AnyRule, so that these rules do not exceed a particular number. Another solution can be to set the priority on these rules and to consider only high priority AnyRule while doing AnyAssessment of dignity.
Challenge ID	0002
Challenge Title	Balance between AnyActor/AnyParty
Scenario	Dignity In Animals
Description	Consider a scenario where, in certain species of animals, male members fight with each other in order to impress female members of the same species. If there are only 10 female members in the species and there are thousands of male members, then only 10 male members would win the championship and the others would die in the fight. This might lead to extinction of the species.
Solution	There should be a balance between the number of female members and the number of male members in the species. To generalize this, there should be a balance between number of actors/parties that are being accessed and the number of actors/parties that accesses the dignity of the former Actors/Parties.
Challenge ID	0003
Challenge Title	Increase in number of AnyDomain
Scenario	Dignity In Sports
Description	Consider a scenario where, a certain sports person is good in a particular field or Domain. For Example—A sports person can be good in basketball, but not in football. Similarly, within a particular field like in cricket, a sports person can have his expertise in balling, but not in bating. If to be a dignified sportsman, a person needs to be expert in multiple games or in all the fields (AnyDomain) in a particular game, then it would be very difficult to find such a dignified sportsman.
Solution	While accessing dignity of AnyActor/AnyParty, there should be a maximum number of domains that should be accessed. If a person is accessed on thousands of domains then it would be hard to find a dignified person who meets the needs.

- Though there can be more attributes and operations of other BOs like AnyRule, AnyType, etc., but the operations and attributes of these BOs are chosen so that they can help in developing and supporting a generic pattern of EBT.

8.2.5 Solutions

The solution shown here utilizes stability model to explain the concept of dignity as shown in Figure 8.1. The derived solution given below focuses all the aspects of dignity and how a party implements different mechanism in myriad domains to achieve dignity. The proposed solution tries to analyze all the ways by which dignity can be achieved. This dignity pattern is applicable to wide range of domain, but the below solution is implemented in such a way that it extracts out the generality from all the domains and represents it in such a way that it is applicable to any of that wide range of domains without trivializing the concept. Figure 8.1 depicts the class diagram for dignity analysis pattern.

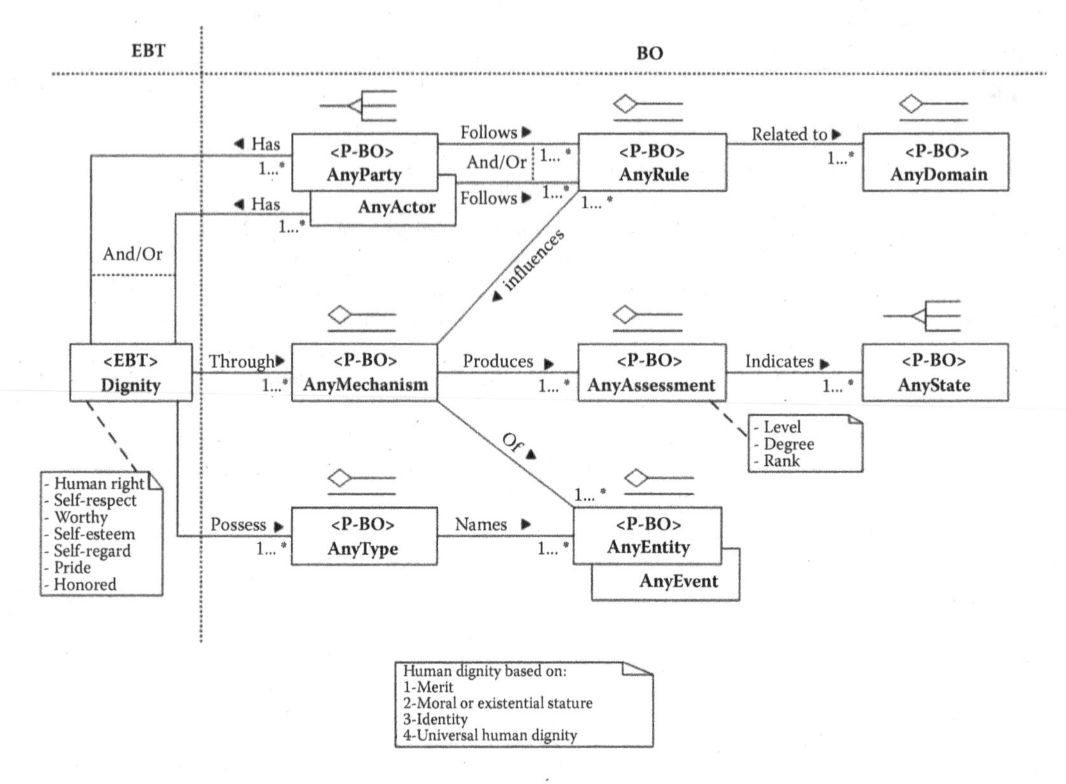

FIGURE 8.1 Class diagram for dignity stable analysis pattern.

8.2.5.1 Class Diagram Description

1. AnyParty/Actor/Domain/Entity has dignity
2. Dignity defines AnyAssessment
3. AnyAssessment asses AnySate
4. AnyState changes form old to new
5. AnyRule influences AnyAssessment
6. AnyAssessment of AnyParty/Actor
7. AnyParty/Actor/Domain/Entity declares AnyRule
8. Dignity possesses AnyType
9. AnyState which changes from old to new meets AnyType

8.2.5.2 Participants

8.2.5.2.1 Class

Dignity: This class is an EBT, which defines self-respect or worthiness for any actor or party.

8.2.5.2.2 Patterns

AnyParty: This class represents an organization or country that has dignity and whose boundary is defined by any assessment.

AnyActor: This class represents any human, plant, or animal, who can attain dignity through their assessment.

AnyDomain: This class represents different domain, where dignity can be applied and has different meaning.

AnyEntity: This class represents anything like road, building, etc, which also has their own dignity.

AnyRule: This class represents set of rules, which are defined to achieve dignity, and it influences evaluation criteria that are defined for dignity.

AnyAssessment: This class represents evaluation of some state to reach a conclusion for a party to achieve dignity. It includes self-assessment of the party also.

AnyState: This class shows a state that needs to be assessed in order to fit in criteria to achieve dignity. It shows two statuses, one of old and another one of new. State changes according to assessment results.

AnyType: This class represents different types of entity, when applied to different context.

CRC Cards: Here are sample CRC cards as shown in Table 8.2.

8.2.6 Consequences

The model is designed with *flexibility and reusability* in mind. It can serve as a reusable component inside many of the applications involving the concept of dignity. BOs and the patterns are generic enough to be suitable for reuse in any application.

The concept of dignity is very abstract and takes different forms based on the context or the application in which it is applied. Hence, it is important to complement the existing model with other domain-specific concepts before it can be integrated into the system. This is a consequence of the concept being abstract and the model being very generic. It is not really a disadvantage of the system, but is a trade off in order to satisfy more applications and domains.

The various business objects like the party, assessment, rule, and state are the core concepts in this model and can easily be adapted into any system. Even these objects are very flexible to make it easier to integrate into any system, where the concept of dignity is relevant. Care must be taken to make sure the domain satisfies all the constraints listed above.

8.2.7 Applicability with Illustrated Examples

8.2.7.1 Application 1: Dignity at Work

8.2.7.1.1 Description

Everyone strives to achieve certain dignity in society. It can be anyone, country, organization, religion, race, or person. Dignity at work gives the concept of how a person doing his or her everyday work at workplace achieves certain esteem through the work. The person's dignity is assessed based on the output of work. The state of work is the criteria for assessing the work .State changes as the on-going work produces some results. Thus, the above concept shows that being skilled in certain area or expertise and using it for productive purposes is definitely a way to achieve dignity and respect.

8.2.7.1.2 Use Cases

Use Case Id: 1.1

Use Case Title: Do work to achieve dignity as shown in Table 8.3

8.2.7.1.3 Use Case Description

1. **Dignity** is possessed by **AnyParty(Employee, Office**) as it gives respect.
   How dignity affects a person's status?
   Why everyone wants to achieve dignity?
2. **AnyParty(Office)** declares **AnyRule(OfficeRule).** Office rule provides guidelines to be followed by employees in order to maintain dignity at work.
   How strictly the rules are implemented?
   Is the rule same for everyone? What are the consequences when the rules are broken?

TABLE 8.2
CRC Cards

Dignity (Dignity) EBT

Responsibility	Collaboration: Client	Collaboration: Server
To define self-respect and worthiness.	1. AnyParty, {OR}	showsWorthiness(), defines(), gives(), respects(), increasesPopularity()
	2. AnyActor,	
	3. AnyAssessment,	meetRules(), typeIndication()
	4. AnyType	determineState()
Attributes	dignityAssessment, possessedBy, stateOfDignity, typeOfDignity, dignityRule, value, context, givenBy	

AnyParty (AnyParty) BO

Responsibility	Collaboration: Client	Collaboration: Server
To achieve dignity	1. Dignity 2. AnyRule	participate(), playRole(), interact(), group(), associate(), organize(), request(), setCriteria(), switchRole(), partake(), join(), monitor(), explore(), receive(), collectData(), integrate(), agree(), disagree(), leave(),
Attributes	id, partyName, address, type, role, skills, designation, affair, activity, partiesInvolved, activity, category (or orientation), purpose	

AnyActor (AnyActor) BO

Responsibility	Collaboration: Client	Collaboration: Server
To achieve dignity	1. Dignity 2. AnyRule	participates(), playsRole(), performs(), interacts(), groups(), associates(), organizes(), requests(), setsCriteria(), monitors(), explores(), receives(0, collectsData(), agree(), joins(), etc
Attributes	id, actorName, address, birthdate, type, role, affair, designation, skills, qualification, activity, category status, workingLocation, workingHours	

AnyEntity (AnyEntity) BO

Responsibility	Collaboration: Client	Collaboration: Server
To specify entity where dignity is required	2. AnyRule	exists(), types(), maintains(), states(), demands(), hasValue(), needs(), represents(), symbolizes()
Attributes	id, entityName, entityType, status, position, states, type, etc.	

AnyRule (AnyRule)BO

Responsibility	Collaboration: Client	Collaboration: Server
To specify standards and to influence assessment.	1. AnyParty, 2. AnyActor, 3. AnyDomain, {OR} 4. AnyEntity, 5. AnyAssessment	states(), influences, imposes(), specifiesStandard(), restricts(), confines(), constraints(), etc.
Attributes	Id, ruleName, typeOfRule, stateOfRule, specifiedBy, numberOfRules, effectOfRule, purpose, description, sternness, impactOnFollower	

TABLE 8.3
Do Work to Achieve Dignity Use Case

Actors	Roles
AnyParty	1. Office 2. Employee

Class Name	Type	Attributes	Operations
Dignity	EBT	1. possessedBy 2. stateOfDignity 3. dignityAssessment 4. dignityRule 5. typeOfDignity	1. defines()
AnyParty	BO	1. designation 2. workHours 3. location 4. skill 5. population	1. declaresRule() 2. assesses() 3. possessesDignity()
AnyRule	BO	1. typeOfRule 2. specifiedBy 3. numberOfRule 4. effectOfRule 5. implementedFrom	1. influences()
AnyAssessment	BO	1. formatOfAssessment 2. resultOfAssessment 3. assessmentCondition 4. doneBy 5. reviewedBy	1. assessesState() 2. providesSelfAssessment()
AnyState	BO	1. typeOfState 2. statusOfState 3. frequencyOfChanges 4. resultOfState 5. factorsAffecting	1. changes()
AnyType	BO	1. factor 2. name 3. numberOfType 4. influencedBy 5. status	1. meets() 2. classifies()
Employee	IO	1. nameOfEmployee 2. workplaceOfEmployee 3. designation 4. salary 5. skillOfEmployee	1. doesWork() 2. followsOfficeRule()
Office	IO	1. locationOfOffice 2. nameOfOffice 3. numberOfEmployees 4. focusOfExpertise 5. numberOfOffices	1. setsOfficeRule() {Should be according to Office type}
OfficeRule	IO	1. definedBy 2. implementedBy 3. followedBy 4. numberOfRule 5. targetedAt	1. formsPartOfRule() 2. providesGuidelineToEmployee() {True, only when employee knows about rules}

(Continued)

Table 8.3 (*Continued*)
Do Work to Achieve Dignity Use Case

Class Name	Type	Attributes	Operations
Appraisal	IO	1. formOfAppraisal 2. basedOn 3. reviewedBy 4. resultOfAppraisal 5. appraisalTime	1. stemsFromWork() 2. partOfAssessment()
Work	IO	1. typeOfWork 2. skillsRequired 3. amountOfWork 4. allottedTo 5. outputOfWork	1. leadsToAppraisal() {True, only when work is worth of it} 2. formsState()

3. **AnyRule(OfficeRule)** influences **AnyAssessment(Appraisal)**
   Do changes in rule affect AnyAssessment and in what way.
4. **Office** sets **OfficeRule**.
   How the office defines rule?
   Is there any limit on number of rules?
5. **Employee** does **Work**.
   Does the employee possess enough skills to complete the work?
6. **AnyAssessment(Appraisal)** asses **AnyState(Work)**.
   In what way AnyAssessment assess sate?
   What are the possibilities for AnyState?
   What are the factors that can affect AnyState?
7. Based on assessment **AnyState** changes state from **old** to **new**.
   Does AnyAssessment lead to change in AnyState?
8. **AnySate(Work)** leads to **AnyAssessment(Appraisal)**.
   What should be the quality of work to lead to appraisal?
   What are the criteria for assessment?
   Does Appraisal assessment always lead to positive result?
9. **AnyType** meets **AnyState**.
   How does type meet state? Does type have any influence on the state?
10. **AnyParty(Employee/Office)** can do self-assessment.
    How can self-assessment be done? How will it affect the party?
11. **AnyAssessment(Appraisal)** is defined by **Dignity**.
    What are the methods for assessment? How does it affects dignity?
12. **Dignity** possesses **AnyType**.
    What are the factors that influence the classification of dignity? How many types of dignity are there in knowledge?

8.2.7.1.4 Alternatives

1. **AnyParty(Employee)** does not get dignified.

8.2.7.1.5 Sequence Diagram

Sequence diagram description

1. Dignity is possessed by AnyParty(Employee, Office).
2. AnyParty(Office) defines AnyRule(OfficeRule).

3. AnyRule(OfficeRule) influences AnyAssessment(Appraisal).
4. AnyAssessment(Appraisal) assesses AnyState(Work).
5. AnyState changes from old to new.
6. AnyState(Work) leads to AnyAssessment(Appraisal).
7. AnyAssessment is defined by dignity as shown in Figure 8.2.

8.2.7.1.6 Class Diagram

Class diagram description

1. AnyParty has dignity.
2. Employee and office inherits from AnyParty.
3. AnyParty defines AnyRule.
4. OfficeRule forms part of AnyRule.
5. Office sets OfficeRule.
6. Employee follows OfficeRule.
7. AnyRule influences AnyAssessment.
8. AnyAssessment assesses AnyState.
9. AnyState is work.
10. AnyState meets AnyType.
11. Dignity possess AnyType.
12. Work is done by employee.
13. Work leads to appraisal.
14. Appraisal is a form of AnyAssessment.
15. AnyAssessment can be defined by dignity.
16. AnyParty can do self-assessment using AnyAssessment as shown in Figure 8.3.

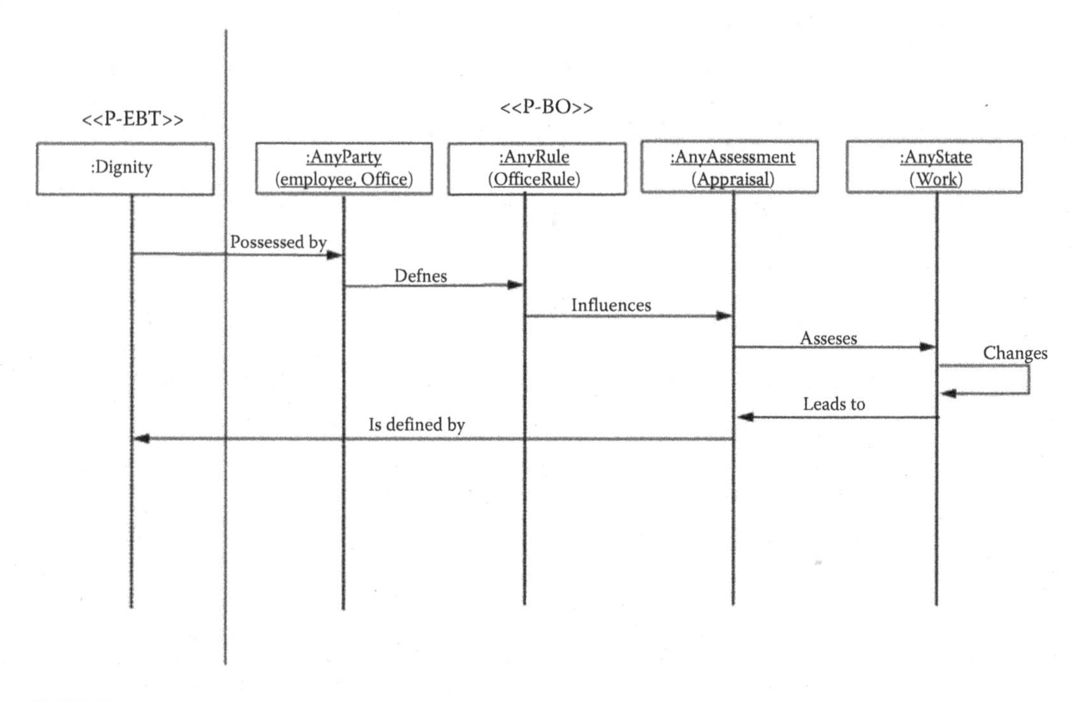

FIGURE 8.2 Sequence diagram of dignity at work.

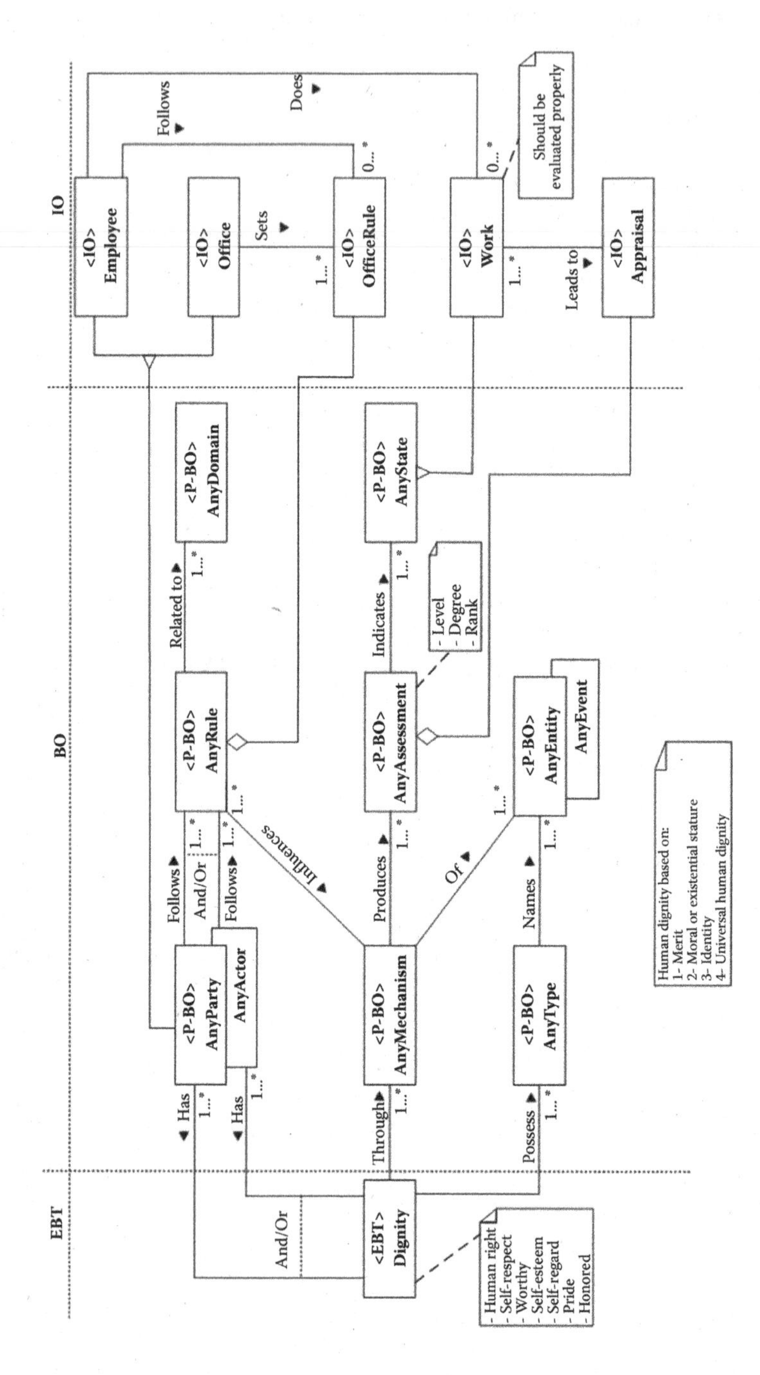

FIGURE 8.3 Class diagram of dignity at work.

8.2.7.2 Application 2: Dignity of Human Rights

8.2.7.2.1 Description

Dignity in terms of human rights is taken into context with reference to the basic rights that a country provides to its citizens. Countries all over the world may face some kind of catastrophe in terms of disaster or disease. The way in which the country tackles this situation dignifies its stand.

Here, catastrophe is taken with reference to public health issues. A WHO official works for a country that needs help and assistance during a critical time. All WHO officials try to resolve and minimize any hard effects that might arise as a result of a disaster or disease. All citizens aspire for medical privileges and help. This system makes it easy to handle all pending health issues. WHO, then prepares health report based on these public health issues by assessing the progress made in tackling the issue. Assessment, then leads to stricter rule or guidelines that are required to bring the situation under control. If assessment yields satisfactory results, then rules remain more or less the same. Finally, a country can self-assess its state of public health based on the report furnished. Thus, providing basic rights to its citizens, emphasizes the dignity it achieves by protecting its people.

8.2.7.2.2 Use Cases

Use Case: 2.1
Use Case Title: Provide and Work for Health Care as shown in Table 8.4

8.2.7.2.3 Use Case Description

1. **Dignity** is possessed by **AnyParty(WHO, Country)**, as it gives respect.
   On what standard is the dignity of country based?
   How dignity classifies the country?
   Why everyone wants to achieve dignity?
2. **AnyParty(WHO)** defines **AnyRule(HealthGuideline)**.
   What are the rules? How strictly the rules are implemented?
   What effect does it have on party?
3. **AnyParty(WHO)** follows **AnyRule(HealthGuideline)**.
   What are the health guidelines?
   Are the health guidelines followed strictly?
   What impact does it create on public health?
   Are the health guidelines standard?
4. **AnyRule(HealthGuideline)** influences **AnyAssessment(HealthReport).**
   How is assessment influenced and in what way?
   Does poor health report lead to change in health guidelines implemented?
5. **AnyAssessment(HealthReport)** assess **AnyState(PublicHealthIssue)**.
   In what way does AnyAssessment assess state?
   What are the possibilities for AnyState?
   What are the factors that can affect AnyState?
6. **AnyState(PublicHealthIssue)** is resolved by **WHOOfficial**.
   What is the approach of WHO officials to tackle Health issue?
   Do they need to take any special precautions?
   How do they reach out to people in need of medical help?
7. **WHOOfficial** belong to **AnyParty(WHO)** and works for betterment of **AnyParty (Country)**.
   How can the condition of country be improved?
8. **AnyParty(Country)** consists of **Citizen** and they need **MedicalPrivilege**
   What kind of privileges do citizens demand? What are the basic rights for citizens in a country? Is medical privilege one of the basic human rights?
   How does country provide this human right to its citizens?

TABLE 8.4
Provide and Work for Health Care Use Case

Actors	Roles
AnyParty	1. Country 2. WHO

Class Name	Type	Attributes	Operations
Dignity	EBT	1. possessedBy 2. stateOfDignity 3. dignityAssessment 4. dignityRule 5. typeOfDignity	1. defines()
AnyParty	BO	1. designation 2. workHours 3. location 4. skill 5. population	1. declaresRule() 2. assesses() 3. possesses() 4. lives() 5. prospers()
AnyType	BO	1. basis 2. number 3. parameterUsed 4. factor 5. basisForClassification	1. classifies()
AnyRule	BO	1. typeOfRule 2. specifiedBy 3. numberOfRule 4. effectOfRule 5. implementedFrom	1. influences()
AnyAssessment	BO	1. formatOfAssessment 2. resultOfAssessment 3. assessmentCondition 4. doneBy 5. reviewedBy	1. assessesState() 2. providesSelfAssessment()
AnyState	BO	1. typeOfState 2. statusOfState 3. frequencyOfChanges 4. resultOfState 5. factorsAffecting	1. changes()
Country	IO	1. nameOf 2. geographicLocation 3. population 4. securityLevel 5. internationalPosition	1. consistsOfCitizen()
Citizen	IO	1. nameOf 2. designationOf 3. worksFor 4. address 5. nationality	1. needsMedicalPriviliges(){ True for every citizen}
WHO	IO	1. numberOfOfficials 2. placeWorkingAt 3. numberOfOffices 4. location 5. headquarterAt	1. includesWHOOfficial() 2. followsHealthGuidelines() 3. preparesHealthReport()

(Continued)

TABLE 8.4 (*Continued*)
Provide and Work for Health Care Use Case

Class Name	Type	Attributes	Operations
WHOOfficial	IO	1. name 2. designation 3. workHours 4. workLocation 5. skill	1. worksForCountry() 2. resolvesHealthIssue()
HealthGuideline	IO	1. numberOfGuideline 2. implementedBy 3. followedBy 4. formulatedOn 5. reason	1. providesGuidelines()
MedicalPrivilege	IO	1. typeOf 2. insuranceType 3. allottedTo 4. benefits 5. aimsTo	1. givesPrivilegeToCitizen() 2. helpsToSolveHealthIssue()
PublicHealthIssue	IO	1. typeOfIssue 2. categoryOfhealthIssue 3. emergencyNeedOf 4. typeofAlert 5. gravityOfEffect	1. formsHealthReportBasis() 2. givesRiseToState()
HealthReport	IO	1. nameOf 2. writtenOn 3. submittedBy 4. analyzedBy 5. reference	1. preparedByWHO() 2. formsAssessmentType() 3. usesPublicHealthIssue()

9. **Dignity** is classified by **AnyType(MedicalPrivilege)**.
   How many types does dignity have? What are the factors on the basis of which dignity is classified?
10. **AnyType(MedicalPrivilege)** meets **AnyState(PublicHealthIssue)**.
    What are the medical privileges? How are they defined and on what basis?
    Does it prove beneficial to help solve public health issue?
11. Based on assessment **AnyState** changes state from **old** to **new**.
    Does AnyAssessment lead to change in AnyState?
12. **AnySate(PublicHealthIssue)** leads to **AnyAssessment(HealthReport)**.
    Does public health provide an insight of country's state in terms of medical facility?
    What are the actions taken to change current state based on assessment?
    What actions are taken if report is satisfactory?
    What actions are required to make the report satisfactory?
13. **AnyParty(WHO)** prepares **HealthReport** based on **PublicHealthIssue**.
    What are the criteria in report formation?
    To whom is the health report submitted?
14. **AnyParty(Country)** can do self-assessment based on **AnyAssessment(Health Report)**.
    Does country's medical provision represent its dignity?
    If its citizens are not provided with medical privileges, does it affect country's status?
    How can the country self-assess its position that leads to dignity?
15. **AnyAssessment(HealthReport)** is defined by **Dignity**.
    What are the methods for assessment? How does it affect dignity?

8.2.7.2.4 Sequence Diagram

Sequence diagram description

1. Dignity is possessed by AnyParty(Country, WHO).
2. AnyParty(WHO) defines AnyRule(HealthGuideLine).
3. AnyRule(HealthGuideline) influences AnyAssessment(HealthReport).
4. AnyAssessment(HealthReport) is based on AnyState(PublicHealthIssue).
5. AnyState(PublicHealthIssue) resolved by WHOOfficial.
6. WHOOfficial works for AnyParty(Country).
7. AnyParty(Country) consists of Citizen.
8. Citizen needs AnyType(MedicalPriviledge).
9. AnyType(MedicalPrivilege) is met by AnyState(PublicHealthIssue).
10. AnyState changes from old to new.
11. AnyState(PublicHealthIssue) assessed by AnyAssessment(HealthReport).
12. AnyAssessment is defined by dignity as shown in Figure 8.4.

8.2.7.2.5 Class Diagram

Class diagram description

1. Country and WHO inherits from AnyParty.
2. AnyParty declares AnyRule.
3. AnyRule includes HealthGuideLine.
4. HealthGuideline are followed by the WHO.
5. WHOOfficial is a part of the WHO.
6. WHOOfficial resolves PublicHealthIssue.
7. Citizen is a part of country.
8. Citizen needs MedicalPrivilege.
9. MedicalPrivilege forms a part of AnyType of dignity.
10. MedicalPrivilege aids in resolving PublicHealthIssue.
11. AnyType meets AnyState.
12. PublicHealthIssue represents AnyState.
13. AnyState can be assessed by AnyAssessment.
14. AnyAssessment is influenced by AnyRule.
15. HealthReport is a form of AnyAssessment.
16. HealthReport is prepared by the WHO based on PublicHealthIssue.
17. AnyParty can do self-assessment through AnyAssessment.
18. AnyAssessment is defined by dignity as shown in Figure 8.5.

8.2.8 Related Patterns and Measurability

8.2.8.1 Related Pattern

Dignity can be used in a wider domain. When we talk about dignity of a person, it is usually coined as Respect/Esteem. A person can gain respect in society through various ways, one way is "Academic Dignity." Based on this definition, following Figure 8.6 traditional model can be modeled for dignity problem.

8.2.8.1.1 Traditional Model (Meta Model) versus Stable Model (Pattern)

- The ***basis*** of traditional model is entirely IOs, which are physical objects and are unstable. On the other hand, the stable model is based on the important concepts—EBT, BO, and IO. The EBTs represent elements that remains stable internally and externally. The BOs are objects that are internally adaptable, but externally stable and IOs are the external interfaces of the system.

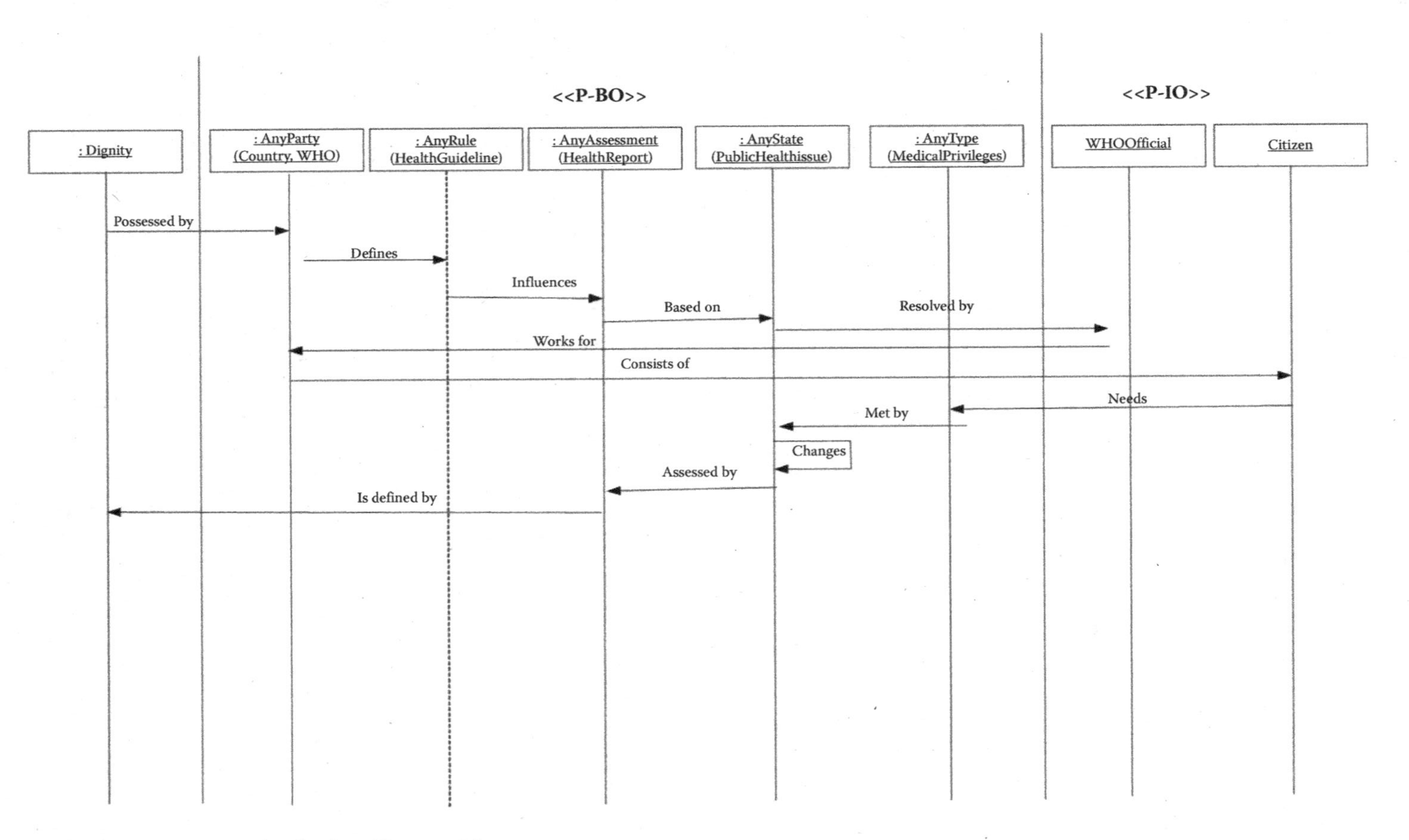

FIGURE 8.4 Sequence diagram for dignity of human rights.

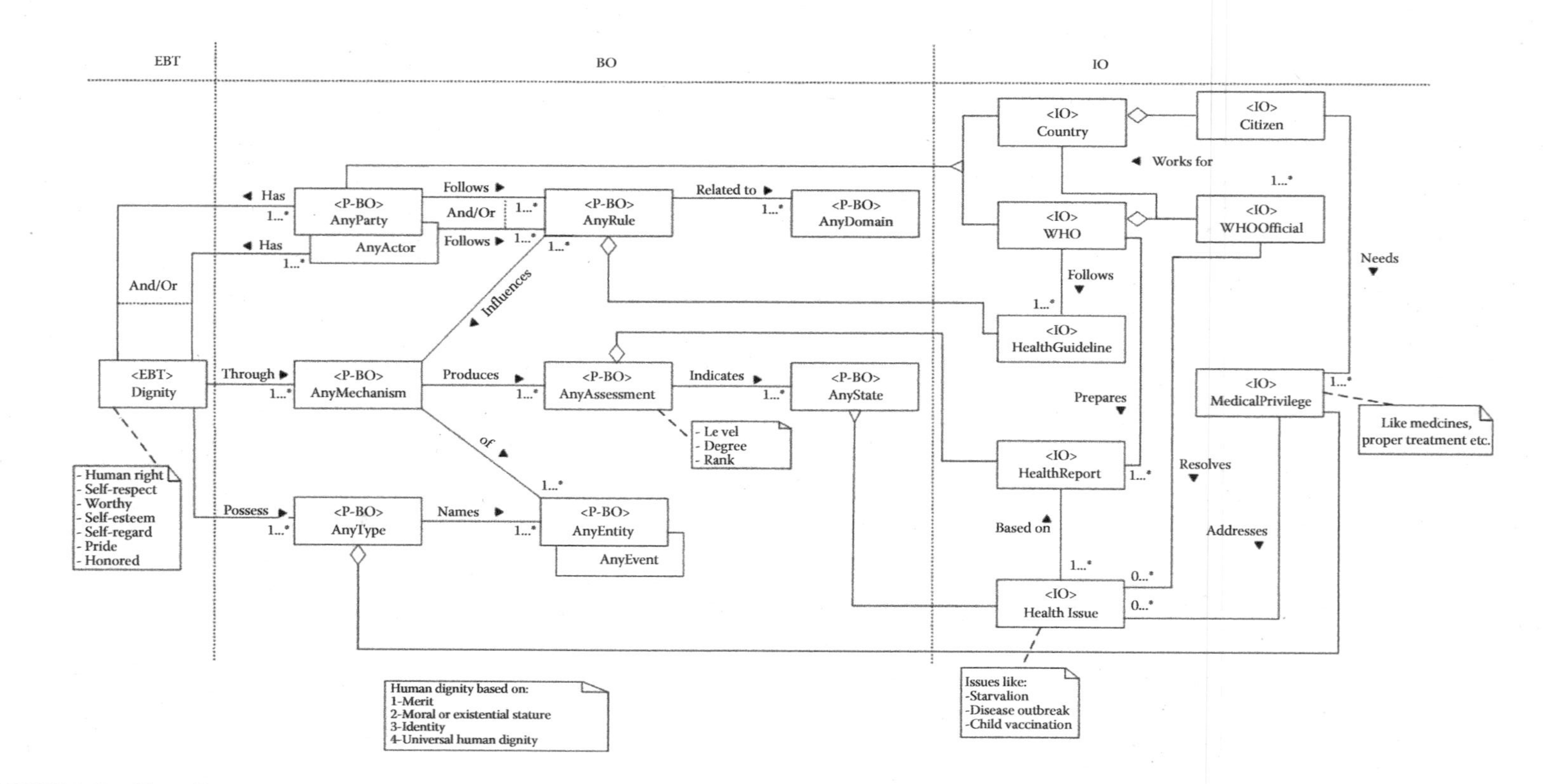

FIGURE 8.5 Class diagram for dignity of human rights.

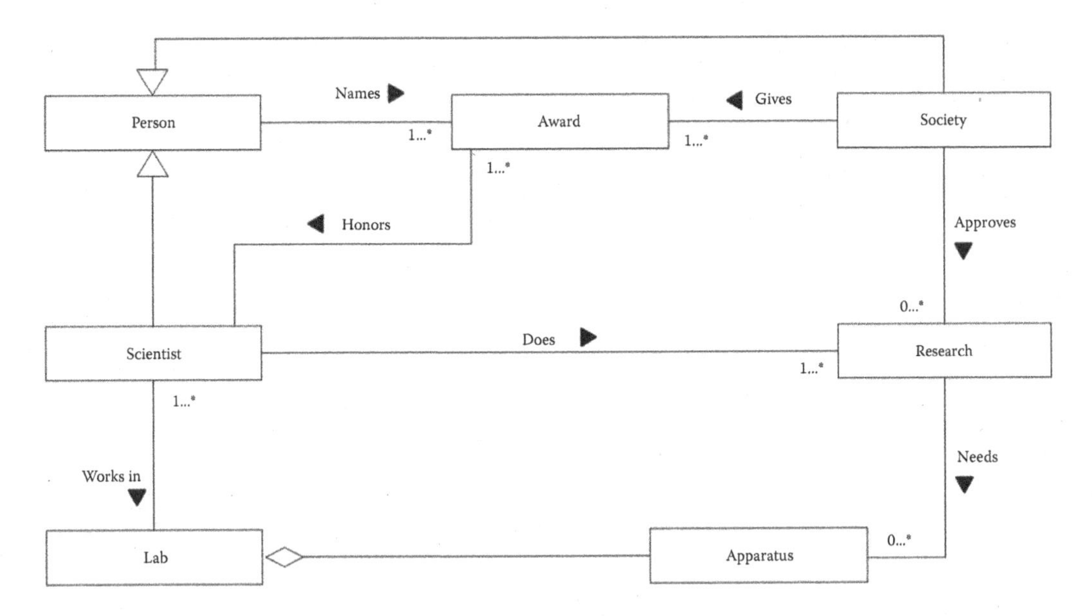

FIGURE 8.6 Dignity traditional model (dignity of person in academics).

- Traditional model is hard to *reuse* if the requirement changes. Any changes in the requirements cause complete re-engineering of project. The stable model is highly flexible and is reusable in wide domains and applications.
- Traditional model requires high *maintenance cost* in terms of time, labor, and money. The system built by using traditional model cannot be extended or adapted. The stable model is easily maintainable and extendable.
- In the above traditional model of dignity, class "Dignity" itself is not used anywhere. As a result, sometimes, it becomes hard to explain the *purpose or goal* of the traditional model. On the other hand, a stability model is developed on the basis of goal or themes that are represented as EBTs. By only looking at the stability model, one can easily define the purpose of the model.
- Division of classes in stability model in categories of EBT, BO, and IO makes it simpler to *understand* and apply as compared to traditional model. This is also one of the reasons for the scalability of stable models as EBT, BOs, and the relationship between them remains stable. IOs are the ones that change with the application. IOs of any application can be easily hooked with the BOs and a new model can be developed very easily and without much effort.
- It is very easy to *identify challenges and constraints* of an application with stability model when compared to traditional model.
- Traditional models are very much susceptible to *dangling*, while dangling is never a problem in stability model because of multiple inheritance and aggregation.
- It is very difficult to define correct *multiplicity constraint* in relationships in traditional model, whereas in stability model, the multiplicity constraints are very much obvious.
- The *inter-dependency* among the classes in traditional model is very high, such that a small change in a single class affects the whole model. While in stability model, the effect of change does not propagate through the whole of the model and hence stability model is more stable than traditional model.
- Limitations of traditional model make it unsuitable for *large projects* having many goals, while stability model because of its stable nature, can be applied to larger projects very easily and their maintenance is also very easy.

- One of the major differences between the two models is the ***cost***. Cost involved in developing project using traditional model is very high, when compared to a stability model. This is because, developing the user requirements keep on changing and a slight change in the requirement results in complete reworking of traditional model, while this is not the case with stability model.

To summarize, the features of stability model such as stability, scalability, understandability, reusability, maintainability, and simplicity makes it far better when compared to traditional model.

8.2.8.2 Measurability

8.2.8.2.1 Quantitative Measure

Quantitative measurability can be applied on the following factors:

1. *Quantity aspect of EBTs, BOs, and IOs:* The more the number of patterns, the more it will result in lines of code during the development of the system. And as lines of code increases, error propagation rate will also increase, and it will be difficult to maintain accuracy in the pattern development. Quantitative aspects shows that EBTs, BOs, and IOs should be selected in such a way that it should cover all the necessary patterns required in modeling, and yet it should be developed in a manageable number of lines of code, which will result in lesser error propagation.
2. *Number of classes:* The second aspect of quantitative metrics is, as compared to traditional model, stability model has lesser number of classes with the focus on explicit as well as implicit factors. Stability model is based on the concept of EBTs, BOs, and pluggable IOs. As a result, the base pattern remains stable and has the capability of representing a large number of applications by just hooking the appropriate IOs with the base pattern. This reduces the number of classes required to represent an application by a drastic amount.
3. *Cost estimation:* In a stable model, as base pattern is known in advance so determining and developing the estimation or measurement metrics is far easier and less time-consuming as compared to that in traditional model.
4. *Coupling among classes:* Coupling represents how tightly the classes are bound together and depend on each other. In a traditional model, coupling among classes is very high. As a result, a small change to any class in traditional model ripples through and affects the whole traditional model. While in the stability model, change in one class does not affect the whole model and it remains restricted to that particular BO.
5. *Constraints:* They represent the multiplicity of the class and are very easy to define in stability model as compared to traditional model.

8.2.8.2.2 Qualitative Measure

Stable model being very generic can be reused to apply to any application, while traditional model is built on application-specific tangible objects and thus cannot be reused. Reusing traditional model requires a lot of re-engineering, effort, time, and cost. This makes stable model more scalable and flexible. Moreover, it is easy to maintain stable model as compared to traditional model as it is easily adaptable.

For software requirement specificity, we can formulate one specific formula. For that, we will need to define few terms. We will use Q1 for specificity of requirements. By specificity of requirements, we mean lack of ambiguity in other words. The second value is completeness. By completeness, we mean how well they cover all the functions of classes to be implemented. We will refer to it as Q2. Therefore, to determine specificity for requirements, we will use the following formula,

$$Ql = Nui/nr.$$

where

1. Ql = specificity of requirements,
2. Nui = number of common requirements identified,
3. nr = total number of requirements, nr = nf + nnf, where nf = number of functional requirements and nnf = number of nonfunctional requirements.

Hence, the lower the value of requirement specificity, the greater will be the degree of ambiguity. So, the value of requirement specificity should always be optimal.

8.2.9 Modeling Issues, Criteria, and Constraints

8.2.9.1 Abstraction

Software stability concept is a multilayered approach for developing robust software systems. In this approach, the classes of the system are classified into three layers: EBTs layer, BOs layer, and IOs layer.

Based on its nature, each class in the system is classified into one of these three layers.

1. EBTs are the classes that present the enduring and core concepts of the underlying industry or business.
2. BOs are the classes that map the EBTs of the system into more concrete objects. BOs are semi-conceptual and externally stable, but they are internally adaptable.
3. IOs are the classes that map the BOs of the system into physical objects. For instance, the BO "AnyParty" can be mapped in real life to "Government/Organization," which is an IO.

Stable analysis pattern is a new approach for developing patterns by utilizing software stability concepts. Stable analysis pattern was proposed as a solution for the limitations of contemporary analysis patterns. The goal of stable analysis pattern was to develop models that capture the core knowledge of the problem and present it in terms of the EBTs and the BOs of that problem. Consequently, the resultant pattern will inherit the stability features, and hence it can be reused to capture the essence of the same problem whenever it appears.

There are two main participants in stable model: (1) classes and (2) patterns. Classes represent the tangible objects, which are unstable, while patterns represent a second level of abstraction to the model, where each pattern is by itself another model that contains classes and, in some cases, other patterns.

The main motive of abstraction is to encapsulate the details of underlying base pattern and expose only the tangible objects to the user along with defining the context of the application. Thus, BOs related to dignity have to be chosen in such a way that they cover all aspects of dignity without exposing themselves.

- Dignity is a very broad concept and is applicable not only to humans, animals, or organizations but also to nonliving things, which distinguish them from each other. For example, a book and a table have their own dignity, which defines their purpose of existence. Hence, BOs like AnyParty or AnyActor were not sufficient for defining dignity pattern applicability, AnyEntity is also required. Similarly, AnyDomain has its own dignity. Any area/domain without dignity cannot exist and its dignity differs from that of AnyEntity. As a result, dignity pattern can be applied to AnyDomain, AnyEntity, AnyParty, or AnyActor and thus, covers all aspects of dignity.

- As dignity has a wider context and is not limited to just living things, it definitely has many types. As a result, it was decided to include one BO called AnyType in the pattern. This BO classifies different forms of dignity and gives more information about it.
- In order to achieve dignity, something needs to be considered. There should be some mechanism through which dignity can be attained. AnyMechanism can be one possible BO for this pattern, but AnyAssessment is a better choice. AnyMechanism relates to some well-defined method, while AnyAssessment is any arbitrary method or task. As dignity can be achieved through any deed or task and no method is defined before hand, so AnyAssessment was given priority over AnyMechanism.
- The next step was to validate AnyAssessment. Some criteria or rules should be defined, as to which assessment can provide dignity and which cannot. As a result, it was required that the party or actor or entity or domain involved with dignity should define some set of rules to validate AnyAssessment. Hence, AnyRule was taken as one of the BOs in the pattern and they influence AnyAssessment done to achieve dignity. Moreover, these rules can be modified from time to time according to the context in which dignity pattern is applied.
- The next part of the debate is how to represent dignity and in what form. Dignity can be given in the form of award or honor or title, but these forms of representing dignity do not fit for AnyEntity and AnyDomain. As a result, AnyState was chosen. This BO can be easily shown as changing from one state to the other and from one time to the other, depending on the assessment done and it goes well with all the aspects of dignity.

Many more BOs can be linked to dignity pattern, but the pattern can lose its meaning and will become unmanageable. In order to develop a good stable pattern, the standard practice is to take minimum number of BOs such that when combined together, they can relate to the correct meaning of the pattern. The BOs defined above are enough to describe dignity pattern and posses the capability to define dignity in any context.

8.2.10 Modeling Heuristics

8.2.10.1 General Enough to Be Reused in Different Application

The stable design pattern so developed can be applied to a wider range of applications. The pattern has been developed keeping generality in mind. Dignity has different meanings under different contexts. The BOs defined for the pattern are general enough such that they can be hooked to IOs of any application and the pattern is capable enough to derive the specific functionality of the application. This part has been well explained in applicability part, where dignity is used to define the importance of human rights and how it can be applied to the domain of knowledge. Similarly, we can use the same pattern to develop a model of dignity for AnyParty (country/organization/government/person), AnyActor, AnyEntity, or AnyDomain.

8.2.11 Design and Implementation Issues

The stability model is based on EBTs, BOs, and IOs. The EBTs used are general enough so that it can be applied in various domains. Nevertheless, there are few implementation issues that we have to deal with and they are given below:

8.2.11.1 Delegation versus Inheritance

8.2.11.1.1 Model Implemented with Inheritance

The above model as shown in Figure 8.7a is very static and fixed. The country and WHO are the subclasses here and they inherit attributes, operations, and methods from AnyParty. If any change

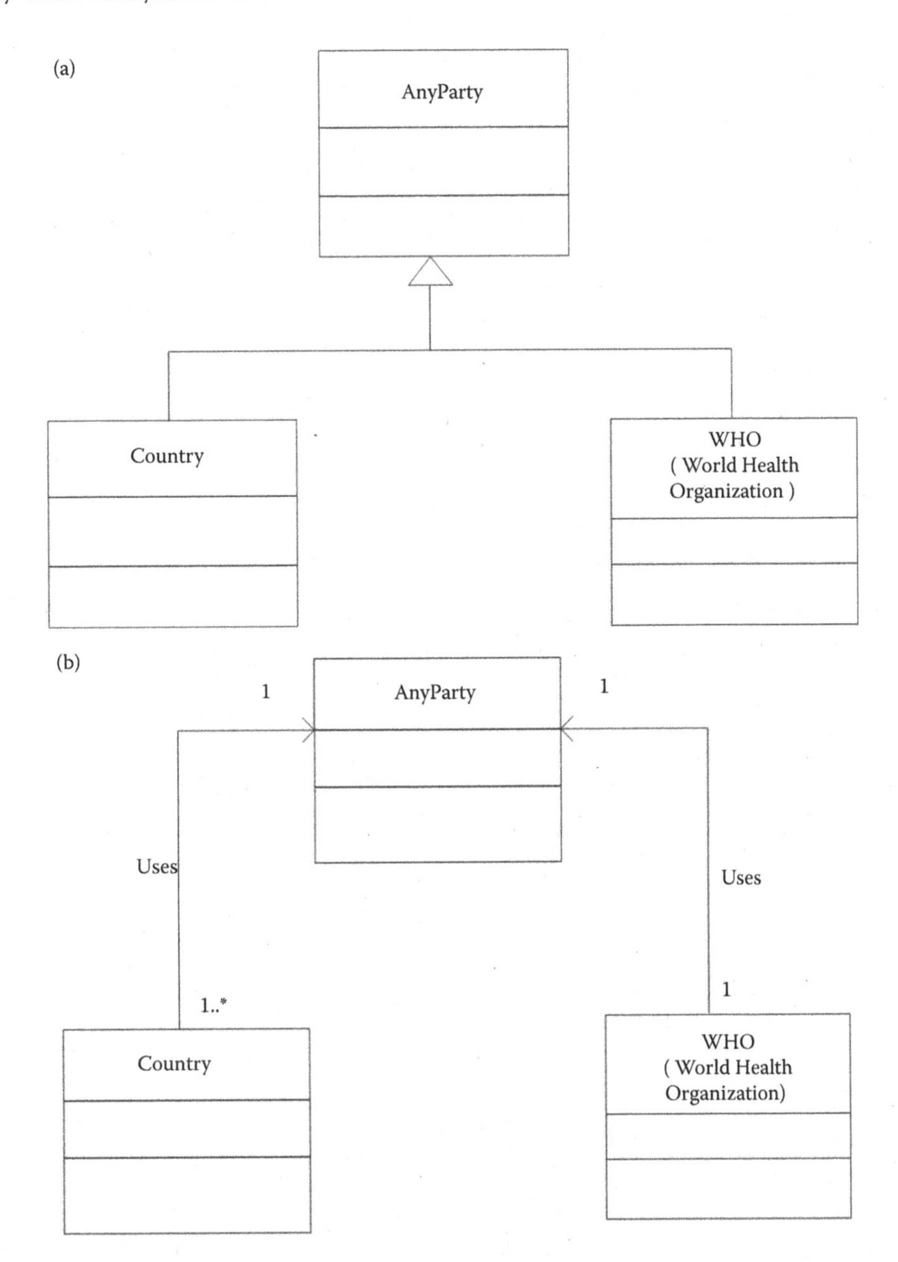

FIGURE 8.7 (a) Inheritance (AnyParty). (b) Delegation (AnyParty).

happens in AnyParty, then it will reflect in all the subclasses, even if that change is not needed for all subclasses concerned here. In other words, superclass will not hide any methods from its subclasses.

8.2.11.1.2 Model Implemented with Delegation

Figure 8.7b shows the use of delegation instead of inheritance. How it affects the modeling pattern is an interesting feature. Delegation provides dynamism, that is, run-time flexibility which is one of the distinct features. The remaining characteristics are similar to inheritance, as it also provides reuse technique. Dynamic coupling between superclass and subclass is the key feature.

In this case, the same submodel is implemented by using delegation instead of inheritance. Now, even if superclass adds some changes, it will not reflect in all subclasses, because of delegation, as it provides dynamic run-time linking by invoking call from one object in superclass to the object in concerned subclass. At this point, if some additional rules need to be implemented for WHO guidelines then, all we need to do is create a separate method for WHO rules and then pass the object to WHO subclass. So, in this case that particular change will not be seen in country subclass. In other words, superclass can hide its methods from subclasses.

The code for the delegation example taken is as below:

```
public class anyparty{
        public void publichealth(system.out.println('public health issue'));
        public void population(system.out.println('number of people in the country'));
}

public class who
{
        anyparty a = new anyparty();
        public void publichealth(a.publichealth());
}

public class country
{
        anyparty b = new anyparty();
        public void population(b.population());
}
```

The above code shows how the class WHO creates an object and delegates the class AnyParty by using that object to invoke the method in class AnyParty. Thus, it will use the relevant methods from class AnyParty for its own class. In this way, class AnyParty can hide its methods from other classes which do not require that method.

8.2.11.1.3 Interface

Interface is a function that would list all operations of BOs in combination that are required to connect BOs to IOs. Thus, BOs will be connected to IOs via an interface. It will increase its functionality. All the links that are used to connect to IOs will be included in interface. To increase functionality is the key feature of interface.

An interface for one EBT is shown below. It takes into account all BOs, which are connected to that EBT. Then, interface for all BOs having their attributes and operations are collected together. This collection of operations is nothing but the interface and then interface is connected to IOs.

Example highlighted below shows interface in Figure 8.8. The box connecting BOs and IOs together is the interface.

8.2.11.2 Testability

In general, patterns designed by using the stability model are more easily testable, when compared to the traditional model. This is because the EBTs and BOs rarely change and can be applied to other applications without any major changes.

In this project, the dignity stability pattern is tested by applying the pattern to two different applications without any changes to the core pattern. This is achieved by plugging in the necessary IOs to the core dignity pattern.

In the same way, other application's IOs from any context can be plugged to BOs. The above pattern so developed will be considered testable only when it can be applied to any scenario/ application and produces the correct output.

Some scenarios in which the above pattern may not give correct output and will fail are

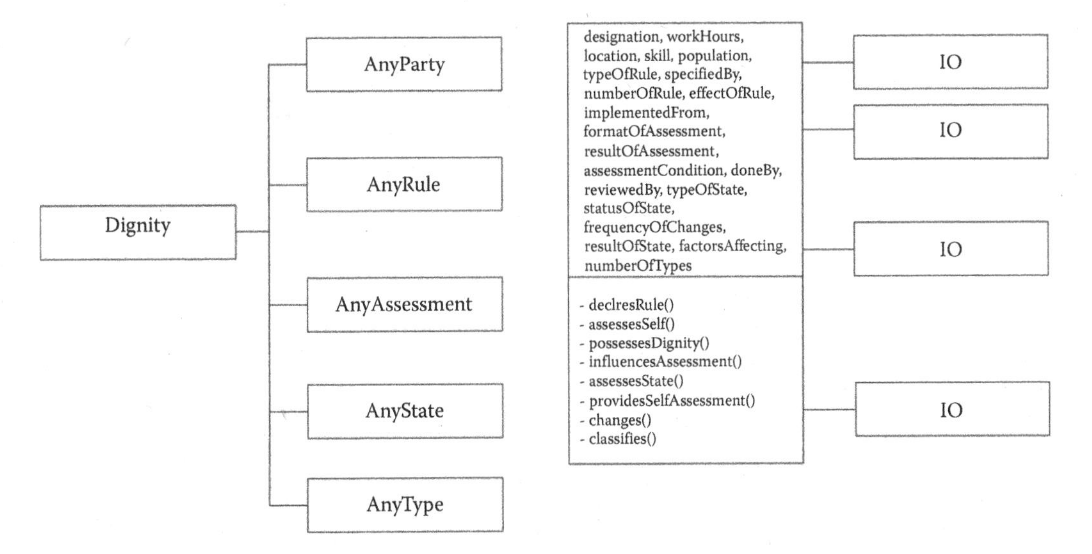

FIGURE 8.8 Dignity patterns' interfaces.

1. This pattern fails, when the assessment defined by dignity is not proper and has no defined boundaries, as then the way to achieve dignity is not known. Moreover, the state change will not be proper as the state depends on assessment.
2. If the party/actor/domain/entity is unable to define proper rules for dignity, they will not create effective influence on assessment. So, this is also one of the cases, where the above-developed pattern will fail.

Dignity has a broader set of definitions and it usually changes with the context. It can be applied to any thing, but the meaning of dignity when applied to some domain may not be true, when used in context of some person or organization and vice versa. All the operations defined for dignity in this pattern may not be valid for an application. For, for example, when dignity is defined for some entity, say a road, then the operations "respects, shows worthiness" does not stand true.

8.2.12 Formalization Using OCL, Z++, Object Z, and/or EBNF

Dignity pattern is formalized by describing in many-sorted first-order languages. This language consists of many sets of sorts and types. Every sort has a universe set and set of function and relations and can have sub-sorts also. Thus, universe of a sort is the union of universe of its sub-sorts.

XML is used for describing formalization because it is a simple text-based language for representation and XML notations can be easily converted into any language such as Java, .net, and C. Syntax of dignity pattern in XML terms has been described below. Only a part of schema has been described, as full description of schema is not possible here and is beyond the scope of the document.

Note: According to mathematical logic conventions, constants over a sort have been defined as constant: sort name.

```
<pattern>
        <title>
                "dignity"
        </title>
        <sort>
                <title>
```

```
        "Dignity"
    </title>
    <sort>
        <title>
            "dignityAssessment"
        </title>
        <sort>
            <title>
                "format"
            </title>
            <type>
                String
            </type>
            <universe>
                {Appraisal,Honor, …}
            </universe>
        </sort>
        <sort>
            <title>
                "type"
            </title>
            <type>
                String
            </type>
            <universe>
                {Promotion,Award,Title,…}
            </universe>
        </sort>
        <sort>
            <title>
                "condition"
            </title>
            <type>
                String
            </type>
            <universe>
                {…}
            </universe>
        </sort>
        <sort>
            <title>
                "result"
            </title>
            <type>
                Action: Nomination
            </type>
        </sort>
        <sort>
            <title>
                "numberOfState"
            </title>
```

```
                    <type>
                         Integer
                    </type>
                    <universe>
                         {1,2,3,...}
                    </universe>
               </sort>
          ...
               <function>
                    <title>
                         "assessesState"
                    </title>
                    <type>
                         State: DignityState
                    </type>
               </function>
               <function>
                    <title>
                         "provideSelfAssessment"
                    </title>
                    <type>
                         Assessment: DignityAssessment
                    </type>
               </function>
               <function>
                    <title>
                         "bringDignity"
                    </title>
                    <type>
                         name?→type?→DignityType→DignityType
                    </type>
                    <description>
                         bringDignity: DignityType=DignityType' ∪ {name?, type?}
                    </description>
               </function>
          </sort>
     ...
</pattern>
```

8.2.13 Business Issues

8.2.13.1 Business Rules

- There should be some assessment defined by dignity, as assessment forms the basis for achieving dignity.
- There should be some party trying to achieve dignity and should have some assessment in accordance to those defined by dignity.
- The party should define some rules, which directly influences the assessment.
- Moreover, the assessment should be able to change states and these changed states should be such that the new states are easily identifiable from the old states.
- There should be an AnyParty that deserves or possess dignity.
- The dignity should be based on AnyRule that are decided on by AnyParty.

- AnyRule influences AnyAssessment and leads to changes in AnyState.
- Dignity is assigned based on the assessments, which are influenced by the rules.
- State changes are triggered by the assessments.

Business rules controls the behavior of the system. They impose constraints on the system and tell the system "what" should be done. Business rules are atomic in nature and thus cannot be further broken down into smaller pieces without loss of information. They are defined prior to defining requirements of the system.

Elements of business rules are

1. Business Items: This element corresponds to different classes forming the pattern. Stable pattern consists of classes at various levels: EBT, BO, IO and they have different functions and responsibilities at each level. Some of the business rules defined for business items are:
   a. Each class should be capable of at least one function
   b. Classes should be able to work independently
   c. IO classes should interact with BO classes only
   d. EBT classes should interact with BO classes only
   e. Classes should be able to reflect the specificity of application
   f. BO classes, when combined, should be able to represent a pattern depicting the meaning of EBT/s involved.
2. Properties: Properties in business language corresponds to attributes and operations of classes in stable language. Business rules related to properties are
   a. The operations defined for the class should be unique and generic, such that they can be used to represent any application
   b. The class should be able to carry out the responsibility assigned to it.
   c. The attributes of the class must cover all the distinct aspects of the class.
   d. The operations defined for the class should be such that the class is able to perform them independently as well as in cooperation with other classes.
3. Relationships: It presents the interdependency among classes and how one class is related to each other. Business rules defined at this level are
   a. One relation can connect only two classes.
   b. Every class should be related to another class through some relation. No classes in a pattern can be standalone.
   c. Relation can be simple relation, connecting two classes, that is, association or it can be "a kind of" or "a part of" relation.
   d. Every associative relation has some multiplicity. The default is one to one.
   e. Every association relation has some name, which represents the type of connection between two classes.
4. Facts: These represent business or common terms, which can occur in the form of EBT or BO. Some of them are
   a. Dignity: represents state of being worthy.
   b. Appraisal: represents the means by which to gain dignity.
   c. Honor/Title: represents forms in which dignity can be given.
   d. Deed: represents task that keeps on changing and leads to dignity.
   e. Knowledge: represents domain that has some dignity.
   f. Office: it is an entity that has some dignity and maintains it.
5. Constraints: This represents the restrictions imposed on pattern. Some of the constraints are
   a. AnyDomain/AnyEntity should define at least one rule for accessing dignity.
   b. AnyAssessment needs to be validated by AnyCriteria.
   c. AnyState cannot be static. It should keep on changing.
   d. AnyAssessment should be based on at least one state.

Based on the above elements some generic business rules for the pattern are as below:

1. AnyAssessment should have defined boundaries.
2. Only bounded assessments should be validated by AnyCriteria.
3. AnyAssessment for dignity should be defined properly.
4. AnyCriteria should be validated for dignity.
5. AnyState should not remain static.
6. Dignity should be associated with some type.

8.2.13.2 Business Integration

Integration means gathering business components and tying them together. This process forms a very important part of any application, as application system is developed as separate components. Developing any application by using stability model approach simplifies the things. This is because the core concept of the problem always remains stable, that is, the pattern developed using EBTs and BOs remain stable and it can be very easily extended to incorporate any application. BOs act as the extension points, where IOs for the particular application can be hooked to make a final product. Hence, it is much easier to integrate the pattern in any business model.

8.2.13.3 Business Enduring Themes

EBT for the dignity pattern is dignity itself. It represents the goal of the business. It answers the question, "what is the main and unique goal of the pattern?" This pattern can be used in any domain that involves the concept of dignity. The pattern models the concept of dignity in a stable way so that it can be used in many applications.

8.2.14 Known Usage

The various scenarios, where the dignity pattern is known to be useful are listed below:

- The pattern can easily be applied to human society, where humans communicate with each other and possess some level of dignity and use it to promote their ideas and satisfy their needs.
- The pattern also applies to dignity in an international scenario, where the countries command certain level of dignity.
- This also can be applied to a corporate culture, where different roles have different levels of dignity.
- Dignity pattern can be applied to any nonliving thing such as road, hospital building, or office. Every nonliving thing has its own purpose of existence and thus try to maintain its dignity.
- The pattern is also applicable in different domains. Every domain has some rules and thus possesses dignity.

1. *Dignity in human society:* The pattern can easily be applied to the human society, where humans communicate with each other and possess some level of dignity and use it to promote their ideas and satisfy their needs.
2. *United Nations:* The United Nations (UN) is an international organization, which aims to facilitate cooperation in various sectors and promote world peace. The dignity pattern can be seen playing an important role in an international scenario, where the countries command certain level of dignity and work toward solving international problems and resolving issues.

3. *Corporate environment:* Any corporation has a set of levels and a tier of leadership that goes from the lower level employees to the higher level management team. Typically, different levels/roles have distinct responsibilities and the nature of work varies. The dignity pattern can easily be applied to such a corporate culture, where different roles have different levels of dignity.
4. *Dignity of profession:* Every profession has some dignity associated with it, in the form of designation, respect in society, honor, and worthiness. Every individual strives hard in his or her profession to reach the topmost position. Some professions even provide some titles to persons such as Sir or Dr. The dignity pattern so developed can be easily applied to this domain, where every profession has some dignity and rules defined by every profession are different.

As explained above, the dignity pattern can easily be applied to any domain or any sphere of human interaction. Besides above usage, dignity also applies to animal interaction of all kinds.

8.3 TIPS AND HEURISTICS

- Aggregation should be preferred over delegation and delegation should be preferred over inheritance.
- It is important to make the pattern as generic as possible to encourage reuse in many domains and applications.
- EBTs should interact only with BOs.
- BOs must interact with EBTs and IOs.
- IOs should interact only with BOs.
- BOs should be kept generic, so that they can be applicable in many different domains and applications.
- Number of interactions between the objects in the sequence diagram should be within the range of interaction in the class diagram.
- Aggregation, delegation, and inheritance should be used diligently in the class diagram.
- Snake patterns in sequence diagrams are easy to follow and understand. Hence, the flow in the sequence diagram should follow the snake pattern.
- The pattern contains EBTs and BOs, which should be generic enough to be easily reused in other applications. The pattern should be reusable after the relevant IOs for an application are identified.
- The IOs in the application should be connected to the BOs in the pattern.
- There should not be any direct connection between the actors or roles in the system.
- BOs and EBTs should be generic, so that they can be easily reusable in other domains.
- Dignity pattern can be applied to any context and is not limited just to humans and animals.
- Every type of dignity has some rules associated with it.
- Dignity can be achieved through some assessment, only which serves as mechanism.
- The state of assessment should not be static, as it is the one which determines dignity.
- Assessment should have defined boundaries.
- Rules defined for dignity should be laid down clearly without any contradictions.
- There should be some rules to distinguish old state from new state.
- The state of assessment that can lead to dignity should be defined properly.

8.4 SUMMARY

The dignity pattern is designed based on the principles of stable analysis pattern. Stable analysis patterns help us to analyze the problem under consideration by extracting the core concepts involved in the problem. As a result, the core concepts of dignity are modeled in the form of BOs like

AnyDomain, AnyParty, AnyAssessment, AnyState, etc. These BOs then helped us in developing a stable pattern for dignity to be applicable in any context.

This method of analysis generally leads to a reusable, extendable, and scalable model that is much more stable than the traditional approach. As a result of this, the pattern can be hooked up to many other applications, where the core concepts are applicable.

In the stability model, the classes are classified into EBTs, BOs, and IOs. The BOs and EBTs should be generic enough to be reusable in other domains. Designing the BOs in such a reusable manner is very challenging. Any application should be able to reuse the pattern without many changes by plugging in the necessary IOs. In this project, we have applied the dignity stable pattern to two different applications without making any major changes, unlike a pattern that is designed using the traditional model. Two different applications could reuse the same pattern just by supplying their own IOs. This demonstrates the power of stability model and the advantages it provides over the traditional mode.

8.5 OPEN RESEARCH ISSUES

1. *Software stability and knowledge maps:* Software stability and knowledge maps usually leads us to many dignity frameworks that include the generation of problem space patterns (analysis) and ultimate solution patterns (design), redefinition of knowledge of dignity, discovery of many possibilities of architectural patterns that are generated from the knowledge map of any domain and used as foundation bases of millions of applications.

   Software stability and knowledge maps allow the development of many meaningful patterns. Developing meaningful patterns is a thing of art and a system of perfect skills; improving the overall quality of patterns is never easy, and quick; more often, developers take an inordinately long time to design perfect and meaningful patterns. To develop meaningful and robust patterns, a developer may need to design them in a phased manner. The most important and critical of all these phases are the diagnostic phase, using which, one can understand and comprehend the main problems that come in the way of development of today's patterns.

   Once a pattern developer identifies and notes all the bottlenecks, it becomes very easy to explore the causes of pattern immaturity and their subsequent usability. In addition, software stability and knowledge maps provide simple and clear guidelines for choosing the appropriate patterns from a large inventory of alternatives and distinguishing clearly between analyses, design, architectural patterns. In addition, software stability and knowledge maps overcome all the pitfalls discussed at [10]
2. *Dignity unified software engine (D-USE):* Utilize the stability model or knowledge map methodology as a way for developing **D**-USE. The engine mainly focuses on several patterns: **Dignity**, Analysis, **Recording, Legality**, etc. The proposed solution attempts to extract the commonality from all the domains and represent it in such a way that it is applicable to a wide range of domains without trivializing or generalizing the concepts. The engine is a stable structural pattern and provides a generic engine that can be applied and/or extensible to any application **of dignity. The D**-USE can be applied to unlimited context and reusability.

REVIEW QUESTIONS

1. "Generally speaking, dignity has many shades." What are the shades of dignity discussed in this chapter? Try to find four more shades of dignity.
2. What is the difference between dignity of a country and dignity of a woman?

3. What are the similarities and differences between dignity and respect?
4. In spite of having the same meaning as dignity, why is worthiness rarely used as a substitute of it?
5. Find out all such terms, which mean exactly same as dignity that can be used interchangeably?
6. Why is pride not a proper way of referring to the dignity of a person?
7. What do you mean by the term "dignity"? Can you use the term dignity in any other context than what you thought of?
8. Explain dignity in the context of
   a. Dignity of country
   b. Dignity of person
   c. Dignity in sports
   d. Dignity in profession
   e. Dignity in animals
9. Rewrite each of the scenarios with EBT and BOs in mind. Make sure that each BO is attached to one or more IOs except AnyParty is attached to two or more roles.
10. List and explain two scenarios of dignity that are not covered in this chapter.
11. Define internal requirements. List and explain the internal requirements for dignity.
12. Define external requirements. List and explain the external requirements for dignity.
13. Examine attributes and operations of dignity. Are there any missing attributes and operations? Discuss them.
14. Examine the functional requirements of dignity pattern—Are there any missing requirements? Discuss them.
15. Examine the nonfunctional requirements of dignity pattern—Are there any missing requirements? Discuss them.
16. Draw and explain the stable analysis pattern for dignity.
17. Write 10 new more challenges following the template used in challenging section.
18. List and explain the participants in stable analysis pattern of dignity.
19. Write the rest of the CRC cards for the classes used in stable analysis pattern of dignity.
20. Explain the consequences of modeling dignity by using stability model in terms of flexibility and reusability.
21. List any four advantages of modeling dignity by using stability model that are not discussed in this chapter.
22. Explain the applicability of dignity in sports with the help of use cases, sequence diagrams, and class diagram.
23. Explain the applicability of dignity in animals with the help of use cases, sequence diagrams, and class diagram.
24. Explain the applicability of dignity of person with the help of use cases, sequence diagrams, and class diagram.
25. Think of any two other applicability of dignity and explain with the help of use cases, sequence diagrams, and class diagram.
26. Explain what EBT, BO, and IO are, and how they form the basis of stability model.
27. Why is it hard to reuse traditional model in the world of changing requirements.
28. "Traditional model requires high maintenance cost." Explain why.
29. Why is stable model easier to understand as compared to traditional model?
30. Why is it difficult to understand the purpose/goal of traditional model?
31. Explain quantitative aspect of EBTs, BOs, and IOs.
32. Explain number of classes as an aspect of quantitative metrics. Explain how stability model reduces number of classes.
33. Compare traditional model and stable model on cost estimation quantitative measure.
34. Compare traditional model and stable model on coupling of classes.

35. Explain software requirement specificity as a qualitative measure with the help of a formula.
36. Explain stability model.
37. What are the main participants in software stable model? Explain them.
38. What is stable analysis pattern? What is its goal?
39. What is the main goal behind abstraction? Explain the abstraction process of dignity stable analysis pattern.
40. Write two scenarios, where dignity can be applied to nonliving things.
41. Explain how dignity can be applied to AnyDomain.
42. Justify the existence of AnyType in dignity stable analysis pattern.
43. Why AnyAssessments is given priority over AnyMechanism in dignity stable analysis pattern?
44. Justify the existence of AnyRule in dignity stable analysis pattern. How are they used to validate AnyAssessment?
45. Justify the existence of AnyState in dignity stable analysis pattern.
46. "The model should be general enough to be reused in different applications." Explain with the help of an example.
47. What implementation issues are related to dignity stable analysis pattern?
48. What is the difference between inheritance and delegation? Why is delegation considered better than inheritance? Explain with the help of an example.
49. What is an interface? How are interfaces implemented for stability model?
50. Why is it easy to test stability model when compared to traditional model?
51. What are business rules? List some business rules for dignity pattern.
52. What are the key elements of business rules?
53. What is meant by business integration? Why is it important?
54. "Developing any application using stability approach simplifies things." Explain and justify.
55. "It is much easier to integrate the dignity pattern in any business model." Explain and justify.
56. What is EBT for dignity pattern? Justify your answer.
57. Explain some of the known usages of dignity pattern in day-to-day life. List a few that are not covered in the chapter.

EXERCISES

1. Dignity in animals
    a. Explain dignity in animals.
    b. Draw a class diagram based on the dignity pattern to show the application of dignity in animals.
    c. Document a detailed and significant use case as shown in Case Study 1
    d. Create a sequence diagram of the created use case of c.
2. Dignity of culture
    a. Explain dignity of culture.
    b. Draw a class diagram based on the dignity pattern to show the application of dignity of culture.
    c. Document a detailed and significant use case as shown in Case Study 1
    d. Create a sequence diagram of the created use case of c.
3. Dignity of woman
    a. Explain dignity of woman.
    b. Draw a class diagram based on the dignity pattern to show the application of dignity of woman.

   c. Document a detailed and significant use case as shown in Case Study 1
   d. Create a sequence diagram of the created use case of c.
4. Dignity of religion
   a. Explain dignity of religion.
   b. Draw a class diagram based on the dignity pattern to show the application of dignity of religion.
   c. Document a detailed and significant use case as shown in Case Study 1
   d. Create a sequence diagram of the created use case of c.
5. Dignity of student in a class
   a. Explain dignity of student in class.
   b. Draw a class diagram based on the dignity pattern to show the application of dignity of student in class.
   c. Document a detailed and significant use case as shown in Case Study 1
   d. Create a sequence diagram of the created use case of c.
6. Think of a few scenarios, where dignity pattern is applicable and create a corresponding class diagram, use case, and sequence diagram as shown in the solution and applicability sections for each of the scenarios.
7. *Dignity in human society:* The pattern can easily be applied to the human society, where humans communicate with each other and possess some level of dignity, and use it to promote their ideas and satisfy their needs.
   a. Draw a class diagram based on the dignity pattern to show the application of dignity in human society.
   b. Document a detailed and significant use case as shown in Case Study 1
   c. Create a sequence diagram of the created use case of b.
8. *United Nations:* UN is an international organization that aims to facilitate cooperation in various sectors and promote world peace. The dignity pattern can be seen playing an important role in this international scenario, where the countries command certain level of dignity and work toward solving international problems and resolving issues.
   a. Draw a class diagram based on the dignity pattern to show the application of UN
   b. Document a detailed and significant use case as shown in Case Study 1
   c. Create a sequence diagram of the created use case of b.
9. *Corporate environment:* Any corporation has a set of levels and a tier of leadership that goes from the lower level employees to the higher level management team. Typically, different levels/roles have distinct responsibilities and the nature of work always varies. The dignity pattern can easily be applied to such a corporate culture where different roles have different levels of dignity.
   a. Draw a class diagram based on the dignity pattern to show the application of corporate environment.
   b. Document a detailed and significant use case as shown in Case Study 1.
   c. Create a sequence diagram of the created use case of b.
10. *Dignity of profession:* Every type of profession has some dignity associated with it, in the form of designation, respect in society, honor, and worthiness. Every individual strives hard in his or her profession to reach the topmost position. Some professions even provide some title to the person such as Sir or Dr. The dignity pattern so developed can be easily applied to this domain, where every profession has some dignity and rules defined by every profession are different.
    a. Draw a class diagram based on the dignity pattern to show the application of dignity of profession.
    b. Document a detailed and significant use case as shown in Case Study 1.
    c. Create a sequence diagram of the created use case of b.

PROJECTS

Develop the following system by using the dignity analysis pattern

1. *Human rights* (http://en.wikipedia.org/wiki/Human_rights) are "basic rights and freedom to which all humans are entitled [11]." Examples of rights and freedom that have come to be commonly thought of as human rights include civil and political rights, such as the right to life and liberty, freedom of expression, and equality before the law; and economic, social and cultural rights, including the right to participate in culture, the right to food, the right to work, and the right to education.All human beings are born free and equal in dignity and rights. They are endowed with reason and conscience and should act toward one another in a spirit of brotherhood.
2. *Pride* (http://en.wikipedia.org/wiki/Pride) is, depending on the context, either a high sense of the worth of one's self or one's own or a pleasure taken in the contemplation of these things. One definition of pride in the first sense comes from Augustine: "the love of one's own excellence" [12]. In this sense, the opposite of pride is humility.

   Pride is sometimes viewed as excessive or as a vice, sometimes as proper or as a virtue. While some philosophies such as Aristotle consider pride a profound virtue, most world religions consider it a sin [13].

   According to the Concise Oxford Dictionary, *proud* comes from late Old English *prut*, probably from Old French *prud* "brave, valiant" (11th century) (which became *preux* in French), from late Latin term *prodis* "useful," which is compared with the Latin *prodesse* "be of use" [2]. The sense of "having a high opinion of oneself," not in French, may reflect the Anglo-Saxons' opinion of the Norman knights who called themselves "proud," like the French knights *preux* [14].

   When viewed as a virtue, pride in one's appearance and abilities are known as virtuous pride, greatness of soul, or magnanimity, but when viewed as a vice, it is often termed vanity or vainglory. Pride can also manifest itself as a high opinion of one's nation (national pride) and ethnicity (ethnic pride).
3. *Honor* (http://en.wikipedia.org/wiki/Honour) Create Honor SAP and show the differences between Dignity SAP using the subheader of detailed documentation template of Chapter 3
4. *Liberty* (https://en.wikipedia.org/wiki/Liberty) Design Honor SAP and show the differences between Dignity SAP using the subheader of detailed documentation template of Chapter 3
5. *Discrimination* (http://en.wikipedia.org/wiki/Discrimination) Create Honor SAP and show the differences between Dignity SAP using the subheader of detailed documentation template of Chapter 3
6. Consider **Human Dignity** [15–17] for applications of Dignity SAP and create the following for each application:
   a. Draw a class diagram based on the Dignity SAP to show the human dignity.
   b. Document a detailed and significant use case as shown in Case Study 1
   c. Create a sequence diagram of the created use case of b.

   A human rights strategy for foreign policyJune 2010, www.minbuza.nl
7. Consider a patient's dignity (Baillie 2007) for applications of Dignity SAP and design the following for each application:
   a. Draw a class diagram based on the dignity pattern to show the application of human dignity.
   b. Document a detailed and significant use case as shown in Case Study 1
   c. Create a sequence diagram of the created use case of b [18].

   Consider workplace dignity for applications of Dignity SAP and create the following for each application:
   a. Draw a class diagram based on the dignity pattern to show the application of dignity of workplace.

b. Document a detailed and significant use case as shown in Case Study 1
c. Create a sequence diagram of the created use case of b.

REFERENCES

1. M. Gregor and J. Timmermann. ed. and tr., *Groundwork of the Metaphysics of Morals: A German-English Edition*, Cambridge: Cambridge University Press. ISBN 978-0-521-51457-6 (hardcover), 2011.
2. M. Gregor, J. Timmermann, and C.M. Korsgaard, *Kant: Groundwork of the Metaphysics of Morals* (Cambridge Texts in the History of Philosophy) [Paperback], Cambridge: Cambridge University Press, 1998.
3. tr. Herbert James Paton (1887–1969). *The Moral Law: Kant's Groundwork of the Metaphysic of Morals*, London; New York: Routledge, 1991.
4. H. Armory, What Murphy Knew: His Interpolations in Fielding's *Works* (1762), and Fielding's Revision of *Amelia*, *Papers of the Bibliographical Society of America*, 77, 1983, 133–166.
5. M. Battestin, and R. Battestin, *Henry Fielding: A Life*. London: Routledge, 1993. ISBN 0-415-01438-7
6. L. Bertelsen, *Henry Fielding at Work: Magistrate, Businessman, Writer.* Basingstoke: Palgrave, 2000. ISBN 0-312-23336-1
7. Autonomy in Moral and Political Philosophy (Stanford Encyclopedia of Philosophy). Plato.stanford.edu. Retrieved on December 7, 2013.
8. R. Shafer-Landau, *The Fundamentals of Ethics*, third edition. Oxford, England, UK: Oxford University Press, 2014.
9. S.B. Cunningham, Review of virtues and vices and other essays in moral philosophy, *Dialogue*, 21(01), 2002, 133–137.
10. M.E. Fayad, *Stable Design Patterns for Software and Systems*. Boca Raton, FL: Auerbach Publications, 2017.
11. Universal Declaration of Human Rights adopted by General Assembly resolution 217 A (III) of 10 December 1948.
12. M. Lewis, K. Takai-Kawakami, K. Kawakami, and M.W. Sullivan, Cultural differences in emotional responses to success and failure, *International Journal of Behavioral Development*, 34(1), 2010, 53–61.
13. C. Oveis, E.J. Horberg, and D. Keltner, Compassion, pride, and social intuitions of self-other similarity, *Journal of Personality and Social Psychology*, 98(4), 2010, 618–630.
14. *Article from Free Online Dictionary*, November 9, 2008.
15. M. Lebech, *What is Human Dignity? Maynooth Philosophical Papers* (ed. by M. Lebech, Maynooth). Faculty of Philosophy, NUI Maynooth, pp. 59–69, 2004.
16. J.Q. Whitman, The two western cultures of privacy: Dignity versus liberty, *The Yale Law Journal*, 113, 2004, 1153–1220.
17. Human Dignity and Bioethics, Essays Commissioned by the President's Council on Bioethics, Washington, DC, March 2008.
18. L. Baillie, A case study of patient dignity in an acute hospital setting, PhD thesis, London South Bank, February 2007.

9 Trust Stable Analysis Pattern

Trust is the glue of life. It's the most essential ingredient in effective communication. It's the foundational principle that holds all relationships.

Stephen Covey

In this chapter, we propose the introduction of *trust* stable analysis pattern. The main objective of this pattern is to provide a conceptual model that embodies the main aspects of the trust concept. The level of abstraction of the trust pattern makes it applicable for a wide spectrum of applications.

9.1 INTRODUCTION

Trust is a concept that plays a major role in any interaction between different entities. Trust has several applications in both social and business activities. With the current booming of web-based applications, especially, e-commerce, e-business, and e-negotiation, trust has become a major component for the success of these applications. Therefore, there is a great need for understanding the trust concept. In addition, trust has become an integral component in future network infrastructures. For example, in peer-to-peer networks, users (or nodes) are allowed to share and exchange data, an activity that necessitates the establishment of trust between different users. In addition, the performance of such collaborative network would be greatly degraded, if no effective trust measures were implemented.

Several studies have given different definitions for trust. We put forward different definitions of trust from the available literature. In Reference 1, trust was defined as *"the firm belief in the competence of an entity to act dependably, securely, and reliably within a specified context" (assuming dependability covers reliability and timeliness).*

In Reference 2, trust has been defined in the following way: *"Trust in a system is defined as an individual's belief in the competence, dependability, and security of the system under conditions of risk."*

A large body of research addresses the topic of trust from different angles. Some studies view the problem from a theoretical perspective, and hence, propose a theoretical framework to capture the aspects of trust [2]. In Reference 3, an abstract, logic-oriented notation called SULTAN (**S**imple **U**niversal **L**ogic-oriented **T**rust **A**nalysis **N**otation) has been proposed to facilitate the specification and analysis of trust relationships. We do not intend to survey the literature of research in trust, and hence, for more details, we refer the reader to the article [1].

Another avenue of research has explored the implementation of trust in e-commerce applications. For instance, in Reference 4, a summary of the three trust-building measures for e-commerce applications, namely, *information policy*, *reputation policy*, and *warranty policy*, is given. For more details regarding the ongoing research in building and analyzing trust in e-commerce applications, we refer the reader to reference 4 and the references thereof. We will revisit some of the ideas presented in Reference 4, when we discuss an example of our trust pattern.

In the area of e-commerce, trust plays a very important role in displaying the security level of a website or portal. Trust is built by the end-users based on various security policies adopted during the development of the website. For example, when a user uses his/her credit card to shop online, he/she looks for signs of trust or indications of a certifying authority on the website.

In the area of enterprise software, the client often requests for data from the server and the server accepts the request and services the same. Similarly, the client submits some data to the server and the server processes that data after conversing with the designated database. In this case, the client and the server trust each other in order for the request and the response to flow smoothly between the two.

In the above cases of e-commerce and enterprise software, we notice that there is always a need for trust to be established for accomplishing various tasks. Thus, the trust EBT-based model can be an all-encompassing solution to model both scenarios.

Because different applications may impose different requirements for establishing trust, it may not seem feasible to develop a unified framework that encompasses all the trust aspects within all these applications. Nonetheless, this does not deter the fact that different applications may still share a portion of trust requirements; even though, they may differ in some aspects. Our goal in this chapter is to develop a conceptual model that captures the core knowledge that is common to the trust concept independent of any specific application or domain. If we are able to capture this common knowledge, we can end up with a generic model that can be used as a starting point for analyzing trust requirements in any application and any domain.

9.2 TRUST ANALYSIS PATTERN DOCUMENT

9.2.1 Pattern Name: Trust Stable Analysis Pattern

The term "trust" in its broadest sense, refers to the trait of trusting without fear of dishonesty in another person or thing. It could also refer to the act of a trustee holding a property on behalf of a trustee. It could also refer to a large group of corporations collaborating to gain monopoly over the production of a product. Since, a trust finds diverse applications in areas such as law, commerce, society, etc., it becomes relevant to capture the core knowledge behind trust. Thus, trust means differently based on the people, the roles they play and the scenarios in which they are living. The trust analysis pattern aims at capturing the crux of the concept of trust that is common in all the applications that use trust.

Trust is a very generic concept and is a synonym of beliefs. However, belief is not a term that encapsulates all the domains while trust is a term that presents the true essence of the problem presented in this chapter. Trust is an enduring concept.

9.2.1.1 Known As

This stability pattern is known as trust stable analysis pattern. Trust is popularly known as Reliance. Nevertheless, reliance does not capture the true essence of the modeled pattern. Trust is a much-generalized concept. Hence, trust is the most appropriate name for this stability pattern. Similarly, other names such as commend, entrust, commit, believe, care, depend, and confide cannot replace the all-encompassing meaning of trust that is applicable to a wide variety of domains. Some of the names that could be identical to trust would include bank on, believe, entrustment, give for safety, hope, expectation, assurance, being positive, confidence, give authority, trusteeship, company, board of advisors, etc. All the words namely, bank on, believe, entrustment, give for safety, hope, expectation, assurance, being positive, and confidence have the same meaning as trust. The similarity that exists in all these words is the context. The context involves a believer delegating a responsibility to another party and hoping that the party will keep up to the expectations of the believer.

The synonyms give authority and trusteeship, though they might seem similar to trust; one can easily classify them to an application domain namely, law. When a person offers a trusteeship to another regarding a property or a deed, it means that he is offering a trusteeship. A company or a group of companies could form a trust in order to gain monopoly or semi-monopoly over a certain product or concept. A board of advisors may refer to a council or a brain trust related to a scientific experiment. Thus, one can conclude that though there are many synonyms for trust including belief, company, and responsibility, each one may refer to a specialized domain and is not all encompassing as the word trust itself.

9.2.1.2 Context

Trust, as explained earlier could be applied in various contexts and domains. For instance, a customer prior to going to a store to buy some items makes sure he/she does proper homework regarding the retailer/merchant's trustworthiness. Even after entering the store, he/she needs to get the feeling of trust from the merchant, in order for him/her to go ahead and buy the goods. The feeling comes from many factors including whether the merchant is selling quality items, how friendly the environment within the store is, and whether he/she uses secure modes of payment.

The scenario of e-commerce is no different from the above. For instance, in the context of e-commerce, trust could refer to a metric of how much a user trusts a particular website for carrying out transactions. Gaining the trust of online shoppers is essential for the success of a website given that the shoppers are always concerned about identity theft, credit card fraud, spyware, etc. Trust plays a crucial role in the completion or cancellation of any transaction on the Internet. In this particular domain, trust is a judgment call made by the shopper based on experience and the hearsays about the perception of the website and the seller. Signs of professionalism are vital for the customer to say, purchase goods from the website.

Hence, it is important for the merchant to make sure he/she eases the concerns of the consumer and makes sure the consumer develops the trust in him/her. For this reason, merchants make sure that they design and build the infrastructure for the website for ease of use in order to increase the willingness of the consumer to complete the transactions.

Various applications, in which trust is an enduring concept, are

a. Self trust in which, the shares and voting rights of one or more shareholders of a company are transferred to a trustee for a trust period
b. Self trust, when, an individual trusting himself
c. Trusted network, for example, a network of trusted servers in a client–server system
d. Trusted computing and trusted systems, where a computer consistently behaves in a specific manner, when it is controlled by hardware and software

Thus, one can observe that the stable pattern of trust could be applied to any of the above application domains.

1. *Treaties between countries:* Consider a scenario, where one country (AnyParty) signs (AnyEvent) a treaty (AnyMechanism) for peace (AnyCriteria) with other countries (AnyParty). Treaties involve rules and regulations that both the countries have to follow. This develops a feeling of trust (EBT) between them that none of them will break the rules (AnyRating). The treaty can be a physical paper (AnyEntity) document or can be in form of electronic media (AnyMedia) stored (AnyLog) over Internet.
2. *Marital relationship:* In a scenario, where a person (AnyActor) is married (AnyMechanism) to another person (AnyActor), they share a bond of trust (EBT) that they will stand (AnyCriteria) with each other at times of hardship (AnyEvent). They share each other's feeling and emotions (AnyEntity), when they are sad or happy (AnyRating). They also store (AnyLog) memories of their marriage and other important events in pictures and videos (AnyMedia).
3. *Trust in trade:* When a customer (AnyParty) purchases (AnyEvent) gold (AnyEntity) from a jeweler (AnyParty), he/she has trust (EBT) on the jeweler that they are getting what is paid (AnyMechanism) for and there are no hidden (AnyCriteria) untold impurities in the gold. The customer usually gets a certificate (AnyMedia) with the gold that logs (AnyLog) the information about how much percentage (AnyRating) of gold the metal contains.

9.2.1.3 Problem

9.2.1.3.1 Functional Requirements

What are the essential components that form the core knowledge of the trust concept?

Trust as explained earlier is a term that has a wide applicability in various domains. The focus of the problem is to bring out a stable pattern for trust that is easily adaptable and easily extendable to all kinds of applications of trust. Essentially, the outcome of this problem should be a solution to a stable pattern for trust to which any kind of application involving trust can be plugged in.

1. *Wide occurrence:* Trust is used in many different domains as explained earlier illustrating the need for a generalized model.
2. *Common abstraction:* The terms faith, belief, confidence, etc. are only synonyms of trust and do not have the generalized properties of trust to be applicable across multiple domains.
3. Structural adequacy: It is a term to denote the rating of the member force and strength of the pattern to model the pattern. Trust, because of its property, supports reusability.
4. Lack of stable model: At present, no stable model is available for a problem that can be applied in a variety of domains by adding domain-specific elements. A stable model helps achieve this requirement by supporting reuse of the EBTs and the IOs.
   a. *AnyParty:* Represents the trust handlers. It models all the parties that are involved in the trust process. Party can be a person, organization, a country, or a group with specific orientation. AnyParty has a unique id, a name, an email address, a phone number, and a mailing address. They ask for trust, provide trust, and specify criteria for trust.
   b. *AnyActor:* Also represents the trust handlers. It models all the actors that are involved in the trust process. Actor can be a person, software a machine, an animated character, or a creature. AnyActor has a unique id, a name, an email address, a phone number, and a mailing address. They ask for trust, provide trust, and specify criteria for trust.
   c. *Trust:* Trust is an enduring concept. Represents the trust process itself. Trust is a feeling of having faith or confidence in other person based on a certain criteria. Trust has an objective, is a process, and has consequences and a description. It defines rule and goals that parties or actors should follow. It also helps in taking action and in making decisions.
   d. *AnyEntity:* Represents an entity that plays a role in the application domain. Every entity has certain properties and characteristics that can be quantified and verified against any criteria. AnyEntity can help in building trust. AnyEntity has a type, a name, a description, and a unique Id. It is related to a party or an actor and helps in finding the level of trust.
   e. *AnyEvent:* AnyEvent can lead to building trust among two actors or two parties. AnyEvent can also occur because of formation of trust. AnyEvent has a unique id, a title, a description, date when it occurred, time of its occurrence and a location associated with it. AnyEvent leads to the building of trust.
   f. *AnyCriteria:* Represents the criteria, which are specified by AnyActor or AnyParty and is based on AnyMechanism followed to verify the trustworthiness of AnyActor or AnyParty. It has rules, boundaries, description, and a unique Id. It has to satisfy party/actor to create trust.
   g. *AnyMechanism:* Represents the mechanism followed by the trust EBT based on AnyCriteria specified by AnyActor or AnyParty. AnyMechanism represents the mechanism of forming trust. It has a unique Id, a type, and a description. It provides method, gives result and fulfills purpose.
   h. *AnyRating:* AnyRating represents the level of trust or confidence that actors or parties have in each other. A high rating would mean that the trust level is high. It has criteria, a level and a result. It rates trust and evaluates trust by reading and analyzing logs.

i. *AnyLog:* Represents different records used in the system to conduct the rating and hence the trust process. These records vary from one system to another and they are application-dependent. It has a unique id, a type, a title, and a description. It stores information. It can add, update, and delete the stored information too.
j. *AnyMedia:* AnyMedia is used to store AnyLogs on it. It helps in storing the information so that it can be accessed whenever needed. AnyMedia has a unique Id, a type, a title, and a description. It is used to store logs and record information.

9.2.1.3.2 Nonfunctional Requirements

1. *Faith:* Faith is an essential nonfunctional requirement of trust. If a person has trust in another person, then he/she can have faith that the other person will not break his trust. Thus, trust and faith go hand in hand.
2. *Reliance:* Reliance is another key nonfunctional requirement of trust. If a person has trust in other person, he/she can rely on that person. He/she can be sure that the other person will be there in times of need. Reliance is thus an outcome of trust.
3. *Uniformity:* Trust brings uniformity in the sense that if there is trust in a relationship, then the relationship will be uniform and smooth without any bumps and hurdles and hence uniformity becomes a nonfunctional requirement of trust.
4. *Integrity:* Integrity means quality of being honest and trustworthy. It also means a state that cannot be divided. Thus with trust, comes integrity and hence it plays major role in acquiring trust of a person.
5. *Constructive:* Trust has to be constructive. For example, if a thief trusts another thief that he will not betray him, then trust loses its purpose and cannot work. The other thief always has the potential to break the trust and hence the relationship of trust cannot be established.
6. *Privacy:* Trust demands privacy. The treaties signed by countries cannot be signed in open. There has to be an element of privacy to establish trust. This makes privacy a nonfunctional requirement of trust.
7. *Confidence:* Confidence is a very important nonfunctional requirement of trust. If a person has confidence that other person will help him in time of need, only then he can trust the other person, else he cannot.

9.2.1.4 Challenges and Constraints

9.2.1.4.1 Challenges

Here are a few challenges as shown in Table 9.1.

9.2.1.4.2 Constraints

1. Trust can be established between one or more AnyParty or AnyActor.
2. Trust can be established with the help of zero or more AnyMechanism.
3. AnyActor specifies zero or more AnyCriteria to establish trust.
4. AnyMechanism for acquiring trust is based on zero or more AnyCriteria.
5. AnyMechanism can have zero or more AnyRating.
6. AnyMechanism operates on one or more AnyEntity and/or AnyEvent.
7. AnyEntity and/or AnyEvent relates to one or more AnyMedia.
8. AnyMedia is recorded on one or more AnyLog.

9.2.1.5 Solution

To resolve the forces aforementioned, we will use the concept of stable analysis patterns and software stability. Software stability concepts can be used to identify the core knowledge of the trust concept, by identifying the EBTs and BOs of the trust notion. The stability characteristics of the

TABLE 9.1
Challenges

Challenge ID	0001
Challenge Title	Ambiguous Roles
Scenario	Investigation for Marriage
Description	AnyParty can play many roles in the system that can be bride, investigator, daughter, girlfriend, etc. The roles if undefined properly, may lead to complication.
Solution	Roles should be defined properly and should be within context to avoid confusion.
Challenge ID	0002
Challenge Title	Too many Parties
Scenario	Person Seeking Loan from Bank
Description	If a person is seeking loan from 1000s of banks or if too many banks are doing credit check on a person then it will end up in confusion and the person will not get loan.
Solution	Only the required number of parties should be involved in the process of building trust.
Challenge ID	0003
Challenge Title	Lots of Criteria
Scenario	Person Seeking Loan from Bank
Description	IF a bank would impose too many criteria on approving loan for a person, then it would be really difficult to get the loan approved by meeting all the criteria and satisfying all the conditions.
Solution	There should be an upper limit on the number of criteria that can be imposed.
Challenge ID	0004
Challenge Title	Too many Mechanisms
Scenario	Person Seeking Loan from Bank
Description	If there can be 100s of mechanism to rate the level of trust, then all of them cannot be very accurate, and thus can lead to false or invalid results.
Solution	Only the few most accurate mechanisms should be used to establish trust.

EBTs and BOs provide a systematic way for extending the model to be applied in specific application, by adapting the BOs of the system externally, and add the required IOs that fit the application in hand.

Stable analysis pattern concepts ensure a deep focus on the trust problem and the representation of the core knowledge in the appropriate abstraction level that makes the pattern reusable in different applications and across domains. We, hereby, present a conceptual pattern that can be used to understand and analyze the main requirements of the trust concept. The pattern structure is shown in Figure 9.1.

We will differentiate between two main participants in the pattern model: classes and patterns. Classes are defined as in any traditional object-oriented class diagram. Patterns present a second level of abstraction to the model, where each pattern is by itself another model that contains classes and, in some cases, other patterns. In the above model, we also use the prefix "any" for patterns. For instance, AnyParty is a stand-alone stable pattern that models the party notion and, hence, can be used to model any party in any application. Therefore, AnyParty can be viewed in the second abstraction level of the trust stable pattern as shown in Figure 9.1. Note that the same argument applies to AnyLog [5] and AnyRating.

Figure 9.1 can be viewed as the core requirements of the trust concept. In a generic trust scenario, an entity (a person, organization, or a machine) (*AnyParty*) wants to make a decision of whether to trust another entity based on some information that is stored or obtained in a defined way (*AnyLog*). This information along with defined metrics is used to identify the trust level of the examined entity (*AnyRating*). Based on the *AnyRating* results, *AnyParty* can decide on whether or not the evaluated entity is trustworthy. Based on this decision, certain actions should take place, which are defined in (*trust*). The following defines the participants of the trust pattern.

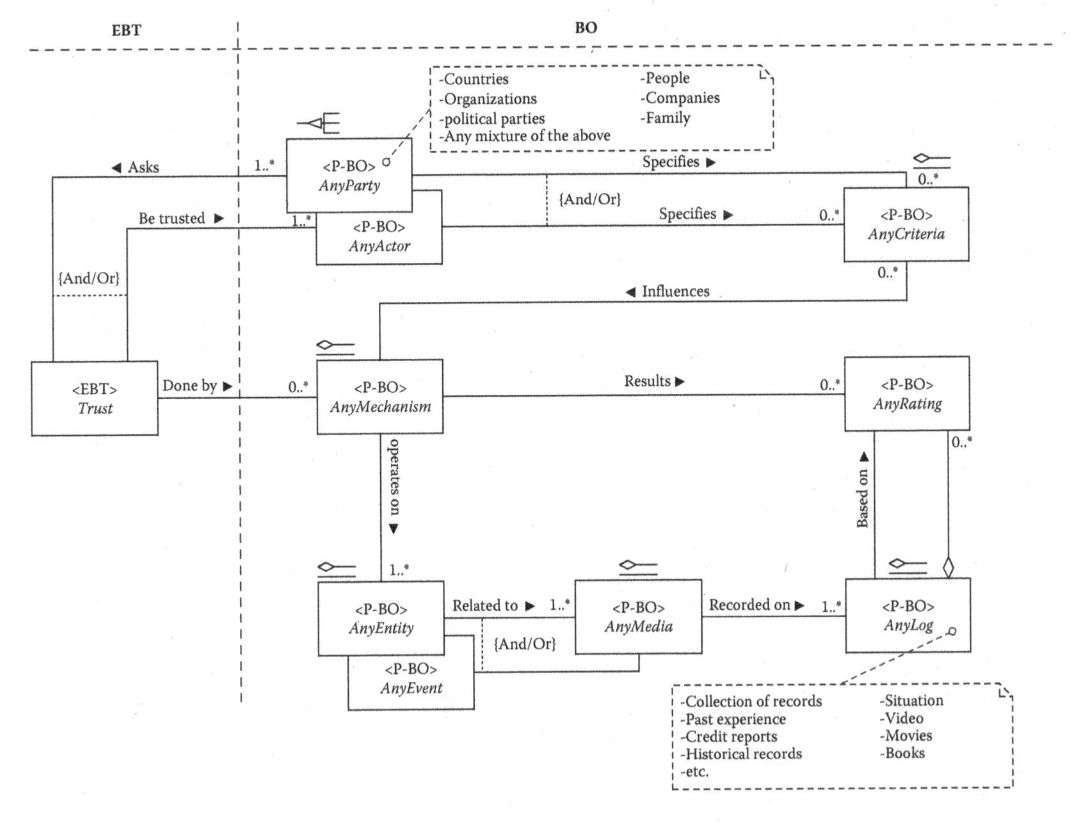

FIGURE 9.1 The trust stable analysis pattern.

9.2.1.5.1 Classes

- *Trust:* Represents the trust process itself. This class contains the behaviors and attributes that regulate the actual trust process.

9.2.1.5.2 Patterns

- *AnyMechanism:* Represents the mechanism followed by the trust EBT based on AnyCriteria specified by AnyActor or AnyParty
- *AnyRating:* Represents the rating mechanism for the evaluation process. The ultimate goal of any trust is to reach a conclusion whether or not the evaluated party should be trusted. Based on the information gathered (e.g., history and recommendations), the AnyRating should be able to identify the trustworthiness of the corresponding party. The AnyRating actual implementation is application-dependent.
- *AnyParty/AnyActor:* Represents the trust handlers. It models all the parties or actors that are involved in the trust process. Party or actor can be a person, organization, or a group with specific orientation. The pattern diagram and detailed pattern description is provided in Reference 6.
- *AnyLog:* Represents the different records that can be used in the system to conduct the rating and hence the trust process. These records vary from one system to another and they are application-dependent. For instance, evaluation for credit cards may use the applicant credit card history as a record for the rating process. However, a customer may decide to buy a certain car model based on a friend recommendation and recommendations in car magazines and so on [5,6].

- *AnyCriteria:* Represents the criteria that are specified by AnyActor or AnyParty and is based on AnyMechanism followed to verify the trustworthiness of AnyActor or AnyParty
- *AnyEntity:* Represents an entity that plays a role in the application domain. Every entity has certain properties and characteristics that can be quantified and verified against any criteria

9.2.1.6 CRC Cards

We will also use the CRC cards format that we proposed in Reference 7. The CRC cards name the class, responsibility, and its collaborations. The CRC cards also name a role for each class, which is useful for identifying the class responsibility. Each class should have only one unique responsibility. The collaboration consists of two parts: clients and server. Clients are classes that collaborate and have relationship with the named class. Server contains all the services that are provided by the named class to its own clients.

- It is worth pointing out that in documenting CRC cards for stable patterns, we deal with any pattern that are included within the main pattern itself as a class. That is, each sub-pattern will be represented by a CRC card that documents its responsibility and collaborations as a black box. To avoid any confusion, and for simplicity, we do not care about how the sub-pattern handle its reasonability according to its internal structure, all what we care about here is that this sub-pattern will perform the task as a black box, leaving the other details to the second abstraction level of the pattern description. For instance, the CRC cards of the sub-pattern AnyMedia will show the details of each class in the black box AnyMedia as shown in Tables 9.2 through 9.8.

9.2.1.7 Applicability with Illustrated Examples

CASE STUDY 1 Credit Report of a Loan Seeker

1. *Scenario Description:* When a named party requests for a loan from a named financial institution, the institution begins a series of checks on the party in order to determine whether the party can be trusted. The NamedFinancialInstt checks

TABLE 9.2
Trust CRC Card

Trust (Trust) EBT		
	Collaboration	
Responsibility	**Clients**	**Server**
Describes the rules and regulations to trust a certain party	AnyMechanism AnyActor AnyParty	defineRules() identifyGoals() takeAction()
Attributes: Process, consequence, objective		

Note:
defineRules(): Define the rules and qualities upon which the AnyMechanism will compute the trust level of a given entity. In addition, this class defines the actions that should be taken based on approving or denying that a given entity is trustworthy.
identifyGoals(): Identify the goals that have to be trusted by AnyActor or which are asked by AnyParty.
takeAction(): Define the actions to be taken by trust in order for AnyMechanism to perform the method in order to achieve ratings.

TABLE 9.3
Anyparty CRC Card

AnyParty (AnyParty) BO		
Responsibility	**Collaboration**	
	Clients	**Server**
Performs and finalizes the trust process	Trust	ask()
	AnyCriteria	specify()
		act()
		provide()
Attributes: name		

Note:
ask(): The evaluating party asks the trust EBT to determine whether the evaluating party should be trusted or not depending on the application.
specify(): The party specifies certain criteria based on which the mechanism carries out operations in order to determine the rating results.

TABLE 9.4
AnyActor CRC Card

AnyActor (AnyActor) BO		
Responsibility	**Collaboration**	
	Clients	**Server**
Performs and finalizes the trust process	Trust	indicate()
	AnyCriteria	note()
		work on()
Attributes: name, address		

AnyRating (AnyRating) BO		
Responsibility	**Collaboration**	
	Clients	**Server**
Describes rating regulations and procedures	AnyMechanism	rate()
	AnyLog	evaluate()
		readLog()
Attributes: result		

Note:
rate(): AnyRating should provide a value because of rating the entity being evaluated. This rate is the result of the evaluation process. This rate depends on the application. For instance, it could be a Boolean function that indicates either trust or do not trust, or it could be a rating scale depending on the nature of the entity being evaluated.
evaluate(): In AnyRating, there should be defined algorithms and approaches by which the rating of an entity will be performed. These algorithms take a defined information regarding the entity being evaluated, and the evaluation metrics as inputs, and process the algorithm to produces the output that can be used to calculate the final rating (by rate()).
readLog(): This is used to read the information required in the evaluation from AnyLog.

TABLE 9.5
AnyLog CRC Card

AnyLog (AnyLog) BO		
Responsibility	**Collaboration**	
	Clients	**Server**
Describes the information used for rating a party	AnyRating AnyEntity	store() add() modify() delete() read() refer()
Attributes: recordName, type		

Note:
store(): The AnyLog stores the information that are supplied from different parties and can be used to evaluate a particular party. For example, recommendation letters from different parties are stored and prepared to be used in the AnyRating of the corresponding party.
add (), modify (), delete (): AnyParty can augment/modify/delete the entries that are used to evaluate a particular party. For instance, a banking enterprise can audit the history of the customer and store its assessment and recommendations regarding this customer in the AnyLog, so that this piece of information can be used to decide in the future whether the evaluated customer qualifies for loans or not.

TABLE 9.6
AnyMechanism CRC Card

AnyMechanism (AnyMechanism) BO		
Responsibility	**Collaboration**	
	Clients	**Server**
Describes the method with which trust is achieved	AnyRating AnyEntity AnyCriteria Trust	nameMethod() result() fulfill() operate()
Attributes: type, name		

Note:
nameMethod(): Names the method by which trust is achieved.
result(): Describes the AnyRating results achieved through AnyMechanism.
fulfill(): Describes the criteria on which the mechanism is based on.
operate(): Describes the operation performed on AnyEntity.

for SpecifiedBackground, which is AnyCriteria and checks for SpecifiedRating and CreditReport and determines whether the NamedParty can be trusted or not.

2. *Stability Model for Credit Report of a Loan Seeker* as shown in Figure 9.2.
3. *Use Case:*
   a. Use Case # 1.
   b. Use Case Name: Request a Credit Report of a Loan Seeker as shown in Table 9.9.

TABLE 9.7
AnyCriteria CRC Card

AnyCriteria (AnyCriteria) BO		
Responsibility	**Collaboration**	
	Clients	**Server**
Describes the rules on which the mechanism of trust is based	AnyActor AnyParty AnyMechanism	satisfyCriterion(criterion)

Attributes: rule, satisfied, criterion

Note:
satisfyCriterion(criterion): Accepts the criterion as a parameter to verify, if that criterion is satisfied or not. This method returns a Boolean.

TABLE 9.8
AnyEntity CRC Card

AnyEntity (AnyEntity) BO		
Responsibility	**Collaboration**	
	Clients	**Server**
Describes the existence of something on which the mechanism operates	AnyMechanism AnyLog	operate() relate()

Attributes: type

Note:
operate(): Describes the entity on which the AnyMechanism operates on.
relate(): Describes the relation between the entity and AnyLog.

4. *Use Case Description*
   a. AnyParty (LoanSeeker) requests a loan from AnyParty (Bank). [Test Case (TC): What is the name of bank? What is the name of the party requesting for loan? What is the email address of bank? What is the phone number of loan seeker?]
   b. AnyParty (Bank) seeks to determine the trust (EBT) of AnyParty (LoanSeeker). [TC: What is the objective behind establishing trust? What process needs to be followed to acquire trust? What are the consequences of this trustful relationship being established?]
   c. AnyParty (Bank) specifies AnyCriteria (SpecificBackgroud) to provide loan. [TC: What is the criterion? What are the rules of fulfilling the criteria?]
   d. AnyParty (Bank) demands (AnyMechanism) AnyEntity (security) to establish trust. [TC: What is the name of the mechanism? How is the mechanism performed?]
   e. AnyEntity (Land) owned by AnyParty (LoanSeeker) can be used for security purposes. [TC: What is the name of the entity? What type of entity is it? Who owns the entity?]
   f. AnyLog (CreditReport) is generated. [TC: What type of log is it? Where is the log stored? What does the log store?]
   g. AnyRating is provided by analyzing AnyLog (CreditReport) and trust is established. [TC: What is the rating of the trust? How is it established? What was needed to get the rating? What is the level of the rating?]
   h. *Sequence Diagram* as shown in Figure 9.3.

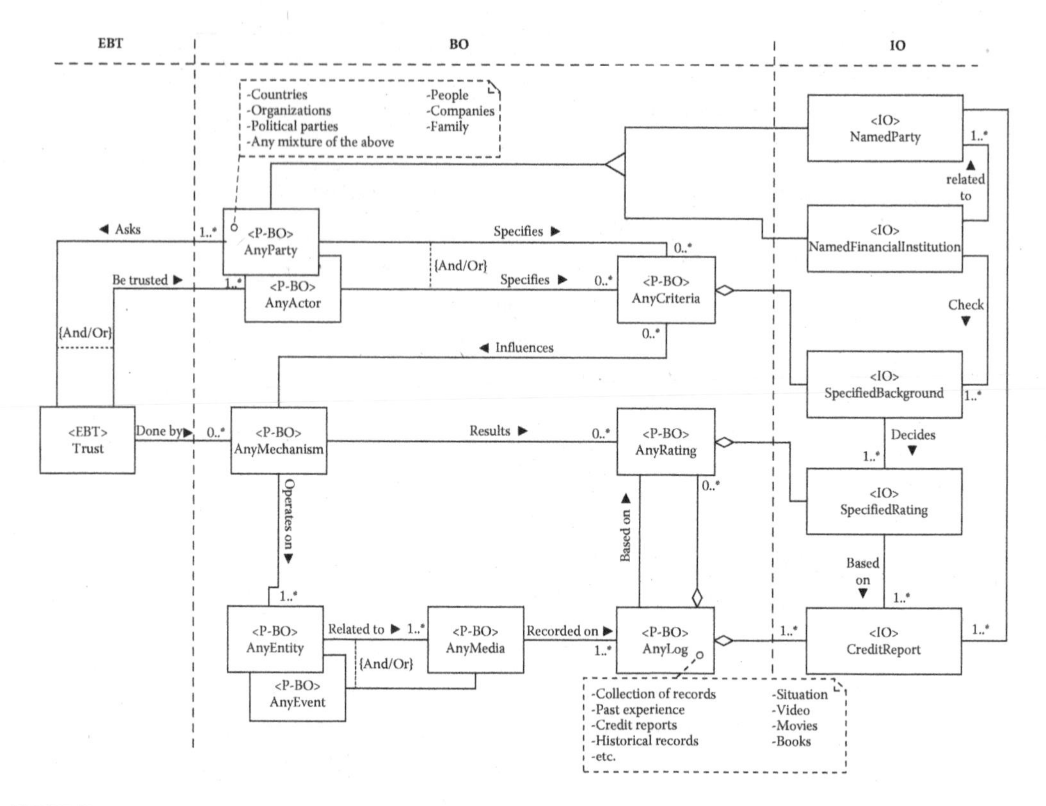

FIGURE 9.2 Trust stable analysis pattern used in Credit Report of a loan seeker.

CASE STUDY 2 Investigation History of a Named Party

1. *Scenario Description:* Consider a scenario where a bride is getting married to a groom. She wishes to check info about the groom and tries to determine the trust level of the groom. The trust EBT determines the rating of the groom by examining various logs of the groom indicating the level of trust of the groom.
2. *Stability Model for Investigation History of a Named Party* as shown in Figure 9.4.
3. *Use Case*
   a. Use Case # 1.
   b. Use Case Name: Investigate the History of a Named Party as shown in Table 9.10.
4. *Use Case Description*
   a. AnyParty (Bride) wants to marry AnyParty (Groom). [TC: What is the name of bride? What is the name of the groom? What is email address of groom? What is the phone number of groom?]
   b. AnyParty (Bride) seeks to determine the trust (EBT) of AnyParty (Groom). [TC: What is the objective behind establishing trust? What process needs to be followed to acquire trust? What are the consequences of this trustful relationship being established?]
   c. AnyParty (Bride) specifies AnyCriteria (SpecificBackground) to establish trust to marry. [TC: What is the criterion? What are the rules of fulfilling the criteria?]
   d. AnyParty (Bride) conducts background check (AnyMechanism) of AnyParty (groom) to establish trust. [TC: What is the name of the mechanism? How is the mechanism performed?]
   e. AnyEntity (BackgroundHistory) of AnyParty (Groom) can be used for this purpose. [TC: What is the name of the entity? What type of entity is it? Who owns the entity?]

TABLE 9.9
Use Case: Request a Credit Report for a Loan Seeker

c. Actors	d. Roles
AnyParty	LoanSeeker (Specify Role) Bank (Specify Role)

Class, Corresponding Attributes, and Interfaces

e. Class	f. Type	g. Attributes	h. Operations
Trust	EBT	process consequence objective	defineRules() defineGoal() takeAction()
AnyParty	BO	name id email phoneNumber	playRole() operate() request()
AnyMechanism	BO	type description id	method() result() fulfill() operate()
AnyCriteria	BO	rule satisfied criterion	satisfyCriteria() defineBoundary() measure()
AnyEntity	BO	name attribute description id	listAttributes() findValue() applyCriteria() assignType() addInterface()
AnyRating	BO	result level description id	rate() evaluate() readLog()
AnyLog	BO	description title id type	create() refer() open() close() edit()

f. AnyLog (Report) is generated. [TC: What type of log is it? Where is the log stored? What does the log store?]
g. AnyRating is provided by analyzing AnyLog (CreditReport) and trust is established. [TC: What is the rating of the trust? How is it established? What was needed to get the rating? What is the level of the rating?]
h. *Sequence Diagram* as shown in Figure 9.5.

9.3 TIPS AND HEURISTICS

9.3.1 Design Heuristics

1. The trust EBT cannot talk directly to the IOs (which are domain- and scenario-specific). The BOs namely, AnyMechanism, AnyCriteria, AnyRating, AnyLog, and AnyEntity help provide coherence between the EBT and the IOs.

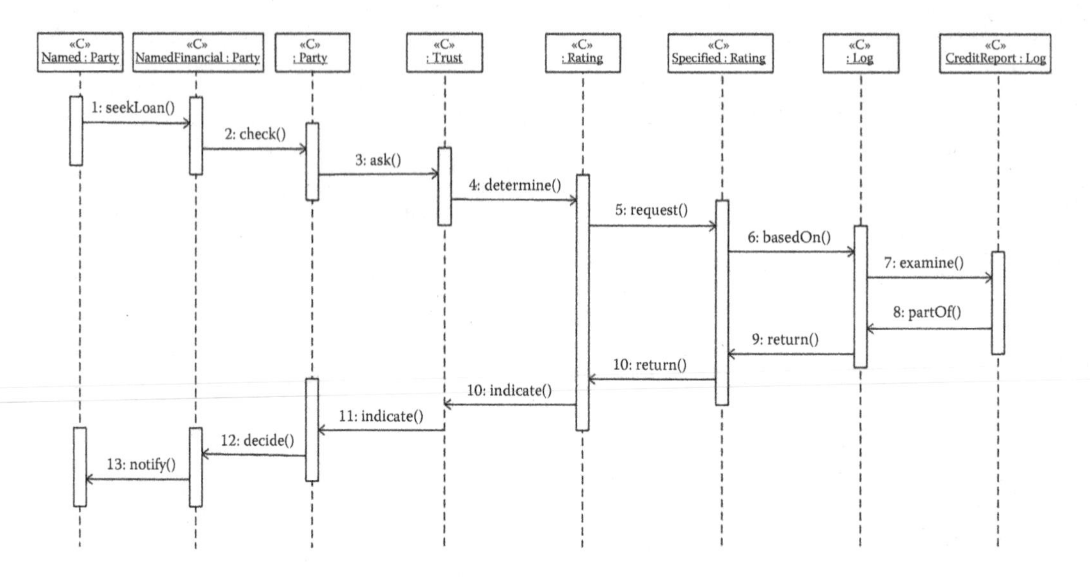

FIGURE 9.3 A sequence diagram for a request of Credit Report for a loan seeker.

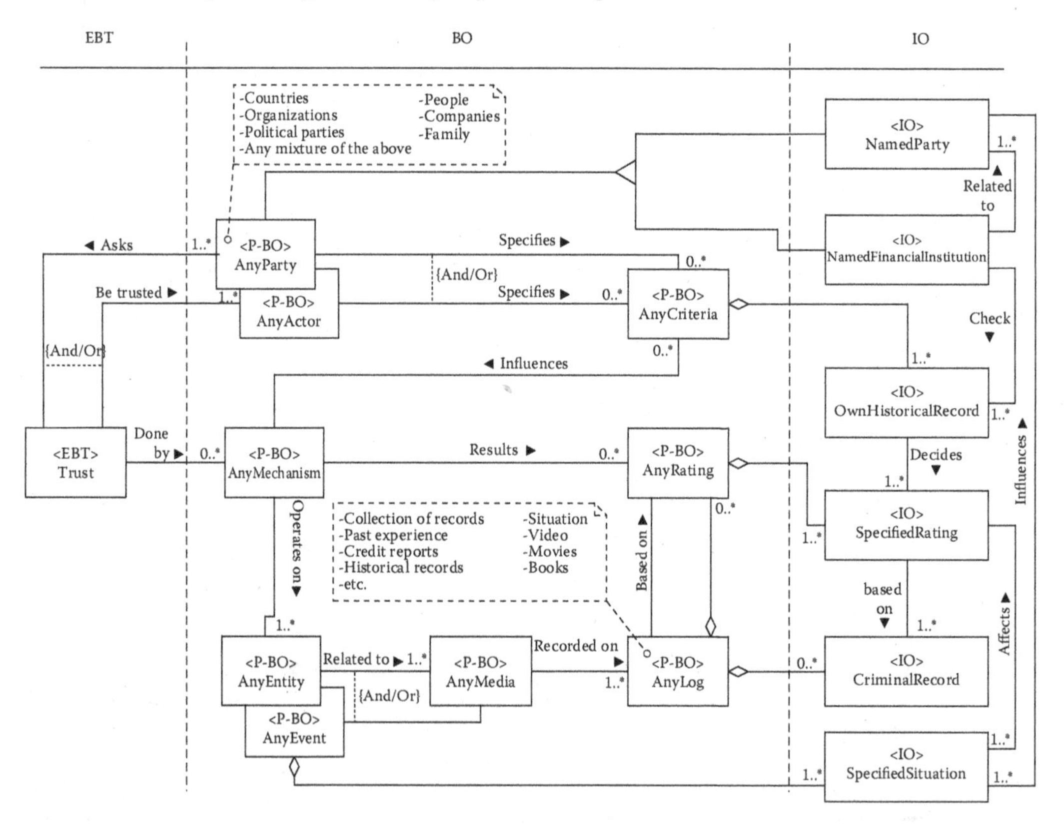

FIGURE 9.4 The application of trust stable analysis pattern for investigating the history of a named party.

2. Reusability is introduced using BOs, as they provide the hooks to which the IOs (belonging to various domains) can be attached.
3. Use of the correct BOs also makes the pattern more adaptable to changes in the IOs.
4. A proper illustration and description for the class diagram and the sequence diagram helps understand the problem in detail.

TABLE 9.10
Use Case: Investigate the History of a Named Party

Actors and Roles

c. Actors	d. Roles
AnyParty	Bride Groom

Class, Corresponding Attributes, and Interfaces

e. Class	f. Type	g. Attributes	h. Operations
Trust	EBT	process	defineRules()
		consequence	defineGoal()
		objective	takeAction()
AnyParty	BO	name	playRole()
		id	operate()
		email	request()
		phoneNumber	
AnyMechanism	BO	type	method()
		description	result()
		id	fulfill()
			operate()
AnyCriteria	BO	rule	satisfyCriteria()
		satisfied	defineBoundary()
		criterion	measure()
AnyEntity	BO	name	listAttributes()
		attribute	findValue()
		description	applyCriteria()
		id	assignType()
			addInterface()
AnyRating	BO	result	rate()
		level	evaluate()
		description	readLog()
		id	
AnyLog	BO	description	create()
		title	refer()
		id	open()
		type	close()
			edit()

9.4 SUMMARY

In this chapter, we propose a possible solution for the trust stable analysis pattern. The pattern forms a conceptual model that can be used to analyze and understand the trust concept. The pattern is generic, since it captures only the core knowledge of the trust concept, which makes the pattern applicable to different domains and applications. In addition, the pattern generalizes the properties of trust, thus making the model a reusable one that can be adapted for any kind of application. The extension of the pattern has been studied and illustrated by using two different scenarios, thus verifying that the pattern is a general one.

9.5 OPEN AND RESEARCH ISSUES

1. *Abstraction:* One of the most complicated issues involved in creating any stable pattern of any BO is finding out the most appropriate EBT for the pattern. Would-be developers and those are just learning to create patterns may find this aspect little complex as different

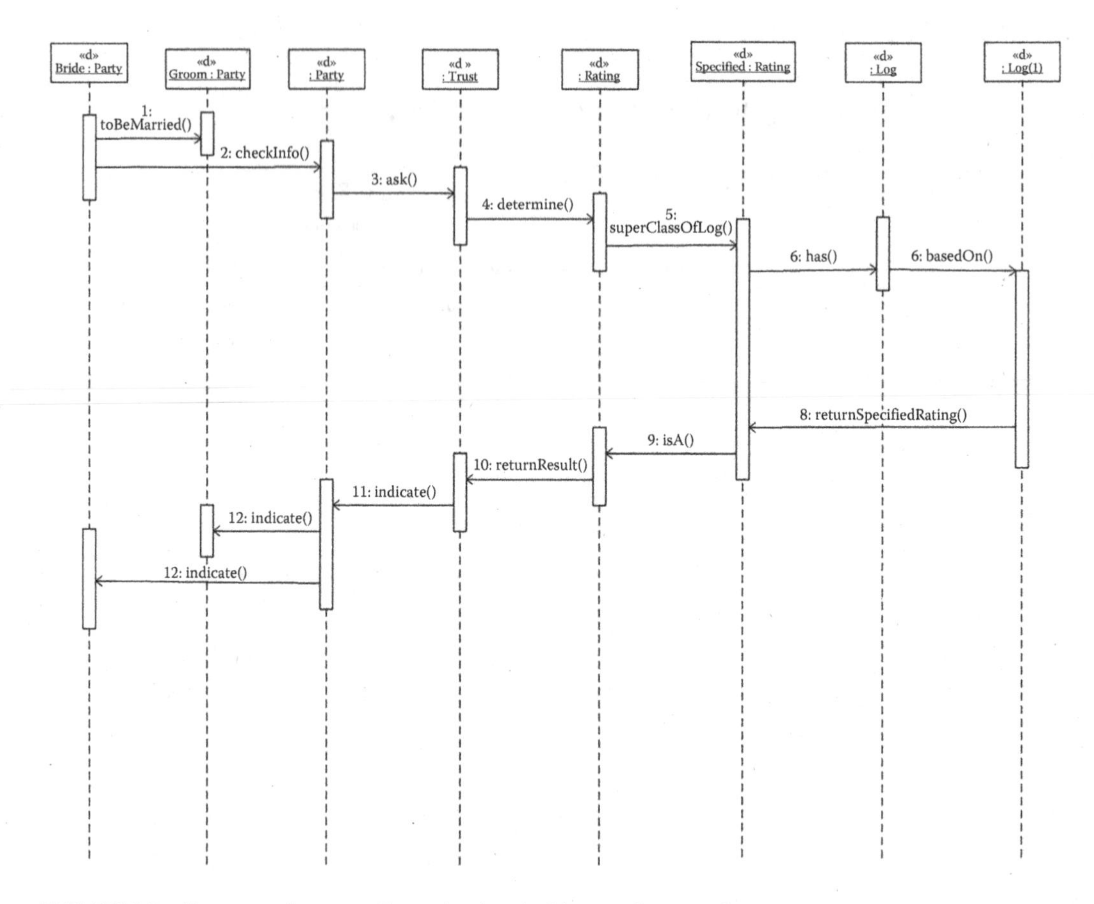

FIGURE 9.5 Sequence diagram of investigating the history of a named party.

meanings for a specific word may confuse and lead them in choosing the wrong EBT. Hence, they would have to gain experience over time to empower themselves in the art of choosing the best EBT. Another important factor that could undermine the effort of creating a stable pattern is establishing a fine coherence between EBT and different IOs. In essence, EBT may not converse directly with IOs. This aspect should be remembered while creating a pattern diagram. Additionally, choosing correct BOs may look tedious at first; however, consistent practice may enable developers with the acumen needed to pull out the most appropriate BOs for the pattern. Another problem that is usually witnessed with traditional pattern is the absence of the factor of extendibility. However, stable pattern when designed correctly, would help pattern developers in making the pattern sweeping and encompassing to all given contexts and situations. In the context of this chapter, trust is a complicated term to understand because of its wider understanding and far-reaching presence in a winder domain. Understanding the core meaning of the problem in depth will help in laying a strong foundation over which several BOs and IOs could be linked to the EBT.

2. *Trust as part of architecture on demand:* Utilize stability model [7–9] or knowledge map methodology [10] as a way for developing architecture on demand using trust and build a trust-unified software engine (T-USE). The engine mainly focuses on several patterns: trust, analysis, adaptability, extensibility, customizability, etc. The proposed solution attempts to extract the commonality from all the domains and represents it in such a way

that it is applicable to a wide range of domains without trivializing or generalizing the concepts. The engine is a stable architecture pattern and it provides a generic engine that can be applied and/or extensible to any application.

REVIEW QUESTIONS

1. What do you mean by the term "trust"? Can the term "trust" be used in any other context than what has already been mentioned in this chapter?
2. Find out all such terms that mean exactly the same as "trust" and can they be used interchangeably?
3. What are the capabilities to achieve trust? Describe each of them.
4. Draw and describe the class diagram for stable trust pattern? Explain the participants and justify their existence.
5. Come up with two scenarios other than those given in this chapter, where trust can be used. Try to fit these scenarios with the trust pattern.
6. Try to create a use case and interaction diagram for each of the scenarios you thought of in the above question.
7. Draw a class diagram for the application of trust SAP in providing trust for a patient in doctor. Describe it. Write a use case for it and its description. Draw a sequence diagram for the use case and describe it.
8. Draw a class diagram for the application of trust SAP in marriage. Describe it. Write a use case for it and its description. Draw a sequence diagram for the use case and describe it.
9. List some differences between the trust pattern described here and the traditional pattern.
10. List some design and implementation issues faced when implementing the trust pattern. Explain each issue.
11. Give some applications, where the trust pattern is used.
12. What lessons have you learnt by studying the trust pattern?
13. Define the trust stable analysis pattern.
14. The trust stable analysis pattern can be applied and extended to any domain (T/F).
15. List some of the domains in which the trust analysis pattern can be applied.
16. List four new applications of the trust stable analysis pattern.
17. List three challenges in formulating the trust analysis pattern.
18. List 10 different constraints in the trust analysis pattern.
19. What are the classes and patterns involved in defining the stable pattern for "trust"?
20. Illustrate by using a class diagram of trust patterns of each of the new applications you listed for question 11.
21. Document the CRC card for the trust EBT.
22. Is the trust pattern incomplete without the use of other patterns? Explain briefly.
23. What is the trade-off of using this stable pattern?
24. Present the sequence diagram for applicability of the trust stable analysis pattern in the e-commerce domain.
25. What are the possible design issues for the trust EBT, when linked to the design phase?
26. What do you think are the implementation issues for the AnySkill BO, when used in the trust stable analysis pattern?
27. List a couple of advantages of using the stable analysis pattern for trust.
28. List two scenarios, which will not be covered by the trust analysis pattern.
29. Describe how the developed trust analysis pattern would be stable over time.
30. List some of the lessons learnt from the use of the stable analysis pattern for trust.
31. List some of the testing patterns that can be applied for testing the trust stable analysis pattern.
32. List three test cases to test the class members of the trust pattern.

33. List some of the related design patterns used in formulating the trust stable pattern.
34. Briefly explain how the trust stable analysis pattern supports its objectives.
35. Assess two different quantitative measures on the "trust" traditional model and trust stable analysis patterns and explain the differences between the each of the measures.
36. What are the consequences of trust SAP?
37. What are the advantages of trust SAP?
38. Draw traditional model for AnyTrust.
39. Compare the tradition model of AnyTrust with the stable model. Which one do you like more and give reasons.
40. List some business models where trust SAP can be applied. Explain them.
41. List some of the properties of trust SAP.
42. List functional and nonfunctional requirements for trust pattern.
43. What did you learn from this chapter?
44. List a few tips and heuristics that you have learnt.

EXERCISES

1. Illustrate with a class diagram the applicability of the trust stable pattern in the following scenario: three or more companies are involved as a trust in monopolizing over a particular product in the market. Draw all the EBTs and BOs involved in the problem.
2. Think of few scenarios, where trust pattern is applicable and come up with corresponding class diagram, use case, and sequence diagram, as shown in the solution and applicability sections for each of the scenarios.
3. Illustrate with a class diagram the applicability of the trust SAP to an ecommerce site such as eBay.
4. Now, adapt the trust SAP to a commerce site similar to Craigslist. Discuss how the different methods of sale between Craigslist and eBay affect the implementation of the trust pattern
5. Illustrate with a class diagram and describe the applicability of trust SAP in following scenario: A person interested in buying a house from a property dealer.
6. Draw with a class diagram and describe the applicability of trust SAP in a scenario, where a marketing company tries to establish trust of its customer in its product.
7. Identify a few challenges and constraints of trust SAP that are not listed in this chapter.
8. Create a traditional model for trust.
9. Identify a few advantages of stable model over traditional model in 8.
10. Compare the trust traditional model in 8 with trust stable analysis pattern that are discussed in this chapter using the following weighted adequacies by assigning 20% for each adequacy: Descriptive, logical, understanding, systematic, and simplicity.
11. Compare traditional model with stable model of trust based on the following quantitative measurability: Number of classes and number of behaviors and show what the indicator (result) means and the impact of the measurement on three different quality of the system: Reuse, complexity, and coupling between classes.
12. Show how to do qualitative measurements: Reuse, complexity, and coupling between classes for trust SAP.
13. Would you discuss the abstraction process of the trust SAP.
14. Write the business rules of trust SAP.

PROJECTS

1. Use the first question of Section 1.9 to explain the stability of the trust design pattern. Add use cases, use case diagrams, sequence diagrams to add credibility to your explanation.
2. *Trust Management System:* A university would like to develop a system for students to evaluate their peers' performance in some group projects. Students' ratings in various

categories will affect a student's trust rating among other metrics that may be used by instructors for evaluating overall project grades.
   a. Name two to three ultimate goals of the system beyond student evaluation.
   b. List all the functional requirements and nonfunctional requirements of each of the ultimate goals.
   c. List software, hardware, or human challenges that relate to the ultimate goals.
   d. Name different applications for each of the goals. For example, how will users of this system utilize the system?
   e. Add use cases, use case diagrams, sequence diagrams based on your design of the system.
3. *Building trust in potential students to join music classes by a music teacher*
   a. Explain the above scenario.
   b. Draw a class diagram for it.
   c. Document a detailed and significant use case.
   d. Create a sequence diagram of the created use case of c.
4. *Building trust in parents by hostel authorities to state that hostel is safe for students to live*
   a. Explain the above scenario.
   b. Draw a class diagram for it.
   c. Document a detailed and significant use case.
   d. Create a sequence diagram of the created use case of c.
5. *Building trust in a company by an applicant for getting a job*
   a. Explain the above scenario.
   b. Draw a class diagram for it.
   c. Document a detailed and significant use case.
   d. Create a sequence diagram of the created use case of c.

REFERENCES

1. T. Grandison, and M. Sloman, A survey of trust in internet applications, *IEEE Communications and Surveys*, Fourth quarter 2000, http://www.comsoc.org/pubs/surveys.
2. A. Kini, and J. Choobineh, Trust in electronic commerce: Definition and theoretical considerations, *31st Annual Hawaii International Conference on System Sciences*, 1998, Maui, HI, http://ieeexplore.ieee.org/iel4/5217/14270/00655251.pdf.
3. T. Grandison, and M. Sloman, Specifying and analysing trust for internet applications, in *Proceedings of 2nd IFIP Conference on e-Commerce, e-Business, e-Government*, I3e2002, Lisbon, Portugal, October 2002.
4. E.A. Kaluscha, and S. Grabner-Kräuter, Towards a pattern language for consumer trust in electronic commerce, in the *Proceedings of EuroPLoP03*, Irsee, Germany, June 2003.
5. M.E. Fayad, J. Rajagopalan, and A. Arun, The anyLog design pattern. *The 2003 IEEE International Conference on Information Reuse and Integration*, Las Vages, NV, October 2003.
6. M.E. Fayad, H. Hamza, and M. Cline, *Stable Software Patterns: Analysis, Design, and Applications.* (Book in progress)
7. M.E. Fayad, H. Hamza, and H.A. Sánchez, A Pattern for an effective class responsibility collaborator (CRC) cards. *Proceedings of the 2003 IEEE International Conference on Information Reuse and Integration (IRI'03)*, Las Vegas, NV, October 2003, pp. 584–587.
8. M.E. Fayad, and A. Altman, An introduction to software stability, *Communications of the ACM*, 44(9), 2001, 95–98.
9. M.E. Fayad, Accomplishing software stability, *Communications of the ACM*, 45(1), 2002, 95–98.
10. M.E. Fayad, H.A. Sanchez, S.G.K. Hegde, A. Basia, and A. Vakil, *Software Patterns, Knowledge Maps, and Domain Analysis*, Boca Raton, FL: Auerbach Publications, 2014.

10 Accessibility Stable Analysis Pattern

"So you're making demands, are you? Well, let's hear them."
"I want unlimited access."
"Now that sounds interesting. To what, exactly?"

Amy Plum
Die for Me

This chapter explores and probes the *Accessibility* stable analysis pattern. The pattern intends to describe the core knowledge behind the concept of "Accessibility." Accessibility finds an extensive range of usage in various applications. The pattern also gives an excellent start to software developers, by defining the core knowledge of any accessibility problem. Any developer can build on, extend, or reuse the pattern to model any specific application by involving the factor of accessibility.

This chapter also defines the accessibility pattern and demonstrates its utilization by two significant software applications, which have an access to Internet contents and student's information access system.

The main essence of this chapter is to define accessibility stable pattern and to show its applicability in various domains.

10.1 INTRODUCTION

An accessible application is the one that is accessible through many avenues and means. An example is requesting for customer service from a computer company. A customer needs to be able to access customer service in many different ways, so that he or she can gain access to help very easily. Most computer/hardware companies and firms offer phone support, email support, interactive online support, and knowledge-based support to cater to the service demand by clients. Accessibility is very important and critical here, because if users are unable to access the application successfully, then it is almost worthless. Accessibility is indeed a good concept; it forces developers or manufacturers to think of all of the different ways in which a person might try to access their applications. This fact makes an application more versatile and extensible. Accessibility could also be bad; it can cause development time to increase exponentially depending on the accessibility requirements for the application. Accessibility can also cause some problems at times; accessibility requirements for a given day do not help a developer to foresee the ever-changing accessibility of the future. The best way to combat these problems is to use an accessibility analysis pattern as it will help and lead developers to create more stable and reusable code, which will be easily changed to work with any new requirements in the future.

Accessibility essentially means the method or possibility of approaching a place or person, or the right to use or view something. Consider the case of accessing an island or successfully contacting a customer support person, or obtaining access to the content of a website. All these cases are good examples of accessibility in different domains. Sometimes, accessibility associates itself to security parameters in different systems, but accessibility could also represent the availability of something. For example, a student's access to his academic records can be a security concern, whereas a visitor's access to the entry code for a gate can be an issue related to availability. All these examples prove that accessibility is applicable in different ways and over different domains. This diverse versatility of the accessibility concept makes it important to understand the semantics of the concept clearly.

A clear understanding of the concept is possible through a better understanding of software stability patterns. Stable patterns describe and document the core knowledge behind any concept. The accessibility analysis pattern describes the characteristics, behavior, and lists all the key players and their relationships involved in any kind of accessibility application. The pattern created with stability in mind, will also promote greater reuse, as it is extendable to suit applications of entirely different domains.

Stable patterns help us to capture domain-independent core knowledge of many different concepts. Such patterns provide the developer with a head start, by identifying the essential components involved in any accessibility application. The application-specific classes now are easily absorbable into the peripherals of the pattern. Hence, the patterns are highly adaptable and extensible. This chapter aims at building a stable pattern that captures core knowledge of the accessibility concept, so that these core concepts are extendable upon to produce long-lasting, stable programs.

10.2 THE ACCESSIBILITY ANALYSIS PATTERN

10.2.1 Pattern Name: Accessibility Stable Analysis Pattern

Accessibility means reachability or availability. Every day, we may find many different types of applications that use accessibility to accomplish different tasks. For example, accessing a person, via phone or accessing information through the web.

The basic idea behind choosing the term accessibility is the issue of generality. Generality is the driving force for choosing the term accessibility, as this term applies to all fields with its different types, takes different values yet leads to same meaning.

10.2.1.1 Known As

The following terms seem to be a good criterion for achieving accessibility:

- *Availability:* The accessibility of an application also depends on the availability of the application, because one will not be able to access an application, if it is not available. The availability of an application also refers to the uptime of the application. Availability seems to be a similar term to accessibility, but they are not the same, because accessibility is more than just availability.
- *Approachability:* The approachability of an application is the ability of the application to be easy to work with. Approachability and accessibility are two different terms that mean the same thing. They both deal with the way in which the application is usable and accessible.
- *Openness:* Openness is how easy it is to understand the application. It can also refer to, if an application is open source, or the source code of the application is readily available to all. Openness and accessibility seem similar, but accessibility is difficult to understand.
- *Exposition:* The exposition of an application is just the beginning, or the start of an application. Exposition and accessibility are not similar terms, because exposition deals with leading into the application, but does not take into consideration, if the application is accessible or not.
- *Convenience:* Convenience of an application refers to how comfortable the application is to use. Convenience and accessibility seem to be similar terms, but they are not really similar. Convenience is more like usability than accessibility in that it is concerned with how easy the application is to use, and not if it is accessible or not.

10.2.1.2 Context

Accessibility is the quality of being able to meet or deal with somebody or something. Accessibility is a general concept that has applicability in various occupations. For example, consider the accessibility to an island for a vacationer, or accessibility to a bank account for a customer. The accessibility pattern addresses the core concept of any person/group/organization that is able to reach/deal with or meet an entity (something or somebody). Accessibility can take variety of meanings, depending upon the context in which it is used. The core implication remains the same: that is, "the ability to access something". When one wants access to an entity (some particular thing), accessibility is the feature that should come into play in order to gain access. Accessing a particular thing may be referenced in terms of accessing some product, environment, service, documents, or some facility.

The accessibility pattern will throw some light on some of the areas in which accessibility is applied in order to get a few obtainable results. When a person wants to board a plane, he/she has to show his/her tickets to gain access to reach the boarding area. Similarly, when a patient wants to access his/her medical records, he has to follow some mechanism in order to gain accessibility to the medical records. Providing accessibility to person with disability is also one important feature. Similarly, when a community or country faces calamities such as hunger, efforts will be on their way to supply food and aid to them. This is nothing but giving accessibility to the things that the people urgently need. When it comes to information technology, accessing the database and ability to access information on remote machines are some of the accessibility contexts.

Certainly, a number of applications deal with accessibility feature. It is certainly one of the important criteria in software architecture/applications and the need for accessibility cannot be challenged.

1. *Accessibility to bank account:* A customer (AnyParty) wishes accessibility (Accessibility) to his bank account (AnyEntity) from bank (AnyParty). The bank provides accessibility to bank accounts through Internet (AnyMedia) banking. The owner of the account has to establish identity (AnyMechanism) by entering username and password (AnyRule) to get access. This gives the client accessibility to the account.
2. *Accessibility to an island:* A tourist (AnyParty) wishes accessibility (Accessibility) to an island (AnyEntity) for tourism from a country (AnyParty) to which the island belongs. The country provides accessibility to island only if the tourist has required visa (AnyRule). Applications for visa can be made online (AnyMedia). Tourist may need to travel (AnyMechanism) to the island so as to visit it.

10.2.1.3 Problem

Accessibility applies differently in different contexts and scenarios. For example, it means availability in case of customer accessing support personnel, while working late hours. On the other hand, it also means security in the case of a student accessing his confidential records. It can essentially mean the right to access something in other cases like access to a door security code. Such wide applicability and versatility of this concept makes it a big challenge to understand the core knowledge behind the concept.

Major requirements with the accessibility model are as follows.

10.2.1.3.1 Functional Requirements

Entity: To be accessed, such as device, service, environment, database, websites, information, etc.

- Accessibility is gained by using AnyMedia. Entity to be accessed works under a broad spectrum. It can be any device, service, environment, database, websites, or any information. So, the spectrum of AnyEntity should be well understood and well covered.

- Accessed entities should be available after the criteria holds true. If the criteria fail, accessibility should be prevented.
- AnyEntity makes use of AnyMedia. Therefore, the mapping between AnyEntity and AnyMedia should be made clear.
- Operations of AnyEntity should be general enough to make it fit into any application.

Accessibility:

- Accessibility has its own internal mechanisms and criteria to check for the validity to provide access. These mechanisms are distinct and should be included in accessibility only. In other words, it should be hidden from users.
- Criteria within accessibility are distinct from those that are defined by the users. Criteria within accessibility cannot and should not be tampered by the user.
- If un-authorized access for any entity is encountered where the criteria fails, then accessibility should throw an exception. In addition to throwing the exception, accessibility should also handle the responsibility of consistency of the data that is, leaving the accessed entity in its original state and not mirroring back the incomplete update.
- Accessibility should be strictly able to restrict malicious users intending to cause undesired entity manipulation.

Actor/Party:

- AnyActor and AnyParty may request or may provide accessibility to some entity. In the request scenario and in provision scenario, the mechanisms involved may be different and this must be well anticipated and dealt with properly to avoid ambiguity.
- AnyActor/AnyParty must go through proper mechanisms to obtain accessibility.

Constraints:

- Mappings between BOs should be clear and should be implemented by using constraints.
- Requirements for giving access should be chalked out during starting phase itself.
- Accessibility (EBT) cannot directly communicate with IOs to be implemented based on application. Communication must flow from EBTs to BOs and from BOs to IOs.
- BOs must be chosen in a way that it can cover maximum domain of applications.
- All the BOs selected for accessibility stable pattern should act as workhorses to achieve the ultimate goal.
- Many-to-many relational mapping should be avoided.

Media:

- Media that are used to gain accessibility should be well identified and documented.
- Media can be of different types and usage of each may vary.

Approached/Mechanisms:

- Mechanisms should be well defined and implemented to gain accessibility to any entity.
- AnyActor/AnyParty must follow predefined, proper, and incorporated mechanisms to gain obtainability.
- Mechanisms should check AnyCriteria that are required to gain access to any entity.
- AnyCriteria, in turn, validates accessibility giving access to those who holds the criteria true.

10.2.1.3.2 Nonfunctional Requirements

Easiness:

- The accessibility stable pattern should be designed in such a way that it should be able to gain or find larger applicability and is easy enough to use.
- Proper documentation of the pattern to be developed should be devised so that it makes the transition from designing to development an easy task.

- Use of ease from the user's point of view should also be taken into account, as the details of the functioning of mechanisms here are hidden from the user. All the user has to do here is to provide the information required for validation.
- When the pattern appeals to a large application domain, the reusability feature also adds up. Therefore, the pattern should be general enough to be used easily as a reusable pattern for other applications as well.

Effectiveness:

- The pattern should be able to answer the "what" and "how" of the system modeled. Only then will the pattern prove effective in giving the result conclusively.
- The pattern developed should produce desired and a logical result.

Intuitive:

- The pattern developed should also be very clear as to what the requirements and inputs are, and what logical outcomes should be generated. No place for ambiguity should be present here.

The main problem is—can we build a stable pattern for accessibility that can identify the essential components of the concept? Can we bring out a stable pattern for accessibility that is easily adaptable and extendable to any kind of accessibility application?

10.2.1.4 Challenges and Constraints

10.2.1.4.1 Challenges

Here are a few challenges as shown in Table 10.1.

10.2.1.4.2 Constraints

The accessibility pattern does not identify the actual procedure and the complexity involved in accessing the entity. For instance, in the case of a client–customer service interaction, the pattern discusses about various types of media available for the client to access the customer service

TABLE 10.1
Challenge

Challenge ID	0001
Challenge Title	Multiple Parties requesting for accessibility
Scenario	Accessibility for Bank Account
Description	If there are 1000 customers requesting to access their bank account at a time, then the bank would be able to handle the request. However, if there are expectedly high volume of customers requesting to access their account at a certain point of time via Internet, then the server might find it challenging to handle such large volume of requests and may crash at a certain point.
Solution	The maximum number of accessibility requests should be set and the server should be prepared to handle that number of requests.
Challenge ID	0002
Challenge Title	Multiple Rules Imposed
Scenario	Accessibility for tourists to visit an island
Description	A country may enforce some visa rules that should be followed by a tourist, to visit an island that is in its territory. If there are only a few rules such as a person must be an adult, must have a clear background, and must have a valid visa then that serves the purpose. But, if a country imposes hundreds of rules on the visitors, then tourists may get turned off and may find it very challenging to fulfill all the criteria. This would have a bad impact on the tourism of the country.
Solution	Only feasible number of rules must be imposed on the mechanism to make an entity accessible.

department, including Internet or a phone; however, it does not address the actual procedure or constraints involved in connecting to the Internet or making a call.

- Accessibility has different meanings under different contexts. For example, it could mean approachability, right, allowance, achievability. The pattern does not address all these different meanings.
- Accessibility can be requested by or provided to one, or many party.
- AnyActor can also request for accessibility and/or can define criteria.
- AnyParty has to follow one or more mechanisms to gain authorized accessibility.
- AnyParty/AnyActor can define none-to-many criteria for the mechanisms to check for in order to validate accessibility.
- AnyMechanisms are based on none-to-many user-defined criteria.
- Accessibility may be granted or denied to AnyActor/ AnyParty, depending upon mechanism and its criteria.
- AnyMechanism involves at least one or many media to gain accessibility.
- AnyMedia helps to access one or many entities required for accessibility.
- AnyEntity can take one-to-many mapping with IO related to the application.
- Accessibility can utilize one or many mechanisms to carry out the task given by the user.
- One or more criteria are needed for validation in order to gain accessibility.
- No un-authorized entity manipulation allowed. The rule for authorization should be atomic, either TRUE or FALSE.
- Some constraints should exist like how many number/copies of same accessed entities can exist at a given point of time.
- Constraints are also needed, when the mechanism encounters same input information multiple times.

10.2.1.5 Pattern Structure and Participants

The proposed solution here is to focus on the basic concept of accessibility, trying to extract the main components of the accessibility concept, and leaving other domain-specific and/or application-specific components far away from this core model. The basic components are presentable in a generic way that allows the developer to utilize them according to the needs of his/her applications as shown in Figure 10.1.

Figure 10.2 shows the stable object model of the accessibility pattern. The model has one EBT named accessibility and BOs are AnyParty, AnyMechanism, AnyEntity, and AnyMedia. Accessibility is the facilitator class that helps to achieve the reachability. AnyParty, wishing to access an AnyEntity, requests the accessibility object to invoke the process. The accessibility object finds the appropriate mechanism to enable the reachability. The AnyMechanism instance specific to the problem is created. The AnyMechanism helps reaching the entity through AnyMedia. AnyMedia can be air, water, wired, or wireless.

10.2.1.5.1 Class Diagram Description

1. Accessibility is requested or provided by AnyParty.
2. AnyActor requests/ needs Accessibility.
3. AnyParty/AnyActor uses AnyMechanism.
4. Accessibility utilizes AnyMechanism.
5. AnyParty/AnyActor defines AnyCriteria.
6. AnyMechanism is based on AnyCriteria.
7. AnyMechanism makes use of AnyMedia.
8. AnyMedia is used to access AnyEntity.
9. AnyEntity is accessed using Accessibility.

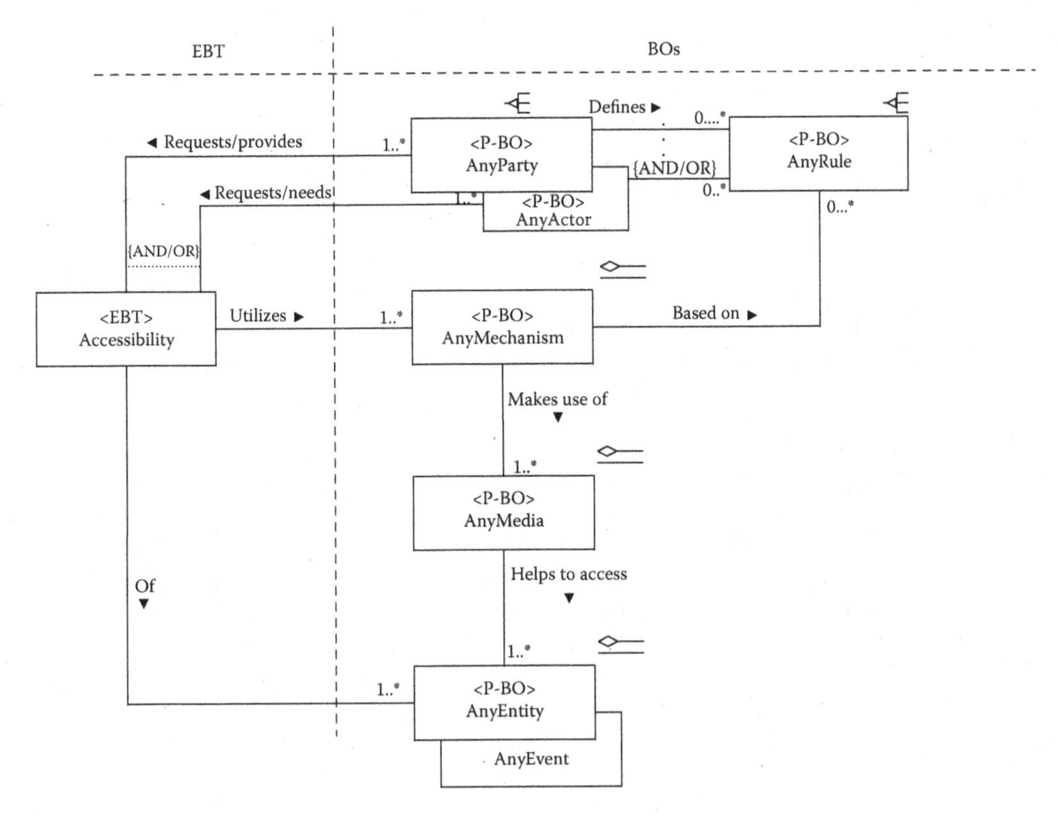

FIGURE 10.1 Accessibility stable analysis pattern.

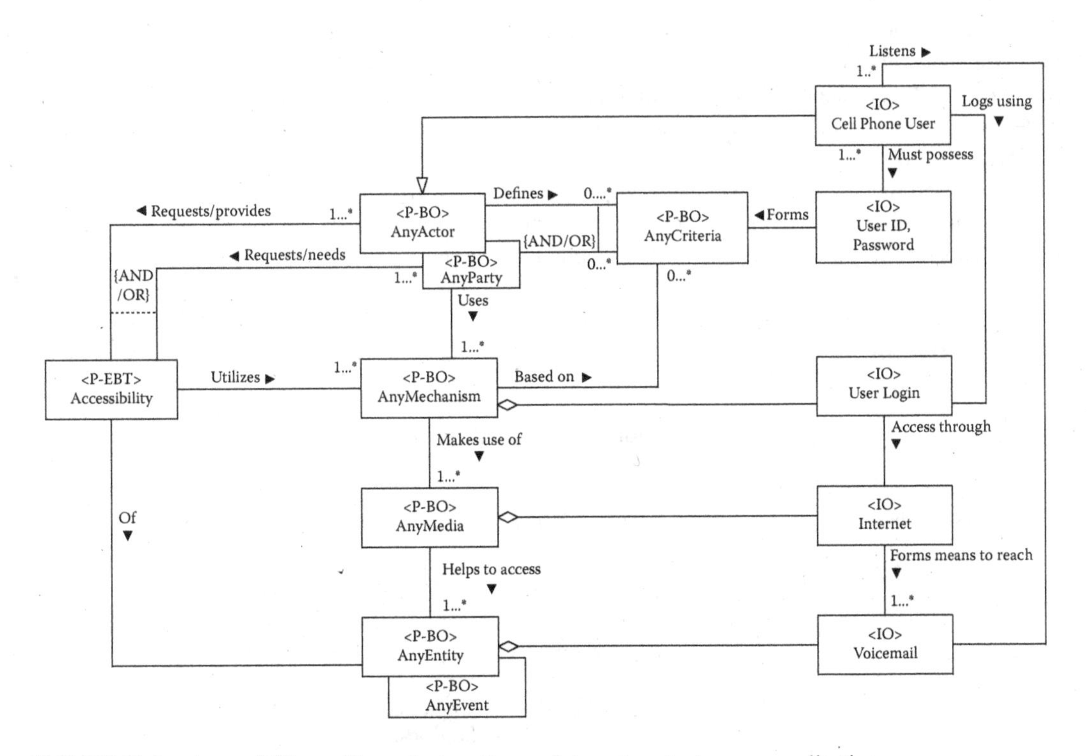

FIGURE 10.2 Accessibility stable analysis pattern—Internet content access application.

The participants of the accessibility pattern are as follows:

Classes

- *Accessibility:* Represents the accessing process itself. The class contains the behaviors and attributes that regulate the actual accessing process.

Patterns

- *AnyActor:* Represents the actor performing the access operation. It models all the parties that are involved in the accessibility process. An actor can be a person, creature, a specific software package, a hardware machine, or robot. The pattern diagram and detailed pattern description is provided in Reference [1].
- *AnyParty:* This class represents a party, an organization, governmental organization, political party, or a country. They may request access to or provide access to some entity.
- *AnyMechanism:* Represents the mechanisms that are usable in performing the accessing process. It deals with different mechanisms that are involved related to access request or access provision. For example, entering the required format of data in case of web meeting.
- *AnyCriteria:* This class represents the rules that should hold true, in order to gain accessibility to some entity. For example, authentication should hold TRUE to gain access.
- *AnyMedia:* Represents the media through which the access of an entity takes place. It is possible to have multiple mediums to access a single entity. For example, the Internet is usable to access a database, and a phone to access a person.
- *AnyEntity:* Represents the entity that needs to an access. The entity defines the media, through which it is accessible as shown in Table 10.2.

10.2.1.6 Consequences

The pattern depicted clearly identifies core knowledge of accessibility by defining the properties, behavior, and relationships. Such a pattern is always adaptable to any accessibility application; however, it is still generic enough to highlight only the domain-independent components of the concept. It gives a head start to a developer to design an application involving accessibility, instead of starting from scratch.

It is important to note that at first glance, it would be hard to capture all of the underlying issues, which relate to the different patterns that form the composite of the main *Accessibility* pattern. For instance, issues regarding which mechanism to choose or what media to choose, various

TABLE 10.2
CRC Cards

Accessibility(Accessibility) EBT		
	Collaboration	
Responsibility	**Client**	**Server**
To access or to make something obtainable	AnyParty AnyActor AnyMechanism AnyEntity	providesAccess(), takesRequests(), utilizes(), specify(), meetAccessibilityStandards(), examine(), denyAccess()
Attributes	accessibilityGivenTo, accessibilityProvidedBy, properties, rules, accessibilityTime, accessibilityDate, accessibilityNeededFor, criteriaForAccessibility, media, conditions, accessibilityGuide.	

complexities involved in them are not addressed in the main pattern. However, these issues need immediate consideration within the AnyMechanism or AnyMedia pattern details, which therefore, present a second level of detail for the original pattern. Nevertheless, the objects' names in the accessibility pattern are chosen carefully, to be both generic and easy to understand. One can understand, to a great extent, the role of each pattern participant, and visualize how he or she interact with each other, even before reading through the examples provided in the applicability section.

10.2.1.7 Applicability

One can apply accessibility pattern to a myriad domain of applications. The significance of this pattern is enormous, when it comes to software architecture/applications.

This pattern proves its worth in software applications, as protecting or restricting the information to limited sources is the prime need today. This pattern is also useful in any application that deals with providing web accessibility to users with some varied and special needs. There are numerous benefits gained by using this pattern in software applications.

- Accessibility pattern allows accessing the information by using proper mechanisms.
- Accessible applications are "viewable/available" only to the person accessing it, thereby securing the data and personal details.
- Accessibility feature is important in applications, where privacy of the user is concerned, as it gives the owner of the application the right to restrict his/her application view.
- Accessibility helps to create a proper view of shared data in large organizations.

All the abovementioned points can be implemented successfully by using the accessibility stable analysis pattern.

CASE STUDY 1 Access to Internet Content

We fully know that any website hosted on the Internet is viewable using a wired (network) or a wireless (handheld devices, etc.) media. Many web users may be operating in various contexts. They may have a disability, be using a different operating system, or have a slow Internet connection. One would have to ensure that the website is accessible in almost all of these contexts. For example, blind people use speech synthesis software to read aloud the content of a web page. Computers can use speech synthesis to read text and screen contents, by giving visually impaired and blind users access to this information. Similarly, the website should use a platform-independent technology, to display the contents, so that a person using different operating systems can view it.

This application demonstrates how a cell phone user is able to listen to his voicemails through the use of the Internet. A user can access voicemails by using the user's login details and if the login is correct, he/she is given access to his cell phone voicemails.

Use Case Id 1.0 as shown in Table 10.3.

Use Case Description

1. Accessibility is requested by AnyActor(Cellphone User).
   For what purpose do users need accessibility?
2. AnyActor/AnyParty(Service provider) defines AnyCriteria.
   How are criteria defined? How are they validated?
   Do the same criteria apply for every service user?
   Are the criteria same for every plan/membership, which the service provider provides?
3. AnyActor(Cellphone user) uses AnyMechanism(User Login).
   On what basis are the mechanisms devised?

TABLE 10.3
Gain Access to Cell Phone Voicemail Use Case

Actors	Roles
AnyParty	1. Cellphone user 2. Service provider

Class Name	Type	Attributes	Operations
Accessibility	EBT	1. accessibilityGivenTo 2. accessibilityProvidedBy 3. accessibilityProperty 4. accessibilityRule 5. accessibilityDate 6. accessibilityGuide 7. media	1. providesAccess() 2. utilizes()
AnyActor	BO	1. name 2. age 3. location 4. accessFor 5. userOf	1. requests() 2. defines() 3. uses()
AnyMechanism	BO	1. name 2. criteria 3. usedBy 4. wayOfFunction 5. status 6. components	1. makesUseOf() 2. used() 3. status()
AnyCriteria	BO	1. name 2. numberOf 3. checkedBy 4. priority 5. condition 6. property	1. validates() 2. providesBase() 3. isDefinedBy()
AnyMedia	BO	1. mediaName 2. mediaType 3. usedFor 4. usedBy 5. securityLevel	1. helpsToAccess()
AnyEntity	BO	1. nameOfEntity 2. typeOfEntity 3. description 4. position 5. states	1. utilizes()
Cellphone User	IO	1. nameOfCellphoneUser 2. cellPhonenumber 3. birthDate 4. nameOfService 5. typeOfService 6. location	1. mustPossess() 2. logsUsing() 3. views()
User ID, Password	IO	1. givenTo 2. givenBy 3. usedFor 4. usedOn 5. condition	1. forms()

(Continued)

TABLE 10.3 (*Continued*)
Gain Access to Cell Phone Voicemail Use Case

Class Name	Type	Attributes	Operations
User Login	IO	1. usedBy 2. providedBy 3. providedFor 4. components 5. status	1. makesUseOf() 2. used()
Internet	IO	1. mediaType 2. providedBy 3. usedBy 4. status	1. formsMeansToReach() 2. used()
Voicemail	IO	1. name 2. recordedOn 3. recordedBy 4. madeBy 5. heardBy 6. status	1. listened() 2. used()

How does the mechanism behave in case of user using some different service? Can they access or will denial of service be issued?
What happens when AnyMechanism fails?

4. User must possess a user Id and password.
   What happens, if multiple user Ids are created by the same user? Is it allowed?
   What are the conditions/steps for creating user Id and password?
   Under what situations can, user Ids be denied?
5. User Id and Password forms AnyCriteria.
   Are these two components sufficient to form criteria?
   Can there be additional components for AnyCriteria? Can one add it at a later stage?
   What happens, when User Id expires? Alternatively, when User Id is inactive for long time?
   What are the conditions for password?
6. AnyMechanism(User Login) is based on AnyCriteria.
   What happens, when incorrect criteria are entered?
   How many times retry are allowed, if AnyMechanism keep failing consecutively?
7. AnyMechanism(User Login) makes use of AnyMedia(Internet).
   Is there any alternate media available?
   Is AnyMedia easily available?
8. Cellphone user logs using User Login.
   What happens, when user login fails?
9. User login is done through Internet.
   Is there any alternate way for logging?
   What happens, when internet is down? Will any notification message be generated for this?
10. AnyMedia(Internet) helps to access AnyEntity(Voicemail).
    What are the entities accessed?
    What happens, if no entity is present/recorded?
    What are the situations in which AnyEntity(Voicemail) will not be recorded?

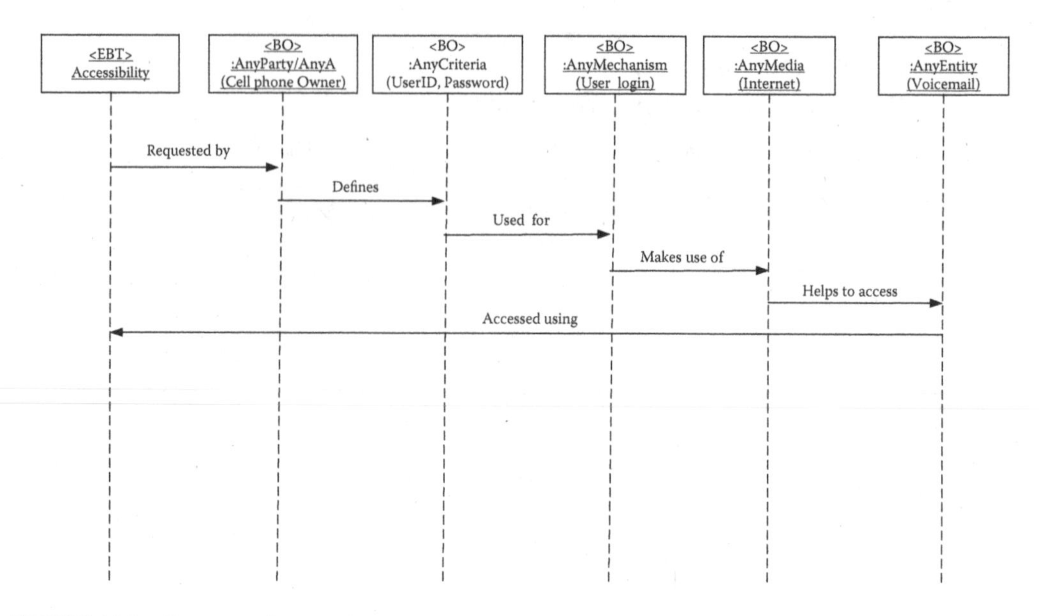

FIGURE 10.3 Sequence diagram for gaining access to voicemails using Internet.

11. Internet allows listening to the voicemails.
    Who has the authority to listen to the voicemail?
    What happens, if the same user has multiple cellphone numbers and is trying to access the voicemails of each?
    Can the service provider give access to user voicemails to someone else? If yes, then under what situation?
12. Cellphone user listens to his/her voicemail as shown in Figure 10.3.
    How many voicemails can be recorded at a given time? Is there any limit for number or duration of voicemails specified?
    What are the allowed actions that the user can execute on his/her voicemails?
    Can the user request any special service on his/her account?

Behavior

CASE STUDY 2 Online Student Information System

To enable students to have an easy access to their confidential information, universities world over have introduced password-protected websites to allow them access to their records. By using a proper user authentication, or by using a user login and a private pin, a student can gain immediate access to his/her records in the college database as shown in Figure 10.4.

Use case Id 2.0 as shown in Table 10.4.

Use Case Description

1. Accessibility is requested by AnyActor(Student).
   What is the need for accessibility? Who controls accessibility?
   For what purpose do you need accessibility?
2. AnyActor/AnyParty(University) defines AnyCriteria.

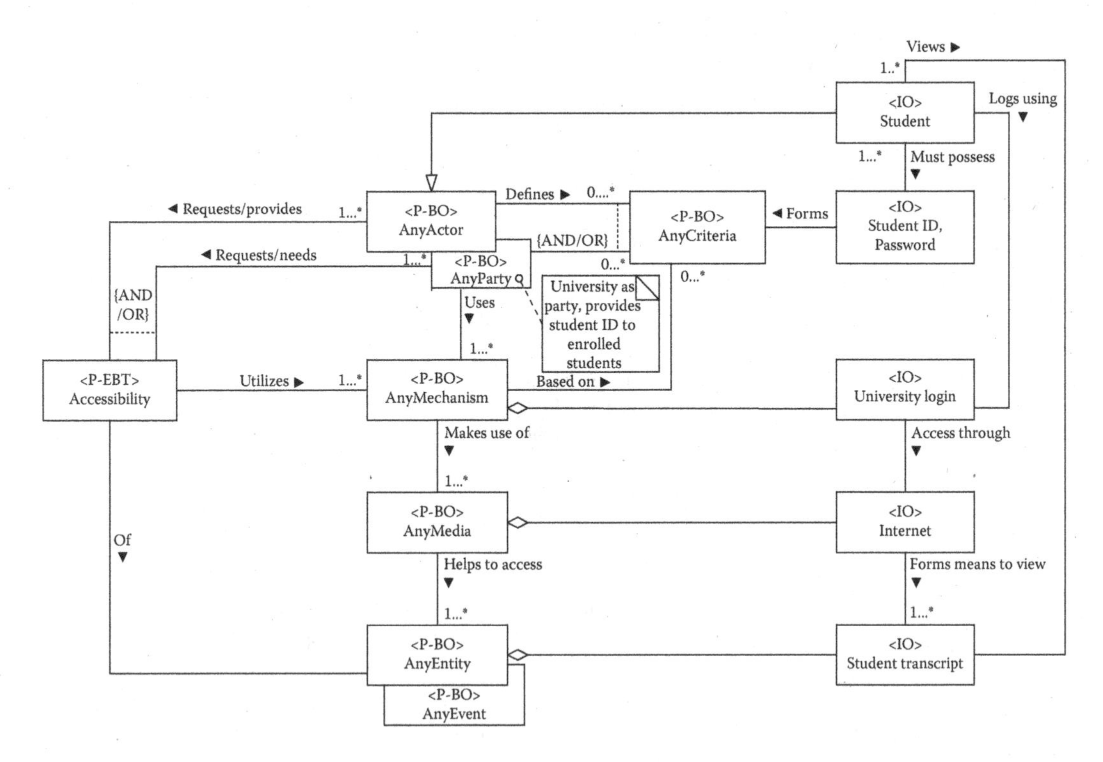

FIGURE 10.4 Accessibility stable analysis pattern—student information system application.

What are the criteria? How are they validated?
What happens, if criteria fail?
What happens, if wrong criteria are entered/given?

3. AnyActor(Student) uses AnyMechanism(University Login).
   What are the components required for AnyMechanism?
   Are components same for every student? How is AnyMechanism used?
4. Student must possess student Id and password.
   What are the criteria for allotting student Id? What happens, if same Id is used multiple times and at the same time?
5. Student Id and password form AnyCriteria.
   Is any other component needed for AnyCriteria? Or, Id and password suffice for restriction of the accessibility?
6. AnyMechanism(University Login) is based on AnyCriteria(Student Id, password).
   What happens, if criteria fail?
   What are the other factors affecting AnyMechanism?
7. AnyMechanism makes use of AnyMedia(Internet).
   What happens if AnyMedia is unavailable?
8. Student logs by using university login.
   Is there any other alternate way for logging?
   What happens, when login fails?
   What should be done, if login expires?
9. University login is through Internet.
   Is there alternate mode for logging?
10. AnyMedia helps to access AnyEntity(Student transcript).
    What are the entities accessed? What is the status of entity?
    Can accessed entity be manipulated?

TABLE 10.4
Gain Access to Student Transcript Use Case

Actors	Roles
AnyParty	1. Student 2. University

Class Name	Type	Attributes	Operations
Accessibility	EBT	1. accessibilityGivenTo 2. accessibilityProvidedBy 3. accessibilityProperty 4. accessibilityRule 5. accessibilityDate 6. accessibilityGuide 7. media	1. providesAccess() 2. utilizes()
AnyActor	BO	1. name 2. Age 3. Location 4. accessFor 5. memberOf	1. requests() 2. defines() 3. uses()
AnyMechanism	BO	1. name 2. criteria 3. usedBy 4. wayOfFunction 5. status 6. components	1. makesUseOf() 2. used() 3. status()
AnyCriteria	BO	1. name 2. numberOf 3. checkedBy 4. priority 5. condition 6. property	1. validates() 2. providesBase() 3. isDefinedBy()
AnyMedia	BO	1. mediaName 2. mediaType 3. usedFor 4. usedBy 5. securityLevel	1. helpsToAccess()
AnyEntity	BO	1. nameOfEntity 2. typeOfEntity 3. description 4. position 5. states	1. utilizes()
Student	IO	1. nameOfStudent 2. studentId 3. birthDate 4. university 5. major 6. location	1. mustPossess() 2. logsUsing() 3. views()
Student ID, Password	IO	1. givenTo 2. givenBy 3. usedFor 4. usedOn 5. condition	1. forms()

(Continued)

TABLE 10.4 (*Continued*)
Gain Access to Student Transcript Use Case

Class Name	Type	Attributes	Operations
University Login	IO	1. usedBy 2. providedBy 3. providedFor 4. components 5. status	1. makesUseOf() 2. used()
Internet	IO	1. mediaType 2. providedBy 3. usedBy 4. status	1. formsMeansToView() 2. used()
Student Transcript	IO	1. name 2. issuedOn 3. madeBy 4. viewedBy 5. year 6. status	1. viewed() 2. used()

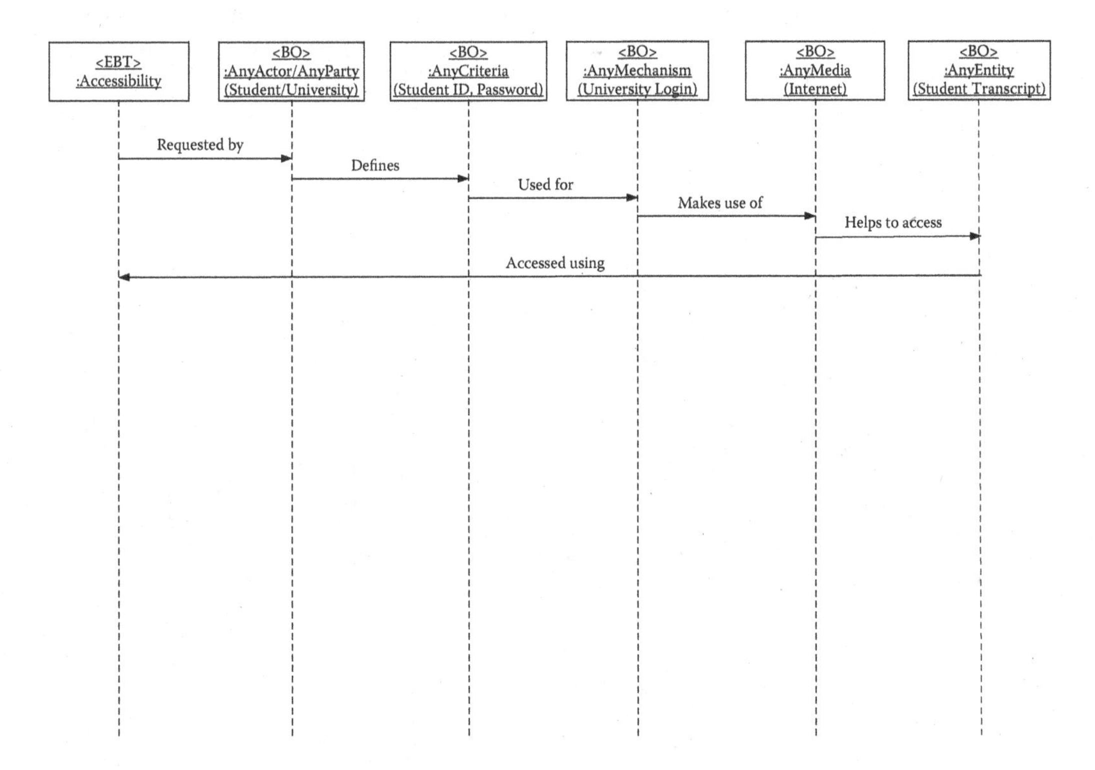

FIGURE 10.5 Sequence diagram for gaining access to view student transcript.

11. Internet provides a means to view Student transcript.
    Who has access to student transcript?
    What are the contents in student transcript?
    Who has the privilege of modifying it?
    Can it be manipulated by student?

12. Student views his student transcript as shown in Figure 10.5.
    Can student view other's transcript as well?
    Can student apply changes to the transcript?
    When does a student need a university transcript?

Behavior

10.2.1.8 Related Patterns and Measurability

10.2.1.8.1 Related Pattern

Here, a related pattern includes traditional models for student information-system access, like viewing student transcript. Related patterns help to compare the pros and cons of the different models that are related to application. The most noticeable difference in a traditional model class diagram given below is the lack of accessibility class. There is no core concept of accessibility in the model given below, which forms a major drawback, as the flow of model does not cover the idea of accessibility related to users and different entities as shown in Figure 10.6. We can draw out comparisons between traditional model and stability model (stable pattern) based on some of their differences as follows:

- Traditional model relies entirely on IOs. No focus is given on relevant concepts and BOs. In the above model, the notion of login is clear in literal meaning, but what exactly login is doing here is too vague. Whereas in stability model, the core focus is on themes/concepts. EBTs and BOs are the backbone of stability model, where EBTs represent the ultimate goal of the pattern and BOs represent and acts as the workhorses to achieve the goal.

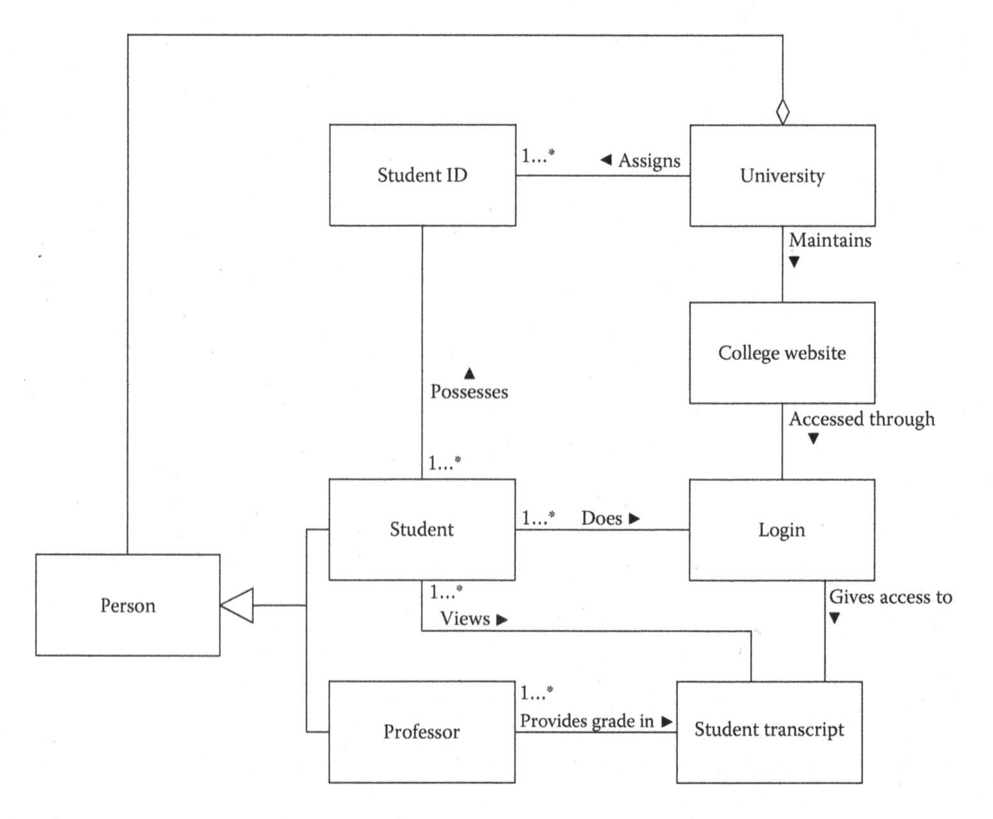

FIGURE 10.6 Traditional model for student information system access.

- Traditional model has nothing to do with the "what" and "how" and it is rather questionable, but stable model tackles both the "what" and the "how' of the system that needs to be developed.
- In a traditional model, the model applies to only that specific application, which is being developed. Its usability is limited to the specific application only.
  Whereas in stability model, the pluggable feature of IOs according to the applications make it easy to reuse for a large number of applications.
  Reusability is also one of the core difference between traditional and stability model.
- In traditional model, if some changes are introduced at some later point in development, all the previous tasks performed are rendered useless. Changes can ripple or reflect back to starting point, also leading to the failure of the system developed. This forms a serious drawback in traditional model. In a stability model, the essence and base work of action is bundled in EBTs and BOs. BOs can contain the objects that are changing internally as per the application. If changes are required at some point, all that is needed is to modify IOs and plug it with the BOs of the application accordingly. Thus, rippling effect is tracked and handled in such a way that it does not lead to serious flaws and alteration of the system developed.
- In a traditional model, the system modeled may be limited in terms of portability as it has minimal interface. Whereas in stability model, interface functionality is implemented by identifying all the operations needed for the compatibility and co-ordination of different elements, which is different BOs to EBTs, thus aiding in portability largely.
- One of the fundamental differences between traditional model and stability model is the implementation of well-defined user input. Stability model has well-defined user input.
- Documentation is much easier in stability model when compared to a traditional model. If any changes occur in traditional model, whole documentation needs to be changed that is an additional overhead burden of cost and time. In a stability model, only changes in specific model can serve the purpose for maintaining up-to-date documentation.
- Cost of development in traditional model is larger when compared to the stability model, as stability model has the feature of reusability and there is least or no maintenance at all involved. Maintainability cost is more in traditional model too.
- Due to its limitability, traditional model is suitable for small projects, whereas stability model can control and handle large projects easily.

10.2.1.8.2 Measurability

10.2.1.8.2.1 Quantitative Measurability Quantitative measures rely on factors such as EBTS, number of BOs and IOs that are required for specific applications, constraints, classes required, division of class functionality, number of operations to be implemented in a class, inter-dependency of classes, and depth of inheritance.

- *Number of classes:* If different models are compared with each other, then we get an idea that as stability model has EBTs, BOs, and pluggable IOs, the number of classes required is less when compared to a traditional model. In addition, the functionality of BOs is divided in such a way that it forms the whole functionality of the application developed. However, each class performs an individually assigned function.
- *Inter-dependency among classes:* If implemented by using stability model, the effect of changes applied in any BOs will be limited to that specific class and other classes will not be directly affected by the rippled changes. Thus, interdependency among classes becomes lesser and this is one of the quantitative measures in a stable model.
- *Constraints:* Constraints represent relational mappings. Relational mapping makes clear the association between classes and shows how many instances of one class can link/associate with the instance/instances of the other associated class.

- *Number of operations in a class:* Each class must perform specific operations required for the functionality of the class. Thus, the operations selected should be general enough to make it fit into any application domain. In a stability model, it is required that there should be 5–7 operations for each class.
- *Depth of inheritance level:* Inheritance does encourage dependency among classes and create child classes that inherit some or all properties of parent class. However, if dependency leads to multilevel inheritance and when any child class fails, then the whole system collapses. In a stability model, this problem is handled by implementing bridges. Thus, the lesser the number of inheritance level, the more will be the freedom of jumping and navigating to number of classes.

10.2.1.8.2.2 Qualitative Measures Qualitative measures depict the features such as reusability, generality, scalability, performance, effectiveness, etc. As we know, stability model relies on EBTs, BOs, and IOs. The idea of pluggable IOs as per the application makes it easy to be reusable for any other application in the related domain. In addition, EBTs and BOs will be the same. Reusability factor enables the use of some set of code to be re-used, instead of writing the code from scratch. Thus, stable software model has the reusability feature.

As reusability factor increases, the generality feature also adds up. The more the general operations selected for a class, the greater its use will be in making it apply to other applications. In stability model, as the classes are assigned some distinct functionality, each class performs its own function, thus generating manageable lines of code per class. Later, final execution of code can be performed by linking all these class functions together. In this way, performance is enhanced. In addition, because of separate functionality, debugging also becomes easy. Scalability is nothing but developing the interfaces in such a way that it provides the base for adding scalable components later in the system without causing much changes in the base code. All these features are implemented in software stable model.

10.2.1.9 Modeling Issues, Criteria, and Constraints

10.2.1.9.1 Abstraction

Stability model revolves around EBTs, BOs, and IOs [8–10].

- EBT classes present the enduring themes and goals of the system to be modeled.
- BO classes present the classes, which are semiconceptual and externally stable, but they are internally adaptable and they map EBTs to IOs.
- IO classes represent the classes that map BOs to external physical objects that are required for the application.

These are the three layers of abstraction in stable models. Abstraction allows dealing with the system in a systematic way by detailing out each required factor (here EBTs, BOs, and IOs). Abstraction also helps us to isolate the approaches applied in modeling the pattern. It also helps to distinguish whether a process approach is applied or measurable approach is used.

Here, a process approach is more suitable as compared to measurable approach. The reason is, process approach is used, and when all the activities are being managed as an individual process and identified based on user requirements and its ability to achieve the goal. On the other hand, measurable approach deals with the quantitative aspect in a large manner. It is inclined toward achieving the desired goal without a clear understanding and limitations on the mechanism needed.

The main goal of abstraction is to find out ways to identify and weave together relevant EBTs, BOs are in order to model a system successfully. In the case of accessibility stable analysis pattern, to achieve the goal of accessibility, we will have to identify different mechanisms needed for gaining accessibility and know and understand the conditions in which the whole mechanism becomes valid. So, it becomes necessary to identify BOs that can handle both the functions separately, yet merge, and work in tandem to gain accessibility. This is how both AnyMechanism and AnyCriteria come into picture while modeling this pattern. Both AnyParty and AnyActor need accessibility for some purpose.

- AnyMechanism may consist of a single action or a group of actions. For accessibility, the mechanism may be a single action or a number of steps; this is the reason why AnyMechanism is preferred over AnyAction for accessibility.
- There must be some means to validate the mechanism to reach accessibility and this is the basic and most important part in accessibility. To validate the mechanism, there must be some well-defined criteria that should be satisfied first. Therefore, AnyCriteria is chosen, which will implement the function of validation of mechanism. If the result is TRUE, point of control is passed to accessibility and when the result of AnyCriteria is FALSE, point of control is returned to AnyMechanism for its re-execution.
- If AnyCriteria is not chosen/implemented, the whole purpose of accessibility is beaten.
- After selecting two capabilities for accessibility, the next important task is identifying ways that will be used by AnyMechanism; here, AnyMechanism deals with discovering means through which it will feed. It is AnyMedia that will provide different ways for AnyMechanism to use. AnyMedia handles the functionality of representing any media like Internet, phone, document, any machine, software, hardware, etc. To identify all the possible media that can be used by AnyMechanism, a separate functionality is implemented, which is AnyMedia.
- After combining the functionality of all these capabilities, we will need to ask a critical question: What are the entities that we are trying to gain accessibility for remains unsolved until now? So, to identify these entities and to assign separate functionality for it, AnyEntity is used by developers.
- Accessibility is the EBT here, as the system modeled wants to achieve valid obtainability to some entity.

By using the relevant BOs, we can effortlessly implement the accessibility stable pattern. The benefits reaped here are manageable number of BOs with each one of them dealing with its own unique assigned functionality, yet jointly working together to achieve the desired goal. The central point to be noted here is, each BO is independently performing with its own assigned function. However, they behave as complete functionality, when they work in combination and in synchronization with other relevant BOs; this approach results in gaining accessibility (this is the key idea behind abstraction).

10.2.1.9.2 Modeling Heuristics

Generality: The stable pattern developed here can be applied to a wide range of application domains. This generality is achieved by using suitable EBTs, BOs, and IOs. A good example of generality is discussed by taking AnyMechanism used in the application here. AnyMechanism has the fixed responsibility to incorporate means/methods needed to gain access. The main responsibility remains the same, no matter what the application is. Nevertheless, internally, AnyMechanism keeps changing according to the application. AnyMechanism can be user login for some information-system access, or it can be entering an access code for gaining access to some secured place. Thus, externally BOs are stable, but internally they adapt to requirement changes depending on the application.

No Dangling: The other example is there should be no dangling class. This means that every class should have some input and it should give out some kind of required output. Dangling class is the class that has either only input and no output, or is connected to any one class only. Dangling class indicates a dead end and this issue is of serious concern to developers, particularly when a code executes. There is no point of talking progress in dangling class that leads to a wedged point in the code eventually causing regression in the process. Navigation is obstructed here mainly because of dangling.

10.2.1.10 Design and Implementation Issues

The most challenging phase of designing is how to link this design to the development phase and how to devise proper interface to make the transition from design phase to development phase. The answer to the above problem in stable model is hooks [2,3] and interface [4]. When some function module is created, there needs to be an interface to make it link in the development phase. Hooks enable the system to call appropriate functions and each hook has some predefined set of parameters/operations to link the functions and it has return type, also. There are different types of hooks such as access hooks and database hooks [11,12]. These hooks contain predefined set of operations to be performed during development phase.

Interface is a pattern [4] that would list all operations of BOs in combination that are required to connect BOs to IOs. Thus, BOs will be linked to IOs through an interface. The key feature of interface is to provide functionality. It is also applicable, while linking EBTs to BOs.

The other design and implementation issue is the use of aggregation or delegation in place of inheritance. As explained in the previous section, inheritance can lead to more dependency, and thus when one node in the inherited structure fails, the whole system will also fail. In addition, in inheritance, the child class inherits all the features/functions of parent class, even if it does not require the whole set of parent functions. This problem can be avoided by using delegation pattern [4]. Delegation provides the unique feature of run-time dynamic coupling between parent class (superclass) and child class (subclass). This allows the subclass to inherit only those properties that are relevant during run time, instead of taking on the whole functions. In delegation, even if some changes are made in superclass, it will not be reflected in subclass unless and until it is desired.

10.2.1.11 Testability

Stable model is easily testable, as it has the provision of implementation of separate functionality in each class. In addition, because EBTs and BOs rarely change, the testing of the model becomes an easy task, even when the applicability changes. In this chapter, one can apply accessibility pattern to different applications and it could be tested whether this pattern satisfactorily handles the base purpose of the stable pattern developed or not. The only changing thing in these applications is the IOs. In the same way, this stable pattern can be applied to different applications by changing the IOs.

The million-dollar question is whether there is any such scenario where the stable pattern fails. Of course, we will have to tackle this question also, because finding a scenario and implementing actions to deal with it in the starting phase itself would make it easy to build a more stable and robust model. For example, if the access information needed to gain access gets in a wrong hand, then that Actor/Party will be using it to gain access to the secured entity for some reason. This means, accessibility is being given here now to the wrong person at the wrong time. The accessibility pattern will work fine, as long as it gets all the required valid inputs and will lead ultimately to the access of secured entity. Here, the pattern eventually fails, as it has no way or means to detect whether the right person is being given accessibility or someone pretending to be the right person is entertained here, as long as all the requirements are validated satisfactorily.

10.2.1.12 Business Issues

Business issues involve business rules, business terms and facts, business models, EBTs, and business patterns.

10.2.1.12.1 Business Rules

Business rules are a set of defined constraints that describe some operations that are needed to fulfill the ultimate goal. The rules tell "what" is needed to reach the goal and "how" to do it. Business rules are normally included as part of requirement-gathering process. Some of the business rules that should be included in accessibility stable pattern are

- The most important rule is to provide accessibility to users holding the criteria true for accessing the entities. The criteria validation should be atomic, either TRUE or FALSE.

There are different elements in business rules that are business items, properties, relations, constraints, and facts.

10.2.1.12.1.1 Business items These represent the classes that are present in the pattern. Each class forms one business item. Some of the rules that apply to these items are

- Classes should be general enough to be able to add any changes as per the business needs.
- Classes should be customizable and adaptable as well.
- Classes should not create cycles.
- The relation between classes should be clear whether it "*is a type of*" or "*is a part of*".
- Classes selected should be able to produce a complete business process for which it is intended.
- One of the business rule that can be inferred based on classes is AnyMechanism must make use of AnyMedia.

10.2.1.12.1.2 Properties Attributes of each class must be unique for that class and must be able to cover all distinct aspect of the class.

- For accessibility, AccessibilityGuide forms one attribute that shows the guidelines needed for accessibility. Guidelines may include different ways to gain accessibility and it may include the troubleshooting guidelines as well. Thus, the attributes defined forms one important property for business rules.
- Operations defined for each class should be unique and must be able to carry out the responsibility assigned to that class.
- In AnyMechanism mentioned above, some of the operations are activate, pause, attach, and status. Thus, AnyMechanism has the operational scope of being activated for execution, being paused at some particular point, where some kind of exception is encountered or attaching itself to some subprocess/subroutine mechanisms for complete execution. These properties become clear and lucid, when operations of each class are identified and this helps to formulate the business rules for solving the business issues.
- Taking into account the above point, one of the business rules that can be formulated using properties is AnyMechanism that can pause or lock the present state when an exception is encountered or help regain point of control after exception is handled or unlock and continue execution from the point, where exception was encountered initially.
- AnyMechanism can pause, if it starts executing, when it is still receiving input from other classes.

10.2.1.12.1.3 Relations Relations focus on business strategies based on all the above points and they deal with generating different policies required to be implemented in the system.

- Relations between classes should be clear whether it forms "a part of" or "a kind of".
- In accessibility EBT, if its internal criteria form part of accessibility then, accessibility cannot be part of that internal criterion.

10.2.1.12.1.4 Constraints Constraints define the limit or range on attributes, operations, classes, etc.

- Accessibility has to utilize AnyMechanism.
- AnyCriteria needs to be validated for accessibility.
- In case of operations, say in AnyMechanism, when the operation of pause is the state then, activate function cannot be invoked at that given instance.
- In case of attributes, say in AnyEntity, there should be limit on how many copies of those particular entities exist at a given point of time.

10.2.1.12.1.5 Facts Facts deal with the general business terms that can occur commonly in EBTs and BOs.

Some of them are

- Accessibility: Represents access to be given to some entity.
- Login: Represents mechanisms needed for accessibility.
- Retry: Shows login/AnyMechanism failed or some interrupt is encountered.
- Success: Shows login has passed the required criteria and can proceed to access the requested entity.
- Authorization: Indicates requested entity is protected/restricted.
- Out of service: May represent that the means, which are used by AnyMechanism, is temporarily out of service.
- Not found may show that the requested entity is not present at the requested place.

These are some facts that are explicitly present, when you use accessibility patterns. There are certainly many additional facts that can be derived based upon the application.

The above important elements together help to formulate distinct, efficient business rules that are required for doing a business. Specific business rules vary as per the application.

10.2.1.13 Known Usage

Accessibility pattern is developed in a generic way, so it applies to myriad domains of applications. Some of the known usages for accessibility are as follows:

- *Access given to person with disability in a public transit facility*: For persons with disability, some seats are reserved in public transits; say for example, a bus. To give them access, certain mechanisms are involved like helping them to board the bus and giving access to the seat reserved for them. This forms one example of access as the person with disability is being given access to something that he/she deserves or has the right to avail. The thing to be observed here is whether the criterion is fulfilled for accessibility and that the criterion for securing the reserved seat is disability of the person. If the person has disability, seat is accessed or else, that service is denied.
- *Access to record in tables in database applications*: When any software application needs to fetch a data from the database to populate its requirement fields, the application needs to have valid database access to gain accessibility to the records stored in that database. For this, specific mechanisms are involved for accessing records and leaving it in a consistent state. Input/output are the mechanisms, validation of the scope of program requesting access form the criteria and the records accessed forms the entities.
- *Web access standards for disabled person*: Many software and hardware accessibility products are developed for a person with some disability to enable them able to use the computer. Many people have disability problems related to vision, hand movements, physical disability, hearing impairment, etc. They may be less in numbers individually,

but together they form a potentially large number of users. To take into consideration the special requirements of these types of computer users, web accessibility software/hardware were designed. These tools help the user with disability to use computers in a better and efficient way, which otherwise would not have been possible. This leads to accessibility for the person with disability to the use of computers.

- *Access to library*: Universities and schools offer services of online libraries for the students to avail their needs for different books and journals. If we take into account the online service of universities as an example for accessibility, the student login feature forms the mechanism and validation of student Id and password is the criteria. After the validation of the criteria, the student is given access to catalog of the books from where he/she can search the database and request the required book or he/she can avail the access of online reading of the book, which otherwise is restricted for accredited users.

10.2.1.14 Tips and Heuristics

- EBTs should represent ultimate goals. It should not focus on subgoals alone.
- There should be a clear distinction between the properties of the pattern and goals of the pattern.
- EBTs can communicate with BOs and can interact with IOs through BOs only.
- BOs chosen should be generic to be able to apply in different domains.
- IOs connect to the BOs by using a proper interface.
- Delegation should be preferred over inheritance in order to *increase run-time linking efficiency*.
- When dependency is less, navigation among the different functions will be easy.
- Use of hooks for BOs for linking it in development phase should be implemented efficiently.
- Requirement gathering is an important process and it should be done in the starting phase itself.
- Flow of the system can be understood easily by looking at the sequence diagram (behavioral diagram).
- Pattern developed should be stable and robust. It should be able to withstand various testing techniques.
- Because of generic EBTs and BOs, the reusability for other applications could be extended.
- If proper interfaces are defined at base level, then it will be easy for adding the scalable components at a later stage.

10.3 SUMMARY

This chapter identifies and describes the underlying components of accessibility that is common across all domains. The AnyParty, AnyEntity, AnyMechanism, and AnyMedia are the fundamental properties for accessibility. These BOs are the workhorses that help achieve accessibility. These core components are inherent in any accessibility applications and a developer cannot create an accessibility application, without these classes. Hence, the chapter successfully provides a solution for the fundamental problem of developing a pattern that captures the atomic accessibility notion. The stable accessibility pattern may serve as a base for modeling any kind of accessibility application.

10.4 OPEN AND RESEARCH ISSUES

1. *e-Accessibility*: Utilize stability model or knowledge map methodology as a way for developing e-Accessibility Engine. Building this engine by using traditional development approaches is not an easy exercise, specifically when several factors can undermine their quality success, such as cost, time, and lack of systematic approaches.

2. *Accessibility unified software engine (A-USE)*: Utilize stability model or knowledge map methodology as a way for developing A-USE. The engine mainly focuses on several patterns: accessibility, control, identity, analysis, adaptability, extensibility, interactive, mobility, security, privacy, etc. The proposed solution attempts to extract the commonality from all the domains and represents it in such a way that it is applicable to a wide range of domains without trivializing or generalizing the concepts. The engine is a stable structural pattern and it provides a generic engine that can be applied and/or extensible to any application and can access Any Entity, Any Event, Any Media, or Any Log and will be applicable to Any Domain by plugging application-specific features. The A-USE can be applied to unlimited context and reusability.

REVIEW QUESTIONS

1. Define accessibility SAP.
2. What does accessibility mean?
3. "Accessibility is also bad." Explain.
4. Describe stable analysis pattern.
5. Describe the characteristics of EBT, BO, and IO.
6. What are the other terms used for accessibility? Explain them.
7. Describe a few contexts in which accessibility is used in day-to-day life.
8. Write two to three scenarios where accessibility SAP is applicable.
9. Discuss the functional requirements for accessibility SAP.
10. Discuss the nonfunctional requirements for accessibility SAP.
11. Write four challenges of using accessibility SAP.
12. Write a few constraints of accessibility SAP.
13. Draw class diagram for accessibility SAP.
14. Explain and justify the participants in accessibility SAP.
15. Write CRC cards for accessibility SAP.
16. What are the consequences of accessibility SAP?
17. Explain the application of accessibility SAP in accessing Internet content. Draw a class diagram for it. Write any one usecase for this application and draw a sequence diagram for the same. Write test cases for the same.
18. Explain the application of accessibility SAP in accessing online student information system. Draw a class diagram for it. Write any one usecase for this application and draw a sequence diagram for the same. Write test cases for the same.
19. Draw traditional model for accessibility.
20. Compare the traditional model and stable pattern of accessibility by using the following adequacies:
    - *Descriptive Adequacy [13]:* Descriptive adequacy refers to the ability to visualize objects in the models. Every defined object should be browse-able, allowing the user to view the structure of an object and its state at a particular point in time. This requires understanding and extracting metadata about objects that will be used to build a visual model of objects and their configurations. This visual model is domain-dependent—that is, based on domain data and objects' metadata. Descriptive adequacy requires that all of the knowledge representation is visual:
        a. Visual models are structured to reflect natural structure of objects and their configurations.
        b. All visual knowledge (data and operations) in the visual model is localized.
        c. Relationships among objects in the visual model are well defined.
        d. Interactions among objects in the visual model are limited and concise.
        e. The visual model must transcend objects, and instead highlight crosscutting aspects.

- *Understanding adequacy:* Understanding adequacy relates to be easy to understand.
- *Simplicity adequacy:* Simplicity adequacy relates to how simple your models will be.
- *Extensibility adequacy:* Extensibility adequacy relates to the degree of extensibility, adaptability, customizability, and configurability of your models.

21. Compare the traditional model and stable pattern of accessibility by using the following modeling essentials [14,15] as comparative criteria:
    a. *Simple*: This property covers those attributes of the object-oriented model that present modeling aspects of the problem domain in a most understandable manner.
    b. *Complete (most likely to be correct):* This property determines whether the object-oriented model provides internal consistency and completeness of the model's artifacts. The model must be able to convey the essential concepts of its properties.
    c. *Stable to technological change:* The model should be stable to technological changes and it cannot require any changes or modifications with change of technology, such as change of the media or the mechanisms.
    d. *Testable:* To be testable, the model must be specific, unambiguous, and quantitative wherever possible, so that we can run an infinite number of scenarios with the context of the pattern.
    e. *Easy to understand:* In addition to the familiarity of the modeling notations, the notational aspects, design constraints, and analysis and design rules of the model should be simple and easy to comprehend by customers, users, and domain experts.
    f. *Visual or graphical:* A picture is worth a thousand words! As a user, you can visualize and describe the model. The graphical model is essential for easier visualization and simulation.
22. Compare traditional accessibility model with accessibility SAP. Which one do you like more?
23. Compare both the models quantitatively in terms of number of classes, interdependency among classes, constraints, number of operations in classes, and depth of inheritance level.
24. Compare both the models qualitatively.
25. Discuss the process of abstraction of accessibility SAP.
26. Discuss the design and implementation issues of accessibility SAP.
27. Discuss the testability of accessibility SAP.
28. Write some business rules for accessibility SAP.
29. Discuss some of the known usages of accessibility SAP.
30. What did you learn from this chapter?
31. Write your findings, tips, and heuristics.
32. Name two more qualitative metrics and utilize them to measure accessibility pattern.
33. Discuss the benefits of using accessibility analysis pattern to generate business rules.
34. Give some examples of applications, where accessibility pattern is being used.
35. What are the lessons learnt by you from studying the accessibility pattern [1–7,8].

EXERCISES

1. Describe a scenario of an application of accessibility for documents in a company.
    a. Draw a class diagram of the application.
    b. Document a detailed and significant use case.
    c. Create a sequence diagram of the created use case of b.
2. Describe a scenario of an application of accessibility of email inbox.
    a. Draw a class diagram of the application.
    b. Document a detailed and significant use case.
    c. Create a sequence diagram of the created use case of b.

3. Describe a scenario of an application of accessibility of permission code for a course in a university.
   a. Draw a class diagram of the application.
   b. Document a detailed and significant use case.
   c. Create a sequence diagram of the created use case of b.
4. Describe a scenario of an application of accessibility of course material of a course on canvas.
   a. Draw a class diagram of the application.
   b. Document a detailed and significant use case.
   c. Create a sequence diagram of the created use case of b.
5. Think of few scenarios, where accessibility pattern is applicable and design corresponding class diagram, use case and sequence diagram as shown in the solution and applicability sections for each of the scenarios.
   a. Building access
   b. Access to reading library materials
   c. Access to gardening resources
   d. Access to public transportation

PROJECTS

1. Model an application of accessibility SAP for accessing a safety deposit box in a bank.
2. You are a part of a team that is developing an e-commerce website. Model the application of accessibility SAP for customers to access their order status on the website and to track their orders.
3. You are tasked with creating a system that tracks access to a secure building. Access should be granted based on the security clearance of the employee. Each employee has an account associated with them that track doors they access and the time of these events. Use the accessibility pattern from this chapter along with the trust pattern, and account patterns mentioned in this textbook. Create the corresponding class diagram, use case, and sequence diagrams required to design this project.
4. Create an architecture on demand (i.e., stable architecture pattern consists of two or more EBTs, or BOs, or a Mix of EBTs and BOs) using accessibility pattern for each of the following contexts:
   a. *System access:* Ability or authority to interact with a computer system, resulting in a flow of information; a means by which one may input or output data from an information source. Access implies authorization or proper clearance. Interaction with a computer or information system without authorization is termed hacking (see hacker) or cracking (see cracker), and is a criminal offense in many countries.
   b. *Room access:* The entrance (the space in a wall) through which you enter or leave a room or building; the space that a door can close.
   c. *Internet access:* Connects individual computer terminals, computers, mobile devices, and computer networks to the Internet, enabling users to access Internet services, such as email and the World Wide Web. Internet service providers (ISPs) offer Internet access through various technologies that offer a wide range of data signaling rates (speeds).

REFERENCES

1. M.E. Fayad, H. Hamza, and M. Cline, *Stable Software Patterns: Analysis, Design, and Applications.* (Book in progress)

2. G. Froehlich, H.J. Hoover, L. Liu, and P. Sorenson, Hooking into object-oriented application frameworks, in Fayad, M.E., Schmidt, D. C, and Johnson, R.E., eds. *Building Application Frameworks: Object-Oriented Foundations of Framework Design*, New York, NY: John Wiley & Sons, Inc., 1997.
3. G. Froehlich et al., Hooking into object-oriented application frameworks, International Conference on Software Engineering, Boston, MA, 1997.
4. M.E. Fayad et al., *Building Application Frameworks: Object-Oriented Foundations of Framework Design*, 1st ed., New York, NY: Wiley, 1999.
5. H. Hamza A foundation for building stable analysis patterns. MS thesis, Department of Computer Science, University of Nebraska-Lincoln, 2002.
6. H. Hamza. Building stable analysis patterns using software stability, 4th European Conference on GCSE Young Researchers, Erfurt, Germany, 2002.
7. H. Hamza, and M.E. Fayad, Model-based software reuse using stable analysis patterns. 16th European Conference on Object-Oriented Programming, Malaga, Spain, 2002.
8. M.E. Fayad, and A. Altman, Introduction to software stability, *Communications of the ACM*, 44(9), 2001, 95–98.
9. M.E Fayad, Accomplishing software stability, *Communications of the ACM*, 45(1), 2002, 111–115.
10. M.E. Fayad, How to deal with software stability, *Communications of the ACM*, 45(4), 2002, 109–112.
11. M.E. Fayad, and P. Strivastava, The hook facility, MS thesis, Department of Computer Engineering, San Jose State University, San Jose, CA, 2005.
12. M. Grand, *Patterns in Java I—A Catalog of Reusable Design Patterns Illustrated with UML*, New York, NY: Wiley, 1998.
13. Chomsky, Noam., Current issues in linguistic theory, in Fodor, J.A. and Katz J.J., eds. *The Structure of Language: Readings in the Philosophy of Language*, Englewood Cliffs, Prentice Hall, 1964, 50–118.
14. M.E. Fayad and M. Laitinen, *Transition to Object-Oriented Software Development.* New York: John Wiley & Sons, 1998, ISBN# 0-471-24529-1.
15. M.E. Fayad, W.T. Tsai, and M. Fulghum, Transition to object-oriented software development, *Communications of the ACM*, 39(2), 1996, 108–121.

Part III

SAPs

Mid-Size Documentation Templates

Part III contains two chapters that are listed below:

Chapter 11 titled "Reputation Stable Analysis Pattern"
Chapter 12 titled "Temptation Stable Analysis Pattern"

11 Reputation Stable Analysis Patterns

You can't build a reputation on what you are going to do.

Henry Ford

Reputation is the opinion (more technically, a social evaluation) of the public toward a person, a group of people, or an organization. In other words, reputation is the general estimation that the public has for a person or an institution. It is an important factor in many fields, such as business, online communities, or social status. It is also a subject of study in social, management, and technological sciences. Its influence may range from competitive settings such as markets to cooperative ones such as firms, organizations, institutions, and communities. Furthermore, reputation also acts on different levels of agency, individual and supra-individual. At the supra-individual level, it concerns groups, communities, collectives, and abstract social entities (such as firms, corporations, organizations, countries, cultures, and even civilizations). It affects phenomena of different scale, from everyday life to relationships between nations. Reputation is a fundamental instrument of social order, based upon distributed, spontaneous social control.

This chapter provides reputation stable analysis pattern without any IO's connected to it. Additionally, it also provides two scenarios (detailed applications) in which the pattern can be used.

11.1 INTRODUCTION

The idea of reputation is commonly used in social life and economy, and there exists a common opinion on its general meaning. When it comes to a person, reputation is described as "a characteristic or attribute ascribed to one person (organization, community, etc.) by another person (or community)." On the other hand, the reputation of say, a service provider, can be formed by means of a collection of ratings by different users; each such rating is intuitively equivalent to user satisfaction. Higher the rating from a user, higher will be the reputation of the service provider.

Reputation is considered to be very relevant to systems, where there is information asymmetry about quality and trust, due to the large number of players involved. Reputation can also be seen as a state variable that gives evidence about the missing information; thus, reputation offers numerous incentives to providers and consumers to behave properly. Reputation provides a suitable mechanism to consumers to identify quality service providers/sellers. A reputation mechanism is quite successful, when a steady-state market situation can be achieved and maintained.

The last decade has witnessed an explosive growth in Internet connections around the globe. Online communities are gaining more popularity, as they neither limit nor restrict human interactions by insisting on geographic constraints; instead, they bring together people of varied backgrounds, ethnicities, and nationalities. EBay, the largest person-to-person auction site, is an excellent example of such a grand community. Selling a product through such a community and becoming very successful entrepreneur depends largely on the reputation of a person or an organization.

Reputation is a must in all types of businesses including online and e-commerce ventures. For example, Apple acquired considerably good reputation by selling a high-quality music player called "IPod" which eventually helped them to gain an eventful entry into the global cell phone market, when they introduced the iconic "IPhone."

Traditional approaches to software design and development may not yield a stable and reusable model for gaining reputation. However, by using the stability model [1–4], it can be represented in any context by using a single model. The stability model requires the creation of knowledge map, by identifying underlying EBT and BO. By hooking IOs, that is specific to each application, the model can be applied to any application domain. The resulting reputation pattern is quite stable, reusable, extendable, and highly adaptable. Thus, any number of applications can be built by using this common model. The reputation analysis pattern attempts to capture the core knowledge of reputation, which is common to all application scenarios that are listed above to emerge with a stable pattern. The overall objective is to conceive and design a stability model for reputation by creating the knowledge map of reputation. This knowledge map or core knowledge can then serve as building block for modeling different applications in diverse domains.

11.2 REPUTATION ANALYSIS PATTERN DOCUMENT

11.2.1 Pattern Name: Reputation Stable Analysis Pattern

Reputation is the opinion of public toward a person, a group of people, or an organization. In other words, reputation is the general estimation that the public has for a person or an institutions. Reputation is very important in every field, be it politics, business, or selling anything, ranging from a coffee to a digital camera. The reputation a person/organization has built in his/its life time determines the quality of the business and sustainability in today's scenario of fierce competition. The overall objective of pattern is to generalize the idea of reputation, so that it can be used as foundation for managing day-to-day activities of business.

The reputation analysis pattern abstracts this concept that can be applied to any party based on any mechanism. The reputation can be of any type and kind. This pattern also depicts the effect of reputation on the user. It is based on the principles of EBT of reputation and is thus stable. It can be used to model any related scenario and is thus not restricted to any one scenario or situation. Hence, the name reputation is a stable analysis pattern.

11.2.1.1 Context

Individuals or any organization tries to build a good reputation because of numerous corporate and business needs. A good reputation differentiates them from other organizations and creates more business. Further, an individual sees immense pride in attaining a good reputation. A bad reputation may be fatal. For example, the bad reputation attained by ENRON made them lose their business eventually leading to filing bankruptcy. Hence, reputation is as crucial and important factor for any business.

Reputation may relate to various organizations and institutions; web portals such as eBay(shopping), Google(search engine), personalities such as Roger Federer (tennis), George W Bush, or companies like Apple computers, CISCO, and countries like Switzerland(for banking), Middle east(for oil) are some of the well-known examples.

Listed as follows are some of the contexts in which reputation is used:

- In online business portals such as eBay, where a seller builds a reputation via the feedback mechanism provided by the buyer and portal. The reputation a seller has gained in the past is displayed in the form of a rating list, which helps a buyer to gain confidence about a seller and cajoles him or her to buy a product from a particular seller eventually increasing the sellers' business.
- In politics, President Bush has gained a bad reputation by waging a long and never ending war with Iraq.
- By selling an MP3 player like "IPod," when there was a Market demand, Apple computers has gained a very good reputation for selling high-quality music players.

- American Energy Corporation gained a bad reputation for creatively planning an accounting fraud.
- Swiss tennis professional Roger Federer gained a good reputation by being the World's No. 1 tennis player for a very long time.

As illustrated above, reputation analysis pattern can be applied to numerous scenarios and in diverse disciplines. In later sections, application scenarios to illustrate the applicability of reputation pattern are shown in full detail. Reputation concept is illustrated in two distinct scenarios such as a seller building a good reputation by selling a product through eBay and Apple computers establishing a good reputation by selling IPod's.

11.2.1.2 Problem

Today, global competition has increased tremendously. Right from selling a small paper clip to selling an airplane, there are numerous players competing with each other to sell their products. In such a scenario, it is very much required that a person/organization set a reputation for them in order to compete effectively with others. This can be done in a variety of different ways, right from selling quality, low-cost products to developing the skills that others like to possess. In an online business, there are numerous ways to develop this reputation like the online feedback and rating mechanism. Building a generic pattern that covers all such cases of reputation development is a challenging and humongous task.

Reputation can be applied to different domains such as politics, business, and to different parties such as an individual, an organization, or a country. Hence, it is very much essential to model a generic pattern. The following criteria must be satisfied before developing a generic reputation pattern.

- The pattern must be so generic and should be reusable to model any reputation application/ scenario.
- A thorough understanding of the core concepts of reputation is absolutely essential, so that the core knowledge can be properly captured.

Since reputation is used in different context and in different domains, building a generic model without loss of functionality is uniquely challenging. By using stability model, this problem is solved and a generic model is modeled for different domains. This model is illustrated and described in the solution section.

11.2.1.3 Solution

The solution shown here utilizes stability model to explain the concept of reputation. Figure 11.1 depicts the class diagram for personalization pattern.

11.2.1.3.1 Pattern Structure and Participants

The following are the participants of the reputation pattern.

11.2.1.3.1.1 Classes

Reputation: This class represents reputation. It is an EBT, which presents the enduring and business knowledge that discloses relevant information based on the attributes of the user.

11.2.1.3.1.2 Patterns

AnyParty: This class represents any person or individual, an organization or group for whom the reputation is associated with.

AnyFactor: This class denotes the factors that affect the reputation of a particular individual or organization.

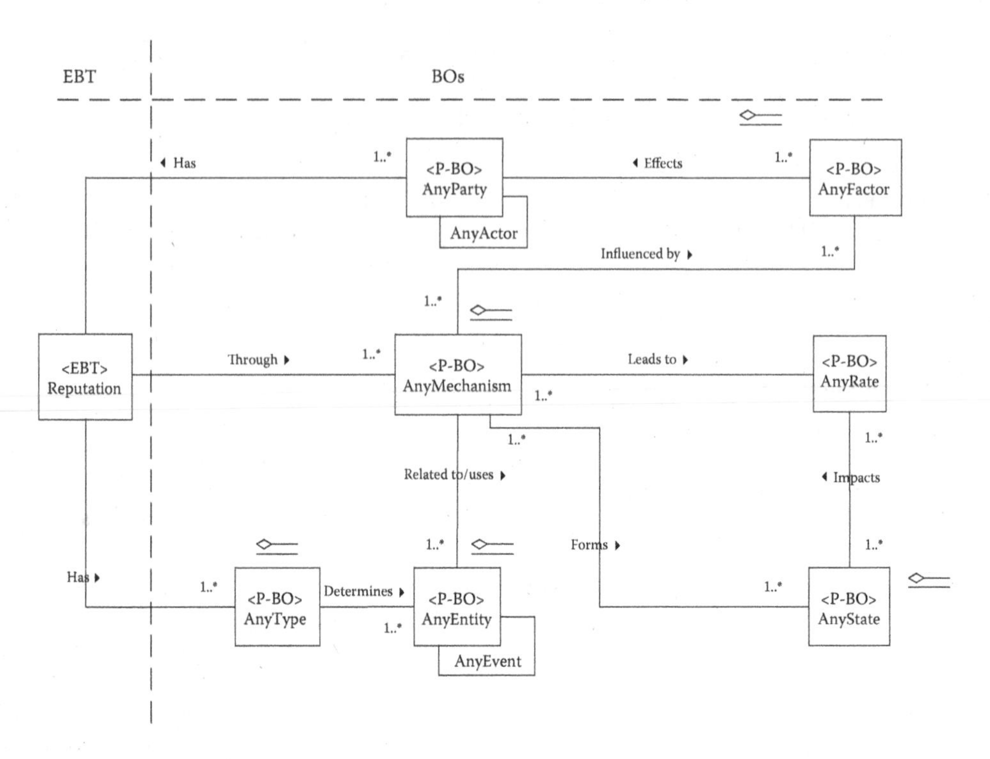

FIGURE 11.1 Class diagram for reputation analysis pattern.

AnyEntity: This class denotes the characteristic or the product for which an individual or organization is reputed for.

AnyMechanism: This class denotes the methodology through which the individual or the organization achieves the reputation.

AnyState: This class denotes the position achieved by any individual or an organization by applying the mechanism.

AnyRate: This class represents the status that was achieved by applying the mechanism and which was impacted by the state.

AnyType: This class represents the nature of reputation achieved by any individual or an organization.

11.2.1.3.2 Class Diagram Description

The class diagram provides visual illustration of all the classes in the model, along with their relationships with other classes. Description of the class diagram is as below.

1. Reputation is the EBT of this pattern and is associated to AnyParty(BO).
2. Reputation(EBT) that has AnyType(BO) is achieved through AnyMechanism(BO).
3. AnyMechanism(BO) uses AnyEntity(BO) to achieve reputation(EBT).
4. AnyParty(BO) chooses AnyMechanism(BO), because of the influence created by AnyFactor(BO).
5. AnyMechanism(BO) forms AnyState(BO) and leads to AnyRate(BO).
6. AnyFactor(BO) affects AnyParty(BO) leading to reputation (EBT).

11.2.1.4 Applicability with Illustrated Example

In this section, illustrate the use of reputation analysis pattern are depicted, by using the use case description and behavior model like sequence diagram.

1.0 Application: Build a Good Reputation by Selling a Product through eBay.

In context of eBay, selling a product online and obtaining a good opinion from the buyers by using the eBay feedback system is called reputation building. The eBay website allows users to enter their ratings on various categories.

11.2.1.4.1 Model

The model for this application is shown below in Figure 11.2.

11.2.1.4.2 Class Diagram Description

The class diagram provides a visual illustration of all the classes in the model, along with their relationships with other classes. Description of the class diagram is as below.

1. AnyParty(Seller) sells AnyEntity(Product) through AnyMechanism(eBay).
2. AnyMechanism(eBay) is surfed by AnyParty(Buyer), who finds the AnyEntity(Product), sold by AnyParty(Seller) suitable for him.
3. AnyParty(Buyer) buys AnyEntity(Product).
4. AnyParty(Buyer) likes the AnyEntity(Product), which creates AnyFactor(Satisfaction) in AnyParty(Buyer).
5. AnyParty(Buyer) through AnyMechanism(Feedback) provided in the eBay website records his AnyState(Opinion).
6. AnyState(Opinion) impacts AnyRate(PositiveRating) given to AnyParty(Seller).
7. AnyRate(PositiveRating) causes good Reputation to AnyParty(Seller).

Use Case No: 1 (as shown in Table 11.1).

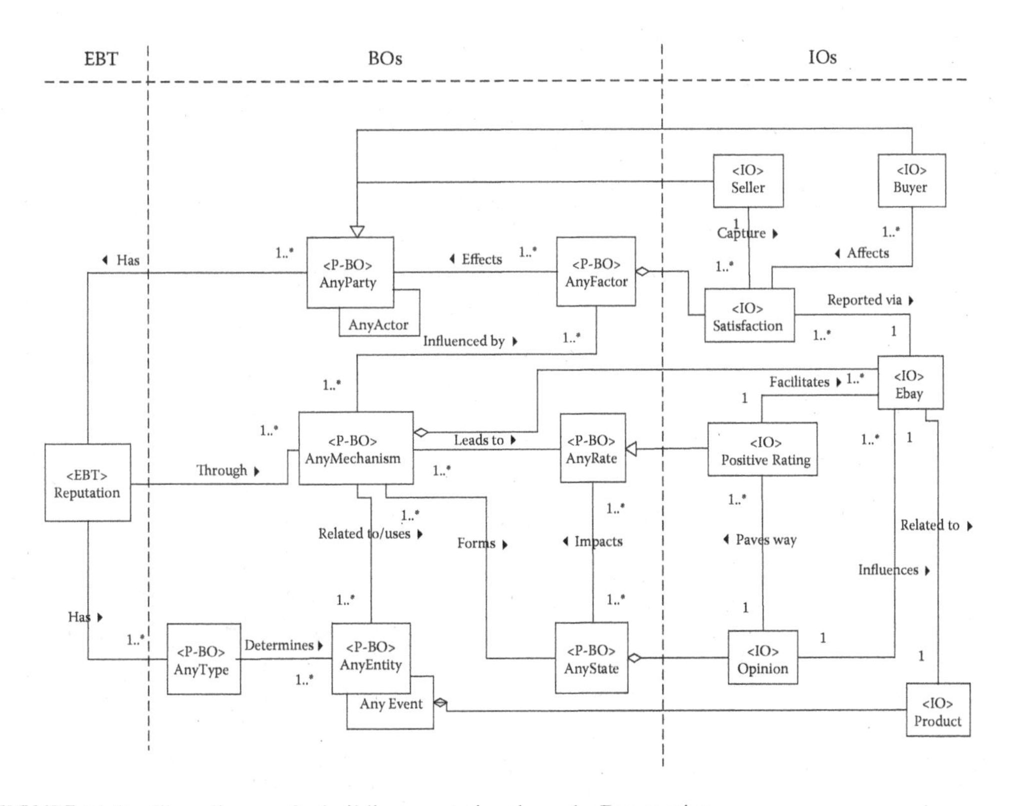

FIGURE 11.2 Class diagram for building reputation through eBay auction.

TABLE 11.1
Use Case Title: Build a Good Reputation by Selling a Product through eBay

Actors	Roles
AnyParty (H)	Seller
AnyParty (H)	Buyer

Class Name	Type	Attributes	Operations
Reputation	EBT	Merit Attainment Estimation Type Credit	holdOpinion() ascribeCharacteristic() seizeDistinction()
AnyParty	BO	Name contactInformation Authenticity	achieveTrait() desireRecognition() interact()
Seller	IO	productArea sellHistory contactInformation Reputation Type	sellProduct() influenceBuyer() useSellingTricks()
Buyer	IO	Requirement Interest Credibility contactInformation	buyProduct() investigateProduct() checkReputationOfBuyer()
AnyMechanism	BO	Procedure Assemble Type Method Perform	execute() build() affect()
Ebay	IO	webAddress Authentication Reputation dealsIn attractUser	assistShopping() authenticateUser() help()
AnyEntity	BO	Name Model Legality Type	facilitateReputation() attractBuyer() profitSeller()
Product	IO	Type Cost Condition Price Quality	renderService() influence() impactFurtherSales()
AnyFactor	BO	Cause Action Type Base Effect	affectEventCourse() impactReputation() causeChange

(Continued)

TABLE 11.1 (*Continued*)
Use Case Title: Build a Good Reputation by Selling a Product through eBay

Class Name	Type	Attributes	Operations
Satisfaction	IO	Reason Percentage Type Reason Effect	level() reportBuyer() capture()
AnyState	BO	Circumstance Rank Credibility Affect Reason	providePosition() mode() relatedTo()
Opinion	IO	Belief Estimate Credibility Reason Affect	expressView() rate() influence()
AnyRate	BO	Progress Grade Value Base Impact	degreeOfSpeed() relativeCondition() facilitate()
PositiveRating	IO	currentScore Grade Credibility Impact Range	classify() affect() helpCustomer()

11.2.1.4.3 Use Case Description

1. AnyParty(Seller) builds a good reputation by selling AnyEntity(Product) through AnyMechanism(eBay). Is AnyEntity(Product) sufficient to build a good reputation?
2. This AnyEntity(Product) creates AnyFactor(Satisfaction) in AnyParty(Buyer). In what way does AnyEntity(Product) create AnyFactor(Satisfaction) to AnyParty(Buyer)?
3. AnyParty(Buyer) develops AnyState(Opinion) about the AnyEntity(Product). Did conduct AnyParty(Buyer) sufficient analysis on AnyEntity(Product) to come up with AnyState(Opinion)?
4. The AnyState(Opinion) is recorded through AnyMechanism(Feedback). Is AnyMechanism(Feedback) useful to record AnyState(Opinion)?
5. AnyMechanism(Feedback) impacts AnyRate(PositiveRating) given to AnyParty(Seller). In what way does AnyMechanism(Feedback) impact AnyRate(PositiveRating)?
6. AnyRate(PositiveRating) increase the reputation of AnyParty(Seller).

11.2.1.4.4 Alternatives

1. AnyParty(Buyer) is not satisfied with the product he bought via eBay (Figure 11.3).

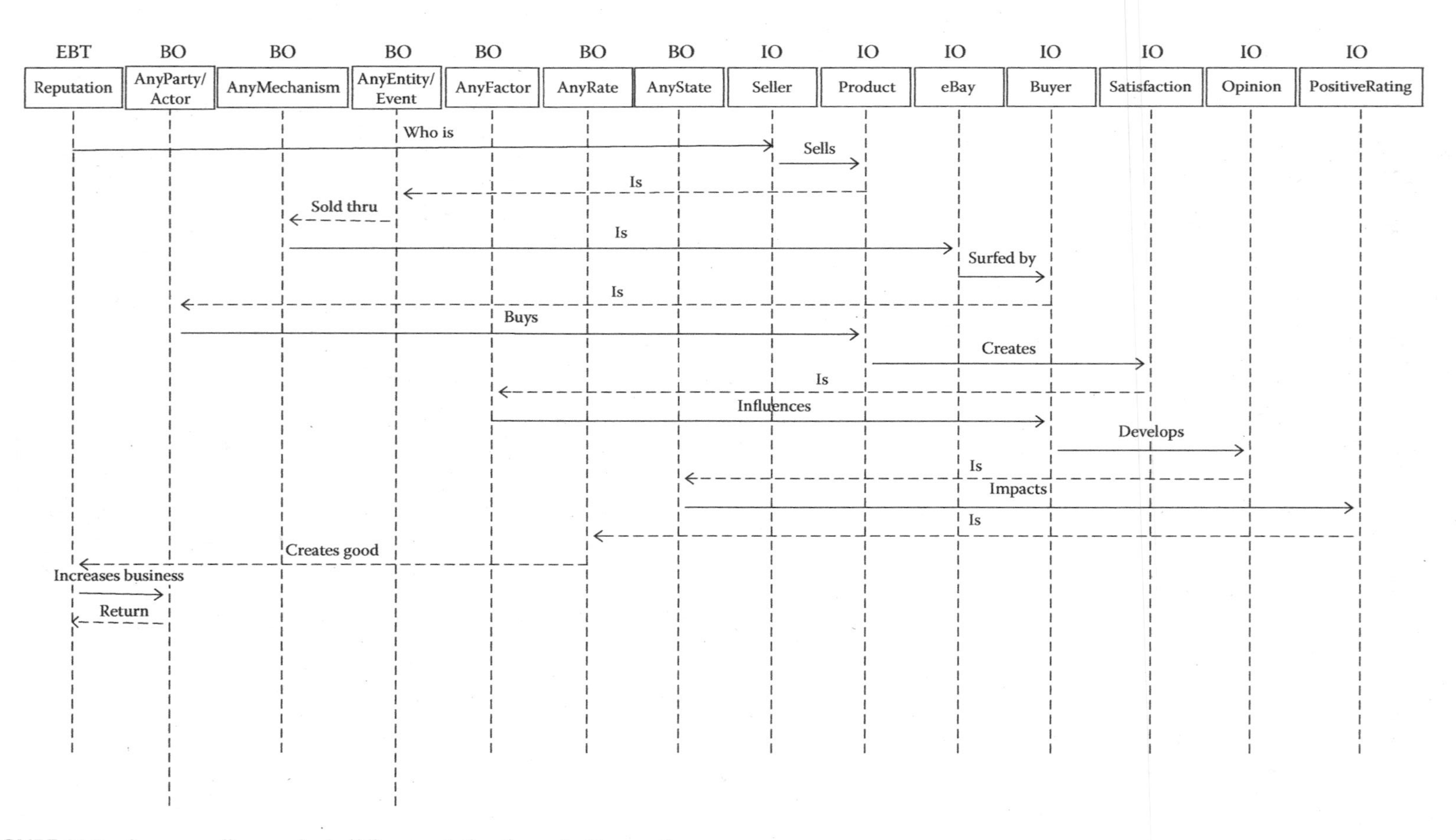

FIGURE 11.3 Sequence diagram for building reputation through eBay auction.

11.2.1.4.5 Sequence Diagram Description

1. Reputation is EBT AnyParty(Seller) sells AnyEntity(Product) through AnyMechanism(eBay).
2. AnyMechanism(eBay) is surfed by AnyParty(Buyer), who finds the AnyEntity(Product) sold by AnyParty(Seller) suitable for him.
3. AnyParty(Buyer) buys AnyEntity(Product).
4. AnyParty(Buyer) likes the AnyEntity(Product), which creates AnyFactor(Satisfaction) in AnyParty(Buyer).
5. AnyParty(Buyer) through AnyMechanism(Feedback) provided in the eBay website records his AnyState(Opinion).
6. AnyState(Opinion) impacts AnyRate(PositiveRating) given to AnyParty(Seller).
7. AnyRate(PositiveRating) causes good Reputation.

11.3 SUMMARY

The Reputation pattern proposed here is based on the principles of stable analysis pattern. The pattern is explained with two applications that perform well based on this model. The depth of this reputation pattern depends on the availability of Any Party's attributes for personalizing the particular application. Each object in the reputation pattern has its own role and play, which is independent of any applications, where this pattern will be applied. More than one mechanism exists to carry out the reputation. Care should be taken, while choosing the appropriate mechanism by utilizing the attributes properly.

One difficult part for modeling reputation problem was finding a good class diagram description. Making the description as clear and accurate as possible, so that it is beneficial in drawing the sequence diagram, is the key for getting good model. The process of creating the sequence then gets much simpler and flexible, as it is just the translation of the class diagram.

Although building a stable design pattern for reputation that can be reused and reapplied across diverse domain is difficult, and requires complete understanding of the problem, it is worth the effort, money, and time. Modeling reputation pattern by using stability model results in reusable, extensible, and stable pattern.

This pattern is so flexible that it can be applied to any type of scenario. IOs can be hooked to BOs to make it more meaningful to the scenario where it is applied. However, the correct identification of EBT and BOs for reputation is the most challenging task and requires some prior experience. Once EBT and BOs are correctly identified, the next challenge is to determine the relationship between EBT and BOs, so that reputation pattern can hold true in any context of usage for reputation. Once this is ascertained, depending on the application, IOs are attached to the hooks, so provided by BOs. Thus, by using reputation pattern as a foundation, infinite number of applications can be built, just by plugging the application specific IOs to the pattern. This results in reduced cost, effort, and a stable solution. Hence, reputation design pattern is very useful and beneficial to developers as well as users.

11.4 OPEN AND RESEARCH ISSUES

1. *e-Reputation:* Utilize the stability model or knowledge map methodology as a way for developing e-Reputation Engine. Building this engine by using traditional development approaches is not an easy exercise, specifically when several factors can undermine their quality success, such as cost, time, and lack of systematic approaches.

REVIEW QUESTIONS

1. What is the need for a stable analysis pattern that describes reputation?

2. What EBTs can be brought under the umbrella of reputation SAP?
3. How does reputation qualify as a subject for a stable analysis pattern?
4. How do we start with the identification of reputation and its related terms?
5. What are BOs that make up the reputation pattern?
6. List several application domains that do not fit the competition SAP.
7. List possible application domains for the SAP competition.
8. Rate the reusability of the competition SAP on a scale of 1–5. Give a short description to support your rating.
9. How sustainable is the reputation pattern?
10. What are the challenges that exist, when trying to utilize the reputation pattern?
11. Are there any constraints, which need to be handled by the reputation pattern?
12. Do the constraints reduce the number of applications for the reputation pattern?
13. What are the constants involved in the reputation pattern if any?
14. What are the limits of scalability that we can achieve with this model? Is it infinitely scalable?
15. Is the reputation model technology independent?
16. Are there any existing anomalies in the reputation pattern? If so, list them, and your argument to support your decision.
17. Are certain modifications required in our stability model, so that even the anomalies can fit in? If so, list the modifications and their effects.
18. In general, anomalies can be dealt with in the application implementation as IOs (T/F).
19. What pre-existing stability models can we take in as inspirations and present our model as an extension or a new dimension to them, or is our model a totally new concept?

EXERCISES

1. Pick an existing consumer product that you are familiar with and apply the reputation pattern to encompass similar products. (Examples: soft-drinks, car manufacturers, etc.)
2. Use reputation pattern to build application to each of the following:
   a. Robin Hood
   b. Your neighborhood
   c. Your own town
   d. Microsoft Corporation
   e. Mercedes

PROJECTS

1. Use the reputation pattern to design a system for rating magazine subscriptions. Develop corresponding class diagrams, CRC cards, uses cases, and sequence diagram.
2. Use the reputation pattern to design a system for tracking user contributions on an Internet forum. Develop corresponding class diagrams, CRC cards, uses cases, and sequence diagram.

REFERENCES

1. M.E. Fayad, and A. Altman, An introduction to software stability, *Communications of the ACM*, 44(9), 2001, 95–98.
2. M.E. Fayad, Accomplishing software stability, *Communications of the ACM*, 45(1), 2002, 95–98.
3. M.E. Fayad, How to deal with software stability, *Communications of ACM*, 45(4), 2002, 109–112.
4. M. E. Fayad, and S. Wu, Merging multiple conventional models into one stable model, *Communications of the ACM*, 45(9), 2002, 102–106.

12 Temptation Stable Analysis Pattern

Every moment of resistance to temptation is a victory.

Frederick William Faber

12.1 INTRODUCTION

Temptation is a universally recognized term that has been prevalent right from the biblical times of Adam and Eve. Temptation is an act that can induce somebody to take an action through manipulation or maneuvering. A desire to achieve something can tempt one to commit certain specific actions. An individual can be tempted to do something by someone using any technique or method. Mostly, the companies and firms market their products by using the art of temptation as the main marketing theme. They make use of enticing advertising techniques to create temptation in the minds of customers, which ultimately results in them buying their products and services. The appearance, look, smell, desire to achieve something, any attraction, are some of the factors that can lead to temptation [1,2].

12.2 REPUTATION ANALYSIS PATTERN DOCUMENT

12.2.1 Pattern Name: Temptation Stable Analysis Pattern

On the whole, temptation analysis pattern can be used in any model that is developed to create temptation. Any party can be tempted by using a particular mechanism based on a specific criterion.

12.2.1.1 Context

Temptation is the desire or craving for something/somebody. Every individual has temptation and nobody can resist it. Temptation is the result of any lure or attraction. Usually, temptation drives us to take some actions, which we may repent in the future. For example, temptation to eat junk food is an irresistible action. Everyone knows the result of eating junk food—that is bad to health. The one who gets tempted forgets to apply judgmental skills and act in haste. Temptation seduces someone to act immediately. For example, when we see a seemingly delicious pizza advertisement on television, which shows slowly melting cheese and its crispiness, our brains immediately start thinking of having one immediately no matter what time it is. Invariably, that pizza advertisement was successful in creating a deep desire in our mind to eat pizza almost immediately. Many stores offer deep discounts on their products to cajole and tempt the people to buy them. Temptation is one important marketing theme that is used by the marketing and sales departments of many businesses to sell their products.

The temptation pattern that we have developed here is based on the stable analysis pattern. Therefore, it is completely stable and can be used for different applications in a variety of domains. This can be achieved by simply connecting industrial objects to the stable business objects. The industrial objects, which we choose depends completely on the application, to which we apply this stability pattern. The biggest advantage of temptation model is that it helps the user to meet his/her objectives.

Some of the contexts, where temptation model can be used are

- A company introduces new products in the market with some lucrative offers in order to tempt its customers, which ultimately results in the sale of those products.
- Projection of sudden growth in the stock market can tempt an individual to invest in the stock market, since it may fulfill his or her desire to get rich quickly.
- To retain employees, they can be tempted with future bonuses and pay increments by the company.

Listed below are two scenarios that are illustrated explaining temptation concept. In the first scenario, chocolate eaters and chocolate retailers are tempted through attractive chocolate advertisements while in the second, obese people and fitness centers are tempted to buy slim and diet products.

1. A Company (AnyParty) introduces (AnyMechanism) new products (AnyEntity) in the market (AnyMedia) with a range of lucrative offers (AnyCriteria) in order to (AnyReason) tempt (Temptation) customers (AnyParty), which ultimately results in the sale (AnyEvent) of those products. The raise (AnyLevel) in sale volume can be measured by calculating the difference from the sales of the previous month. The sale volume is usually calculated on a month-to-month basis (AnyUnit).
2. Projection (AnyCriteria) of sudden growth (AnyEvent) in the stock market (AnyParty) can tempt (Temptation) an individual (AnyParty) to invest (AnyMechanism) money (AnyEntity) in it, since it might fulfill his or her desire (AnyReason) to get rich very quickly. The price quotes of stocks can be checked regularly from Internet (AnyMedia). The investor may earn a lot of (AnyLevel) money by following this approach and one can measure the earning performance by creating a record of monthly earning (AnyUnit) earned on papers.
3. For retaining employees (AnyReason), they (AnyParty) can be tempted (Temptation) with good (AnyLevel) future bonuses and increments (AnyCriteria) by the company (AnyParty). The increment can be in the form of perks (AnyEntity) or a pay rise (AnyEvent) in monthly (AnyUnit) salary. This can be announced in a meeting or through an email (AnyMedia).
4. Chocolate retailers (AnyParty) upload (AnyMechanism) chocolate (AnyEntity) advertisements on Internet (AnyMedia) to tempt (Temptation) chocolate eaters (AnyParty) to buy chocolates to raise (AnyReason) their product sale (AnyEvent). The advertisement should be good enough (AnyCriteria) to cajole and tempt a huge number of buyers, so as to raise the sale volume to a good level (AnyLevel) and to make a good profit (AnyUnit).
5. Fitness Institutes and firms (AnyParty) buy (AnyMechanism) slim fast and diet products (AnyEntity) from Internet (AnyMedia) to tempt (Temptation) obese people (AnyParty) to join their institute and to raise (AnyReason) their monthly sale (AnyEvent). The products should be good enough (AnyCriteria) to show results, so as to tempt a huge number of crowd to raise the sale to a good level (AnyLevel) and to make a good profit (AnyUnit).

12.2.1.2 Problem

Temptation is an act that can induce somebody to take an action through manipulation. A desire to achieve something can tempt one to commit to certain actions. It is necessary to find the level of desire in an individual and that desire needs to be used in the manipulation act.

The process of manipulation invariably involves using some proven techniques. An individual can be tempted to do something by someone by using any technique that is effective. It can be implemented by using different kinds of media such as TV and radio. The act of manipulation returns certain levels and those levels must match the desire of the individual. When the individual sees that those levels are addressing his desires, he is tempted. It is necessary to address the needs or desires of the individuals to tempt them to do anything. The technique used to manipulate these individuals must concentrate on that entity for which those individuals has to be tempted.

The set of requirements for temptation are numerous. Meeting the criteria of any party to create temptation is a must. The process of temptation also requires the use of the right type of mechanism and the media.

Here, we will try to provide a pattern that can be used by different applications, which requires the use of temptation concept.

12.2.1.2.1 Functional Requirements

1. *Temptation*: Temptation describes a feeling that forces AnyParty/AnyActor to perform an action. Temptation has enough strength that can be measured and a unique ID allotted. It seduces AnyParty/AnyActor, creates a desire, induces an action, and appeals.
2. *AnyParty*: AnyParty is the legal user of the system. He/she has a unique ID, a name, and an interest. AnyParty fulfills his/her temptation through AnyMechanism. AnyParty gets seduced, can lose self control, can commit any action, and can resist a temptation.
3. *AnyActor*: AnyActor is the actual user of the system. He or she has a unique IS, a name, an address, a contact number, an email address, and an interest. He or she fulfills his temptation through AnyMechanism. He or she can act, can get tempted, can get seduced, can lose self-control, and can resist a temptation.
4. *AnyMechanism*: AnyMechanism is a way to fulfill temptation. AnyParty/AnyActor uses AnyMechanism to fulfill his or her temptation. AnyMechanism can be of many types. It can have attributes and should be within the context of the system. AnyMechanism is operated, build, and executed by AnyActor/AnyParty.
5. *AnyReason*: AnyReason is the reason for which the temptation arises. It also represents the reason for which the temptation is created and thus impacts on AnyMechanism. AnyReason can have a unique ID, description, type, and attributes. It can be implemented, controlled, reduced, and increased.
6. *AnyLevel*: AnyLevel is the level returned by the technique used to create temptation and it also represents the level that impacts the factors used to win AnyParty/AnyActor. AnyLevel has a unique ID, a description and ingredients. It can be found and measured. It gives the proportion of the existing temptation to the temptation satisfied.
7. *AnyCriteria*: AnyCriteria are the factors that are considered to tempt AnyParty/AnyActor and which can be used to win over AnyParty/AnyActor. AnyCriteria has a unique ID, quality, and a description. It specifies characteristics, rates quality, and decides taste.
8. *AnyUnit*: It represents the unit of measurement, which indicates the value of level in a relevant unit. AnyUnit has a unique ID, a unit name, and a standard. It gives measurement unit, adopts a standard, expresses physical value, and measures quantity.
9. *AnyMedia*: It represents various media used to tempt AnyParty/AnyActor. It also represents the media used to implement AnyMechanism to tempt AnyParty/AnyActor. AnyMedia has a unique ID, a name, type, and cost associated with it. It is used to implement mechanism, convey message, and to communicate with AnyParty/AnyActor.
10. *AnyEvent*: It is the event that tempts AnyParty/AnyActor like advertisements, etc. AnyEvent has a unique ID, an occurrence date, time, and address and a name. It occurs at a specific time, is organized for a specific reason, and it effects AnyActor/AnyParty by increasing or decreasing temptation.
11. *AnyEntity*: It represents anything toward which AnyParty/AnyActor should be attracted to and which impacts AnyMechanism used to create temptation. AnyEntity has a unique ID, a name, an effect, and a type. IT defines characteristics and is used by AnyMechanism.

12.2.1.2.2 Nonfunctional Requirements

1. *Controllable*: The temptation should be controllable. For example: If a person cannot control his or her temptation of eating chocolates, then he or she may suffer from future problems of obesity, and it can have bad influence on the health status. Thus, it is very necessary that the temptation be controllable.
2. *Identifiable*: Temptation should be identifiable. If a company is trying to make products that are successful in the market, then it should be able to identify what tempts people. Also, in a case where a person wishes to fulfill temptation, he or she should be aware of what one needs to fulfill temptation.
3. *Measurable*: Temptation should be measurable. A person should know, if the temptation is high and is uncontrollable, or if he or she can control the temptation. Also, when a company raises salary to retain an employee, then it should know how much salary would be sufficient to retain a particular employee and this is done by knowing the quantum of salary that will be needed to get the employee tempted sufficiently.
4. *Relevant*: The temptation should be relevant enough. If a company is showing an advertisement for a candy, but a person is tempted to buy a blanket by looking at that advertisement, then the temptation being created is not relevant and hence the whole purpose is not fulfilled.
5. *Timely*: The temptation should be timely too. If a person is tempted to eat sweets, when he or she is suffering from diabetes, then this would harm the health and it would not be beneficial for the person. Thus, it is very important that the temptation occurs in a timely fashion.
6. *Feasible*: A person should be able to fulfill his or her temptation. If a person has the temptation to see a black hole or to fly in sky, then this would not be possible. Also, if a company creates a temptation for a thing that does not exist or which cannot be created, then this would harm the reputation of the company.
7. *Legal*: If a person is tempted to kill another person or to steal money from other person's wallet or to rob a bank, then the temptation would not be good for the society as well as for him. Thus, legality is a very important quality factor of temptation.

12.2.1.3 Challenges

Here are a few challenges as shown in Table 12.1.

12.2.1.4 Constraints

1. There should be at least one actor/party, who can be tempted.
2. There should be one or more mechanism that can be implemented to create or satisfy temptation.
3. There should be atleast one criteria defined by AnyActor/AnyParty.
4. AnyMechanism should be based on one or more AnyCriteria defined by AnyActor/AnyParty.
5. One or more AnyMechanism should be there to return AnyLevel.
6. AnyLevel should at least have one AnyUnit.
7. There should be one or more reason behind a temptation.
8. AnyReason for temptation should be for one or more AnyEntity.
9. AnyReason should be a result of one or more AnyEvent.
10. AnyEntity/AnyEvent should be on one or more AnyMedia.
11. AnyMechanism should use one or more AnyMedia.

12.2.1.5 Solution

The solution shown here utilizes stability model [3–7] to explain the concept of temptation. Figure 12.1 depicts the class diagram for temptation pattern.

TABLE 12.1
Challenges

Challenge ID	0001
Challenge Title	Too many criteria
Scenario	Raise in salary and perks to retain employee
Description	If an employee set up criteria that he will stay with a company only if his or her salary is raised by 10%, then the company can do it so as to retain that good employee. But, consider a case, when an employee sets too many criteria like raise in salary by 10%, a car, a house, two months of paid leave, 1000 stocks, etc. In this case, the company might find it easier to hire a new employee than to retain him or her. Also, any other company might not be able to fulfill his criteria. Hence, too many criteria imposed by AnyActor/AnyParty makes it challenging to fulfill the temptation.
Solution	There should be a limited number of criteria imposed or else fulfilling the temptation would be extremely challenging.
Challenge ID	0002
Challenge Title	Too many Actor/Party tempted
Scenario	Advertisement of chocolate
Description	If an advertisement tempts 1000 people to buy the chocolate, then it would be not be a problem, as company will have the capability to satisfy the temptation of all of them. But, in a case where an advertisement tempts × number of people and the company does not have the capability to manufacture, as many chocolates a day so as to satisfy the temptation of all of them, the reputation of the company will go down immediately, as there will be a shortage of chocolate and people will get utterly disappointed.
Solution	The temptation should be created on only to the level that can be satisfied. If too many people are tempted to buy the chocolate, then the company should impose a limit of chocolates that a person can buy and it should try to outsource some of the chocolate manufacturing work.
Challenge ID	0003
Challenge Title	Too many mechanisms used
Scenario	Company doing publicity of a product to tempt people to buy it
Description	If a company uses ten mechanisms like advertisements, hoardings, providing samples of the products, etc. to tempt the people to buy it, then that would be alright, but in a case where a company uses 1000 mechanisms, it would definitely raise the demand of the product, but the company might find it challenging to make profit, as it has invested a lot of money in publicizing the product.
Solution	The company should use only that number of mechanisms that are sufficient enough to create the required temptation and should not waste unnecessary money by implementing too many mechanisms.
Challenge ID	0004
Challenge Title	Multiple Entity
Scenario	Temptations in a child
Description	If a child is tempted for 10 entities like a chocolate, a soda, a toy, etc. then his parents can fulfill them. But, when a child has 1000s of temptations, then it would be challenging to fulfill this temptation for all of the entities.
Solution	A person should have temptation for a realistic number of entities or should learn to compromise on his or her temptations.

12.2.1.5.1 Pattern Structure and Participants

The participants of the temptation pattern are presented below.

12.2.1.5.1.1 Classes

Temptation: This class represents the temptation itself. It is EBT, which presents the enduring and business knowledge that describes an act that induces an individual to perform some action.

12.2.1.5.1.2 Patterns

AnyParty: This class represents any person, individual, an organization, or group that can be tempted.

AnyCriteria: This class denotes the factors that are considered to tempt AnyParty and, which can be used to win over AnyParty.

AnyEntity: This class represents anything toward which AnyParty should be attracted to and which impacts the mechanism used to create temptation.

AnyMechanism: This class represents the various techniques employed to tempt AnyParty.

AnyMedia: This class contains different types of media and any one type is used to show AnyEntity with the help of AnyMechanism.

AnyLevel: This class represents the level returned by the technique used to create temptation and it also represents the level that impacts the factors used to win AnyParty.

AnyUnit: It represents the unit of measurement, which indicates the value of level in a relevant unit.

12.2.1.5.2 Class Diagram Description

The class diagram provides a visual illustration of all the classes in the model, along with their relationships with other classes. Description of the class diagram is as below.

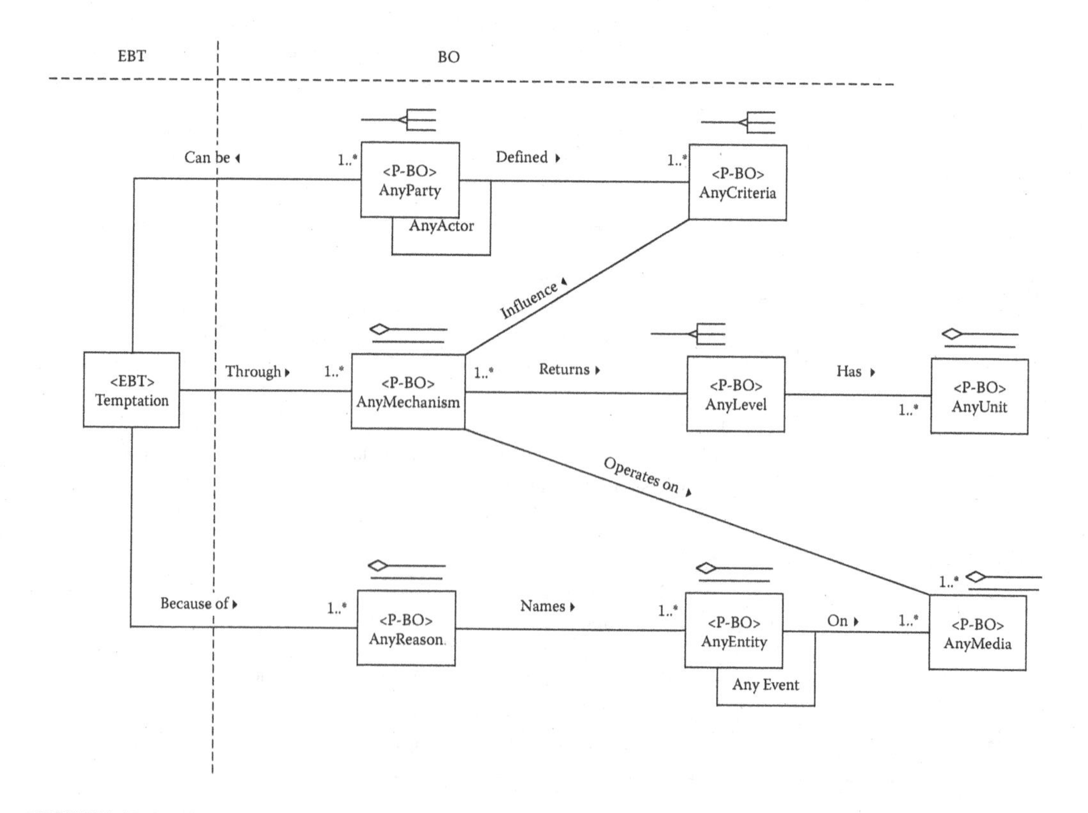

FIGURE 12.1 Class diagram of temptation pattern.

- Temptation is the EBT of this pattern and it describes an act that appeal AnyParty and force them to commit some action.
- AnyParty (BO) can be tempted through AnyMechanism(BO).
- AnyCriteria(BO) influence AnyMechanism(BO) and is considered, while choosing AnyMechanism(BO) to win AnyParty(BO).
- AnyMechanism(BO) returns AnyLevel(BO) and AnyLevel(BO) impacts AnyCriteria(BO) and operates on AnyMedia(BO).
- AnyLevel(BO) has AnyUnit(BO) and AnyUnit(BO) indicates the relevant unit of measurement of AnyLevel(BO).
- AnyEntity(BO) impacts AnyMechanism(BO) and is shown/ implemented through AnyMedia(BO).

12.2.1.6 Applicability with Illustrated Example

In this section, two examples to illustrate the use of personalization analysis pattern are depicted, by using case description and behavior model like sequence diagram.

CASE STUDY: Tempt Chocolate Consumers through Advertisement

In case of marketing chocolates, it is very necessary to attract the attention of the chocolate consumers and the chocolate retailers. This can be achieved by considering a preferential factor that chocolate consumers demand in a chocolate product. To win the confidence of chocolate consumers and chocolate retailers, advertisement can be used as an effective mechanism to attract their attention. Advertisement can be created by recognizing a readily identifiable factor that the chocolate consumers and retailers are finding in chocolates. Advertisement can use different media such as TV and radio or newspapers. One of the prominent among them is the television. Television prime time slots advertise particular chocolates and reveal the level that impacts the factors. When the chocolate consumers and retailers observe that the advertised chocolate meets their criteria, they are tempted to buy or sell it.

12.2.1.6.1 Model

The model for this application is shown below in Figure 12.2.

12.2.1.6.2 Use Case

Use Case No: 3.1 as shown in Table 12.2.

12.2.1.6.3 Use Case Description

1. AnyParty(BO) can be tempted through AnyMechanism. Temptation appeals, creates desire, seduces, and induces AnyParty to take some action AnyParty looses self control, seduced, and commits certain verifiable actions. AnyParty can be ChocolateConsumer and ChocolateRetailingCompany. ChocolateConsumer views chocolate advertisements, buys chocolates, consumes, and enjoys chocolates. ChocolateRetailingCompany do retail business of chocolates, contact manufacture, purchase, and sell chocolates. AnyCriteria(Deliciousness) is used to win AnyParty(ChocolateConsumer). Is AnyCriteria(Deliciousness) sufficient to win AnyParty(ChocolateConsumer)?
2. AnyMechanism is Advertisement. AnyMechanism executes, operates, and builds. Advertisement promotes the sales, encourages buyers, and assigns any identity to the product. AnyMechanism operates Any Media. Is the right AnyMechanism(Advertisement) chosen to tempt AnyParty(ChocolateConsumer)?

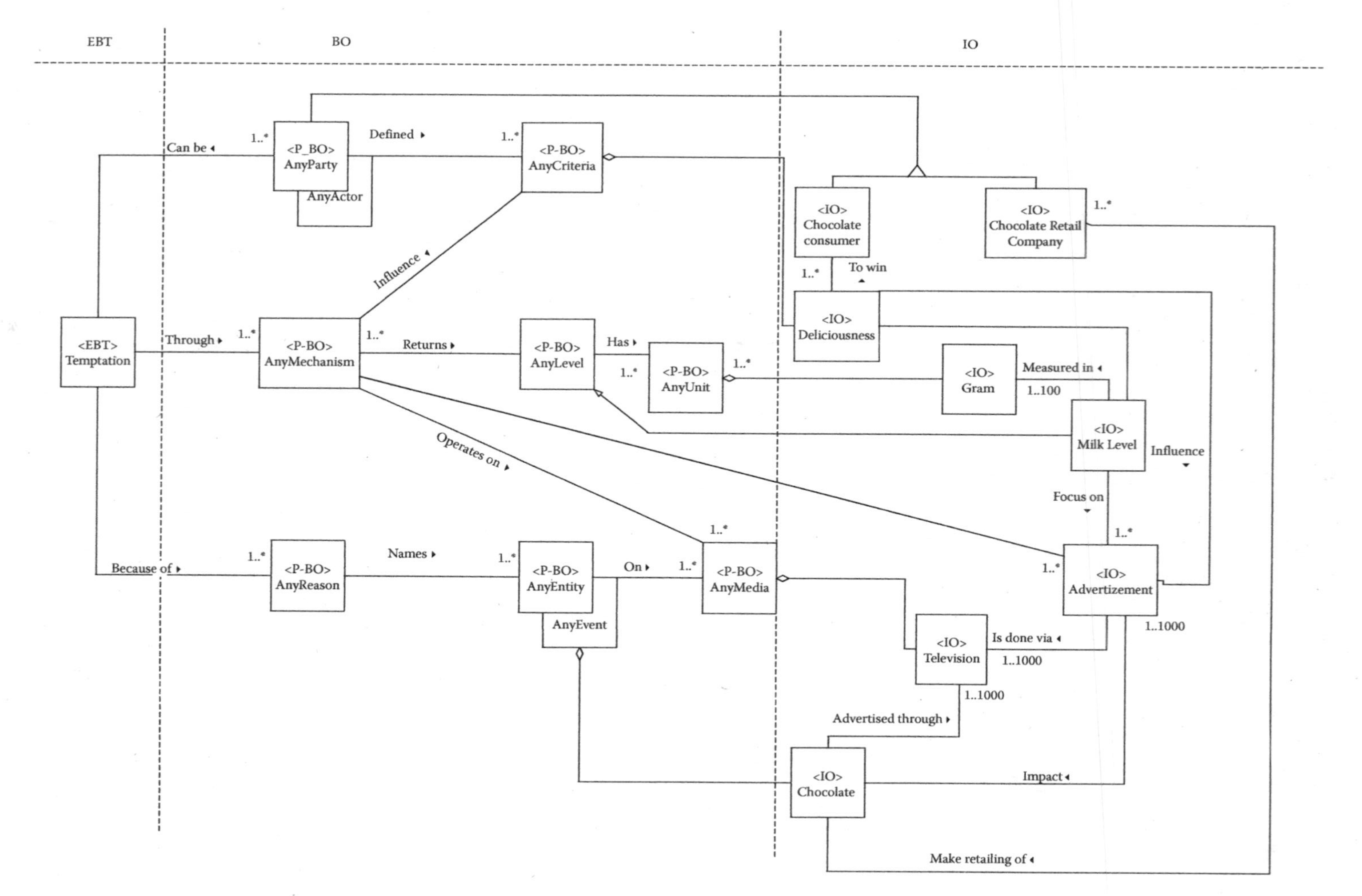

FIGURE 12.2 Class diagram for illustrating how chocolate consumers can be tempted.

TABLE 12.2
Use Case Title: Tempt Chocolate Consumers through Advertisement

Actors	Roles
AnyParty	ChocolateConsumer
AnyParty	ChocolateRetailingCompany

Classes	Type	Attributes	Operations
Temptation	EBT	• TemptationStrength	• appeal() • createDesire() • seduce() • induce()
AnyParty	BO	• Name • Interest	• looseSelfControl() • seduceTo() • commitAnyAction() • resist()
AnyMechanism	BO	• Type • Attribute • Context	• operate() • build() • execute()
AnyEntity	BO	• Effect • Type	• defineCharacteristics() • name() • type()
AnyCriteria	BO	• Quality • ValueDescription	• specifyCharacteristics() • rateQuality() • decideTaste()
AnyLevel	BO	• Ingredient	• find() • measure() • giveProportion()
AnyMedia	BO	• MediaName • MediaType • Cost	• implementMechanism() • conveyMessage() • communicate()
AnyUnit	BO	• UnitName • Standard	• giveMeasurementUnit() • adoptStandard() • expressPhysicalValue() • measureQuantity()
ChocolateConsumer	IO	• Weakness	• buy() • consume() • viewAdvertisement() • enjoy()
ChocolateRetailingCompany	IO	• Name • AreaCovered • ChocolateBrand	• sell() • contactManufacturer() • purchase()
Deliciousness	IO	• Taste	• checkTaste() • increaseAppetite() • matchExpectation()
Television	IO	• ChannelName • Time	• transmit() • repeatAdvertisement() • chargeFees() • time()
Advertisement	IO	• Type • Content • Duration	• promoteSales() • encourageBuyers() • assignIdentityToProduct() • fixDuration()

(*Continued*)

TABLE 12.2 (*Continued*)
Use Case Title: Tempt Chocolate Consumers through Advertisement

Classes	Type	Attributes	Operations
Chocolate	IO	• brandName shape size	• putNutrition () • decideSize() • stateIngredients() • shape() • packagingStyle()
MilkLevel	IO	• quantity	• indicateQuantity() • compare() • increaseTaste()
Gram	IO	• measuredValue	• inputLevel() • expressValueInGram()

3. AnyMedia(BO) is Television. AnyMedia conveys message, communicates, and implements the chosen mechanism. Television transmits the advertisement, shows an advertisement, and charges fees for the advertisement. AnyMedia advertises AnyEntity Verify, if AnyMedia(Television) conveys the right message?
4. AnyEntity is Chocolate. AnyEntity defines the characteristics, name, and type. Chocolate puts nutrition values, decides the size, states the ingredients present in the chocolate, and gives the shape and packaging style. AnyEntity(Chocolate) impacts AnyMechanism(Advertisement). Is AnyEntity(Chocolate) considered while choosing AnyMechanism?
5. AnyMechanism returns AnyLevel. AnyLevel is MilkLevel. AnyLevel finds the level of the chosen ingredient and gives exact proportion. MilkLevel indicates the quantity of milk present and compares with other chocolate products. AnyLevel impacts AnyCriteria(). Check if AnyCriteria is affected by AnyLevel?
6. AnyLevel(BO) has AnyUnit(BO). AnyUnit gives measurement unit, adopts standard, expresses physical value, and measures the quantity. AnyUnit is gram. Gram takes input level and expresses its value in grams. AnyUnit(BO) gives the measurement unit of AnyLevel(BO). AnyLevel(BO) impacts AnyCriteria(BO). Is the correct AnyUnit(BO) chosen to give the measured value of AnyLevel(BO)?
7. AnyCriteria characterizes and influences the mechanism used to create temptation. AnyCriteria is Deliciousness. Deliciousness checks the taste and increases the appetite of AnyParty(ChocolateConsumer). AnyCriteria influences the AnyMechanism used and is used to win AnyParty. How does AnyCriteria(BO) influence AnyMechanism(BO)?

12.2.1.6.3.1 Alternatives AnyParty resists temptation.

12.2.1.6.4 Sequence Diagram

Sequence diagram description as shown in Figure 12.3.

- AnyParty(ChocolateConsumer/ChocolateRetailingCompany) can be tempted(EBT) through AnyMechanism(Advertisement)
- AnyMechanism(Advertisement) operates AnyMedia(Television)
- AnyMedia(Television) advertises AnyEntity(Chocolate)
- AnyEntity(Chocolate) impacts AnyMechanism(Advertisement)
- AnyMechanism(Advertisement) returns AnyLevel(MilkLevel)

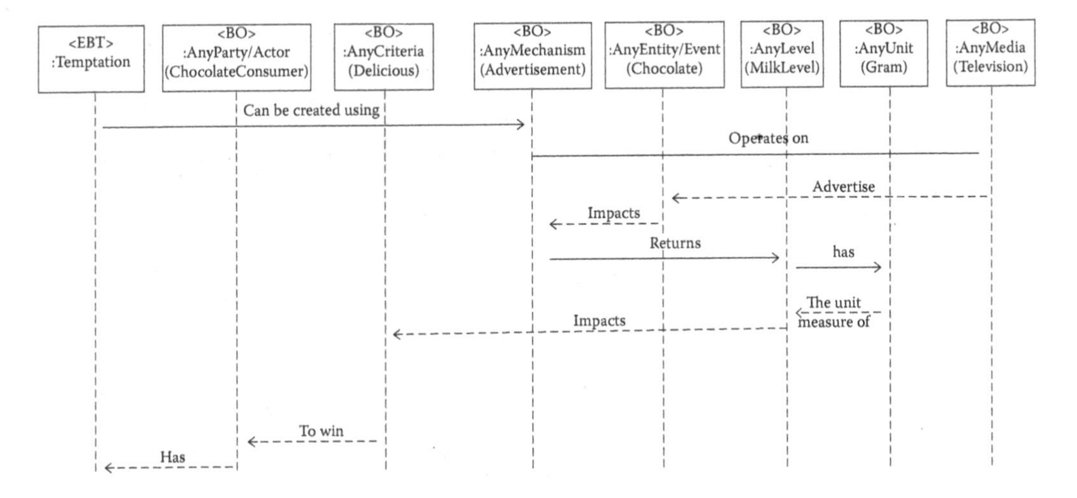

FIGURE 12.3 Sequence diagram for tempting chocolate consumers through advertisement.

- AnyLevel(MilkLevel) has AnyUnit(Gram)
- AnyUnit(Gram) gives the measurement unit used to measure AnyLevel(MilkLevel)
- AnyLevel(MilkLevel) impacts AnyCriteria(Deliciousness)
- AnyCriteria(Deliciousness)is considered to win AnyParty (ChocolateConsumer/ChocolateRetailingCompany)

12.3 SUMMARY

The temptation model is developed with stability in mind. It is stable and complete in itself and can be used in applications based on temptation concept from different domains using different contexts.

REVIEW QUESTIONS

1. What do you mean by the term "temptation"? Can the term "temptation" be used in any other context other than what has already been mentioned in this chapter?
2. Find out all such terms (synonyms), which mean exactly the same as "temptation" and can they be used interchangeably?
3. What are the capabilities required to achieve temptation? Describe each of them.
4. Draw and describe the class diagram for stable temptation pattern?
5. Create two scenarios other than those given in this chapter, where the temptation pattern can be used.
6. Try to create a use case and interaction diagram for each of the scenarios you thought of in the above question.
7. List differences between the temptation pattern described here and the traditional pattern.
8. List some design and implementation issues faced, when implementing the temptation pattern. Explain each issue.
9. Give some applications, where the temptation pattern is used.
10. What lessons have you learnt by studying the temptation pattern.
11. Define the temptation stable analysis pattern.
12. The temptation stable analysis pattern can be applied and extended to any domain (T/F).
13. List some of the domains in which the temptation analysis pattern can be applied.
14. List four new applications of the temptation stable analysis pattern.
15. List three challenges in formulating the temptation analysis pattern.

16. List ten different constraints in the temptation analysis pattern.
17. What are the classes and patterns involved in defining the stable pattern for "temptation"?
18. Illustrate by using a class diagram of temptation patterns of each of the new applications you listed for question fourteen.
19. Document the CRC card for the temptation EBT.
20. Is the temptation pattern incomplete without the use of other patterns? Explain briefly.
21. What is the trade-off of using this stable pattern versus the traditional approach?
22. Present the sequence diagram for applicability of the temptation stable analysis pattern in the e-commerce domain, specifically marketing.
23. What are the possible design issues for the temptation EBT, when linked to the design phase?
24. What do you think are the implementation issues for the AnyLevel BO, when used in the temptation stable analysis pattern?
25. List a couple of advantages of using the stable analysis pattern for temptation.
26. List two scenarios, which will not be covered by the temptation analysis pattern.
27. Describe how the developed temptation analysis pattern would be stable over time.
28. List some of the lessons learnt from the use of the stable analysis pattern for temptation.
29. List some of the testing patterns that can be applied for testing the temptation stable analysis pattern.
30. List three test cases to test the class members of the temptation pattern.
31. List some of the related design patterns that are used in formulating the temptation stable pattern.
32. Briefly explain how the temptation stable analysis pattern supports its objectives.
33. Assess two different quantitative measures on the "temptation" traditional model and temptation stable analysis patterns, and explain the differences between each of the measures.
34. What do you think are the implementation issues for the AnyUnit BO, when used in the Temptation stable analysis pattern?

EXERCISES

1. Illustrate with a class diagram and describe the applicability of the temptation SAP in the following scenario: A marketing company of a movie is trying to tempt people to come and watch a movie.
2. Illustrate with a class diagram and describe the applicability of the temptation SAP in a scenario, where a YouTube channel is trying to tempt people to come and watch the uploaded videos.
3. Draw with a class diagram and describe the applicability of the temptation SAP in a scenario, where the YouTube channel is trying to tempt people to come and watch the uploaded videos, subscribe to the channel and like the videos.
4. Draw with a class diagram and describe the applicability of the temptation SAP in a scenario, where the government of a country is trying to tempt its citizens to use only the products made within that country, so as to increase the economy of the country.
5. Identify a few challenges and constraints of temptation SAP that are not listed in this chapter.
6. Identify a few advantages of stable model over traditional model that are not discussed in this chapter.
7. Illustrate with a class diagram the applicability of the temptation stable pattern in the following scenario: three or more companies are involved in sponsoring a sporting event. Each company would like to maximize their investment in the event by attracting television viewers and spectators.

8. Think of few scenarios where temptation pattern is applicable and create the corresponding class diagram, use case and sequence diagram as shown in the solution and applicability sections for each of the scenarios.
9. Illustrate with a class diagram the applicability of the temptation SAP targeting advertisements to users of a popular webs search engine.
10. Adapt the temptation SAP to an e-commerce site that is similar to eBay. Illustrate with a class diagram how the ecommerce site can advertise user generated sales.

PROJECTS

1. Attracting sponsors to sponsor a movie.
   a. Explain the above scenario.
   b. Draw a class diagram for it.
   c. Document a detailed and significant use case.
   d. Create a sequence diagram of the created use case of c.
2. Tempting people to work by providing a good work environment.
   a. Explain the above scenario.
   b. Draw a class diagram for it.
   c. Document a detailed and significant use case.
   d. Create a sequence diagram of the created use case of c.
3. Tempting people to buy a house by doing a good interior.
   a. Explain the above scenario.
   b. Draw a class diagram for it.
   c. Document a detailed and significant use case.
   d. Create a sequence diagram of the created use case of c.
4. Tempting a company to hire you by showing capabilities and skills.
   a. Explain the above scenario.
   b. Draw a class diagram for it.
   c. Document a detailed and significant use case.
   d. Create a sequence diagram of the created use case of c.
5. School sporting event—A university would like to develop a system for attracting students, alumni, and members of the local community to the school's sporting events. Using an automated email service, the sporting event can be distributed in a generic format. However, specific emails may be tailored specifically to the receiver. Each student has an associated ID and with this a class history, and student profile. Using the information from the profile, class history, generate a system or advertising sporting events using the temptation pattern.
   a. Name two to three ultimate goals of the system beyond sporting events.
   b. List all the functional requirements and nonfunctional requirements of each of the ultimate goals.
   c. List software, hardware, or human challenges that relate to the ultimate goals.
   d. Name different applications for each of the goals. For example, how will the system react to user feedback and email unsubscriptions.
   e. Add use cases, use case diagrams, sequence diagrams based on your design of the system.

REFERENCES

1. R.H. Seiter, and J.S. Gass, *Persuasion, Social Influence, and Compliance Gaining*, 4th ed., Boston, MA: Allyn & Bacon, 2010, p. 33. ISBN 0-205-69818-2.
2. L. Fautsch, Persuasion, *The American Salesman*, 52(1), 2007, 13–16, Retrieved 9 December 2012.
3. M.E. Fayad, and A. Altman, An introduction to software stability, *Communications of the ACM*, 44(9), 2001, 95–98.

4. M.E. Fayad, Accomplishing software stability, *Communications of the ACM*, 45(1), 2002, 95–98.
5. M.E. Fayad, How to deal with software stability, *Communications of ACM*, 45(4), 2002, 109–112.
6. M.E. Fayad, and S. Wu, Merging multiple conventional models into one stable model, *Communications of the ACM*, 45(9), 2002, 102–106.
7. M.E. Fayad, H.A. Sanchez, S.G.K. Hegde, A. Basia, and A. Vakil, *Software Patterns, Knowledge Maps, and Domain Analysis*, Boca Raton, FL: Auerbach Publications, 2014.

Part IV

SAPs

Short Documentation Templates and Future Work and Conclusions

Part IV contains seven chapters that are listed below:

Chapter 13 titled "Analysis Stable Analysis Pattern"
Chapter 14 titled "Deployment Stable Analysis Pattern"
Chapter 15 titled "Change Stable Analysis Pattern" where the change pattern is an enduring
Chapter 16 titled "Propaganda Stable Analysis Pattern"
Chapter 17 titled "Fairness Stable Analysis Pattern"
Chapter 18 titled "Anxiety Stable Analysis Pattern"
Chapter 19 titled "Future Work and Conclusions"

13 Analysis Stable Analysis Pattern

The ultimate authority must always rest with the individual's own reason and critical analysis.

Dalai Lama XIV [1]

Analysis is the process of breaking a complex topic or substance into smaller parts to gain a better understanding of it. It is a careful study of something, to learn about its components, what they do, and how they are related to each other. The process of analysis can also be understood as a method of studying the nature of something or of determining its essential features and their relations. Similarly, dictionary.com defines the word, analysis, as "a process or a method of studying the nature of something or of determining its essential features and their relations." Many online free dictionaries such as the Free Dictionary also define the word analysis as "the separation of an intellectual or material whole into its constituent parts for individual study" and as "the study of such constituent parts and their interrelationships in making up a whole."

This word also encompasses a wide variety of meanings in various circumstances and situations. Its range of use covers logic, philosophy, mathematics, chemistry, computer software engineering, English grammar, political and international affairs, psychology, and physics. In other words, all these domains demand different requirements for engaging someone in analyzing an event or an entity. Thus, it may not be feasible to design and create a unified pattern framework that can contain all aspects and issues of the word "analysis." Yet, all those applications still associate themselves with a small part of analysis in spite of several basic differences that set them apart in how they are used by an entity. Hence, the main goal of this chapter is to create conceptual yet workable software pattern model that is robust and extendible across a number of domains. In fact, this chapter perceives that generated pattern models actually capture the core knowledge that is common to the concept of analysis independent of a specific application.

13.1 INTRODUCTION

The word "Analysis" in a wider context refers to "a systematic examination and evaluation of data or information, by breaking it into its component parts to uncover their interrelationships." However, it could also refer to several meanings each of which could be used as a specific situation or context. In the study of English language grammar, the grammatical analysis of a line or sentence is the method of studying the nature of something or of determining its essential features and their relations. A basketball coach analyses the reasons for his team's humiliating defeat in a league match. A business analyst, who is also a consultant to a business firm, sits down with board members to analyze the reasons for dwindling corporate profits. A philosopher uses a unique analyzing method to exhibit complex concepts as functions of more basic attributes; in fact, he/she is adept in the analysis of complicated ideas and deriving very simple solutions. A research student will conduct an analytical research for more than a year and later publish his/her thesis to announce the results of the entire research. A psychoanalyst, to find out reasons that are causing mental affliction, analyzes a mental patient's medical profile carefully.

Thus, the word *Analysis* finds a diverse number of applications in many areas of life. This necessitates an immediate need for capturing the core knowledge that is hidden behind analysis. In fact, it tries to evaluate how different entities are using its concepts to create an entirely different set of inferences although they signal a similar meaning, which is to analyze something. The analysis stable analysis pattern, as depicted in this chapter, focuses on encapsulating the nucleus of the concept that seems to be common to all applications that meticulously use analysis. Analysis is an enduring and generic concept. Hence, creating a stable software pattern called analysis, might help pattern developers to decode the core problem that links all analysis situations and potential problems that might arise during the pattern-making process. Forthwith, this chapter provides a brief view of a stable pattern for analysis that could act as a common template for all domains, contexts, and scenarios where analysis is used.

13.2 PATTERN NAME: ANALYSIS STABLE ANALYSIS PATTERN

This chapter intends to design and create a stable analysis pattern for analysis—An EBT is taken as analysis itself because of its specific nature. Several BOs, that are the essential parts of a stable pattern, are also identified and included in this pattern. Analysis is performed for some specific reason (*Reason*), which leads to different types of events and entities (*Entity* and/or *Event*) on which analysis intends to work. An event or an entity examines a specific aspect (*Aspect*) very carefully and conducts an analysis on the issue at hand. To analyze an event or an entity, a tool is employed that can carry out the process of analysis (*Mechanism*). Analysis is done depending on some specific context (*Context*), whereas *Mechanism* produces different versions of analysis (*Form*), which are stored and protected in a specific media (*Media*) such as CDs, DVDs, hard discs, or written notes. In nutshell, different BOs as stated here play a vital role in indexing the process of analysis by a *party* or an *actor.*

13.2.1 Context

Analysis is the process of describing the way one understands something by looking at it from different angles and by studying its different parts. It is a method of investigating the component parts of a whole and their relations in making up the detailed examination of the elements or structure of something, typically as a basis for discussion or interpretation. It also involves a set of techniques for exploring underlying motives and a method of treating various problems, subjects, or things. The process of analysis does not occur accidentally and in most of the cases, it is elaborate and systematic. It is continuous because it opens up a number of possibilities and future options to start the entire process once again. In other words, the entire process of analysis is unending and is ever happening. Analysis is a stronger candidate to be graded as an EBT. Analysis always consists of many consequences, results, reasons, actions, and supporting tools that together make up for a set of business objects called BOs. Here are some scenarios where one can find an instance of analysis.

13.2.2 Scenarios

1. *Analysis in chemistry:* The field of chemistry uses analysis in at least three ways: to identify the components of a particular chemical compound (qualitative analysis), to identify the proportions of components in a mixture (quantitative analysis), and to break down chemical processes and examine chemical reactions between elements of matter. For an example of its use, analysis of the concentration of elements is important in managing a nuclear reactor; so, nuclear scientists will analyze neutron activation to develop discrete measurements within a vast number of samples. Analysis can be done manually or with a device.
2. *Analysis in linguistics:* Linguistics began with the analysis of numerous languages that are at least 1000 years old; however, today it looks at individual languages and all languages in general. It breaks the language down and analyzes its component parts: theory, sounds, and their meaning, phonics, utterance usage, the history of words, the meaning of words and word combinations, sentence construction, basic construction beyond the sentence level,

stylistics, and conversation. It also analyzes the above using statistics and modeling, and semantics.

3. *Analysis in engineering:* Analysts in the field of engineering look at requirements, structures, mechanisms, systems, and dimensions. Electrical engineers analyze systems in electronics. Life cycles and system failures are broken down and studied by engineers. It also looks at different factors incorporated within the design.

13.2.3 Problem

13.2.3.1 Functional Requirements

1. *Analysis:* It is the EBT for this pattern. Analysis is to describe the way one understands something by looking at it in different ways and studying its different parts. Analysis is a goal, is a process, and has consequences, aftermath, and a description. It defines different rules that parties or actors should follow and obey. It also helps in taking action and in making decisions.
2. *Party:* It represents the analysis handlers. It molds parties that are involved in the analysis process. Party can be a person, an entity, organization, a country, or a group of people with a predefined orientation toward analysis process. AnyParty has a unique id, a name, an email address, a phone number, and a mailing address. They ask for analysis to be performed, offer analysis, and specify different norms for analysis.
3. *Actor:* It represents the analysis handlers. It also models actors that are engaged in the analysis process. An actor can be a person, an organization, a business, a computer machine, or a creature that can carry out analysis. An actor has a unique id, a name, an email address, a phone number, and a mailing address. They ask for analysis to be performed, offer analysis, and specify different norms for analysis.
4. *Entity:* It represents an entity that plays a vital role in the application domain. Every entity has certain properties and attributes that can be quantified and verified against any criteria. An entity can help in carrying out an analysis. An entity has a type, a name, a description, and a unique id. It is related to a party or an actor and helps in finding the quality of analysis.
5. *Aspect:* It represents the issue, which are specified by AnyActor or AnyParty and it influences mechanism deployed to create a definite form of analysis that can produce a meaningful result. It has rules, boundaries, description, and a unique Id. It has to satisfy party/actor to carry out analysis.
6. *Mechanism:* It represents the means followed by the analysis EBT based on aspect specified by an actor or a party. Mechanism represents the mechanism of carrying out analysis. It has a unique Id, a type, and a description. It provides method, ensures result, and fulfills purpose and goal.
7. *Media:* Media is used to store transcripts of analysis on it. It helps in storing the information related to analysis, so that it can be accessed whenever needed. Media has a unique Id, a type, a title, and a description. It is used to store logs and record information.
8. *Context:* A context refers to a set of facts or circumstances that surround a situation or event. Analysis can be done in any context by a party/an actor and it determines the type of analysis to be performed.
9. *Form:* The way something or someone is shaped or arranged is its form. Mechanism used in analysis can produce various forms of result. An analysis could be deductive or inductive and it can infer something that is meaningful and understandable.
10. *Type:* A type is a category of things distinguished by some common characteristic or quality. Analysis can be of various types depending on a given circumstance. An entity: an entity is something that is perceived, known, or inferred to have a distinct existence, be it living or nonliving. An entity may be subject to or help in doing an analysis.

13.2.3.2 Nonfunctional Requirements

1. *Completeness:* Analysis carried out should be complete and wholesome and it should mean something that is understandable, comprehensive, meaningful, logical, and self-explaining. An incomplete analysis may lead to undesirable consequence and eventually a wasteful exercise.
2. *Consistency:* Consistency in analysis is the conformity in the application of something, typically that which is necessary for the sake of logic, accuracy, or fairness. Any analysis that is carried out should be consistent to avoid lack of proper understanding. An argument made should be analytical in all respect and its tone and tenor should be even, reliable, uniform, and regular.
3. *Understanding:* Analysis is the process of assimilating knowledge and ability to judge a particular situation or subject. For any analysis to be effective, it should be easily understandable. People who want to understand something that is concrete and meaningful should easily understand what is being analyzed and deciphered.
4. *Synthesis:* Synthesis and analysis are different concepts. Synthesis derives something new that is open in the future for further analysis. It makes a new proposition and poses a challenge to prove or disprove it. In other words, analysis, if any carried out, should also help the person who analyze to synthesize something that can lead to fresh propositions.
5. *Criticism:* The entire analytical process should be open to constructive criticism and comments. In essence, the results of any analysis could be further improved and honed if there is a constructive criticism.
6. *Investigation:* Analysis should also focus on investigating something hidden or invisible. A police team that is investigating a crime should produce a detailed analysis of crime committed that eventually helps it solve the entire mystery. In other words, analysis should also support investigation process so that the core of the problem could be understood properly.

13.2.4 Solution

13.2.4.1 Class Diagram Description

1. AnyActor and/or AnyParty do analysis
2. Analysis is done using AnyMechanism
3. Analysis has AnyContext
4. AnyActor and/or AnyParty looks at AnyAspect to do analysis
5. AnyAspect influences AnyMechanism
6. AnyMechanism produces AnyForm of analysis
7. AnyForm of analysis is kept on AnyMedia
8. AnyEntity is also kept on AnyMedia
9. AnyContext determines AnyType of analysis
10. AnyType names AnyEntity
11. AnyEntity is to be analyzed using AnyMechanism Class diagram description as shown in Figure 13.1

13.3 SUMMARY

The pattern developed here is an attempt to capture the real essence of the term, *Analysis*. The software stable pattern designed for this term is conceptual and it tries to present a bird's eye view of how developers can use the concept of analysis to decipher hidden meaning and its associated problems and pitfalls. In addition, the generic nature of analysis concept makes the pattern applicable to many contexts and domains; in fact, the pattern represents different properties of

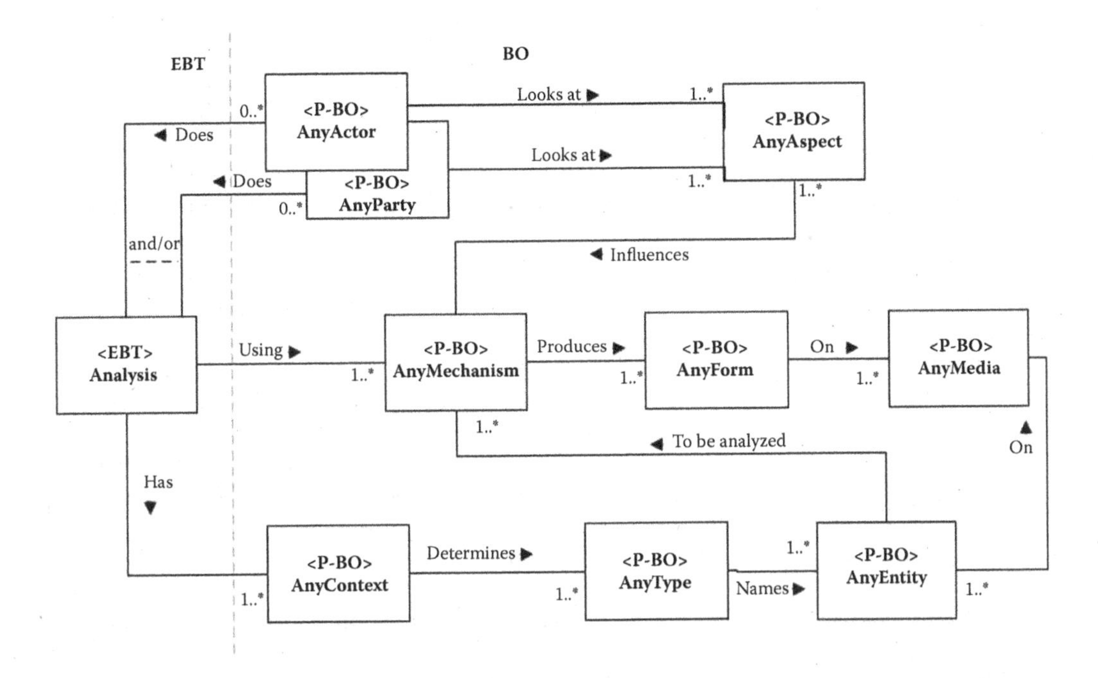

FIGURE 13.1 Class diagram for analysis stable analysis pattern.

analysis concept and the resulting pattern is reusable, applicable to different domains, and extremely stable as it considers different objects that are likely to play their vital role in forming the pattern.

13.4 OPEN AND RESEARCH ISSUES

1. *e-Analysis:* Utilize stability model [2–5] or knowledge map [6] methodology as a way for developing e-Analysis Engine. Building this engine by using traditional development approaches is not an easy exercise, specifically when several factors can undermine their quality success, such as cost, time, and lack of systematic approaches.
2. *Analysis-unified software engine (Analysis-USE):* Utilize stability model or knowledge map methodology as a way for developing Analysis-USE. The proposed solution attempts to extract the commonality from all the domains and represents it in such a way that it is applicable to a wide range of contexts without trivializing or generalizing the concepts. The engine is a stable structural pattern and it provides a generic engine that can be applied and/or extensible to any application and can access to analysis of anything and will be applicable to AnyContext by plugging application-specific features. The Analysis-USE can be applied to unlimited context and reusability.
3. *Analysis as part of architecture on demand:* Utilize stability model or knowledge map methodology as a way for developing architecture on demand, using Analysis and build unlimited architecture patterns that can be applied to many applications within context, such as accounting analysis, accident analysis, contract analysis, algorithm analysis, benefit cost analysis, behavior analysis, capability analysis, code analysis, data analysis, content analysis, empirical analysis, energy analysis, formal analysis, functional analysis, hazards analysis, requirements analysis, etc. The proposed solution attempts to extract the commonality from all the domains and represents it in such a way that it is applicable to a wide range of contexts without trivializing or generalizing the concepts. Each of the stable architecture patterns (architectures on demand) can be used to provide a generic engine that can be applied and/or extensible to any application.

REVIEW QUESTIONS

1. What is "analysis?" List scenarios where you can use this term in your daily life.
2. Find out all terms that are similar to "analysis" and you can use them interchangeably.
3. What type of capabilities and attributes do you use to achieve analysis? Describe each one of them.
4. Draw and explain the class diagram for stable analysis pattern? Identify and highlight all participants and explain why they exist.
5. List two new situations for analysis and fit them with the analysis pattern.
6. Create a use case and an interaction diagram for scenarios that appeared in the above question.
7. Draw a class diagram for the application of Analysis SAP in carrying out medical analysis for a mentally disabled person. Describe and write a use case for it and its description. Draw a sequence diagram for the use case too.
8. Draw a class diagram each for the application of Analysis SAP in philosophy and metaphysics. Write a use case for them and their description. Draw a sequence diagram for the use cases.
9. List some differences between the analysis pattern described here and the traditional pattern.
10. List some design and implementation issues faced, when implementing this pattern. Explain each issue.
11. Give at least four applications where the analysis pattern is used.
12. What lessons have you learnt by studying the analysis pattern?
13. Define and highlight the importance of analysis stable analysis pattern.
14. The analysis stable analysis pattern can be applied and extended to any domain (T/F).
15. List three important challenges in formulating the analysis pattern.
16. List five advantages of analysis pattern.
17. What are the classes and patterns involved in defining the stable pattern for analyses?
18. The term analysis is a generic term, which could be used in different contexts and situations. Do you agree? If yes, give your reasons.
19. Analysis pattern is applicable to many domains (T/F).
20. List some nonfunctional requirements that make up this pattern. Do you agree that they are different from the concept of analysis?

EXERCISES

1. Analysis in chemistry
   a. Explain analysis in chemistry and select any context (scenario) in chemistry.
   b. Draw a class diagram based on the analysis pattern to show the application of selected context in a.
   c. Document a detailed and significant use case as shown in Case Study 1.
   d. Create a sequence diagram of the created use case of c.
2. Analysis in business
   a. Explain analysis in business and select any context (scenario) of analysis in business.
   b. Draw a class diagram based on the analysis pattern to show the application of a selected context in a.
   c. Document a detailed and significant use case as shown in Case Study 1.
   d. Create a sequence diagram of the created use case of c.
3. Analysis in computer science
   a. Explain analysis in computer science and select any context (scenario) of analysis in computer science.

b. Draw a class diagram based on the analysis pattern to show the application of a selected context in a.
c. Document a detailed and significant use case as shown in Case Study 1.
d. Create a sequence diagram of the created use case of c.

PROJECTS

Develop the following systems using the analysis pattern:

1. *Image analysis:* The extraction of useful information from images
2. A *map analysis*, whereby a study is made regarding *map types*, that is, political maps, military maps, contour lines, etc., and the *unique physical qualities* of a map, that is, scale, title, legend, etc.
3. *Market analysis:* marketing research that yields information about the marketplace or the act or process of examining in detail the performance of AnyMarket, such as stock, corporation, special industry, market place, with a view to suggesting future trends
4. *Requirements analysis:* The process of reviewing a business's processes or the user's requirements to determine the business or the user needs and *functional requirements* that a system must meet.
   a. List all the functional requirements and nonfunctional requirements for each area.
   b. List two challenges for each area.
   c. Name five contexts for each area.
   d. Draw the application of the pattern for each context in c.
   e. Select a significant use case per application and describe each one of them with text cases.
   f. Map each use case in e into a sequence diagram.

REFERENCES

1. Dalai Lama quotes Head of the Dge-lugs-pa order of Tibetan Buddhists, 1989 Nobel Peace Prize, b.1935.
2. M.E. Fayad, and A. Altman, An introduction to software stability, *Communications of the ACM*, 44(9), 2001, 5–98.
3. M.E. Fayad, Accomplishing software stability, *Communications of the ACM*, 45(1), 2002, 95–98.
4. M.E. Fayad, How to deal with software stability, *Communications of ACM*, 45(4), 2002, 109–112.
5. M.E. Fayad, and S. Wu, Merging multiple conventional models into one stable model, *Communications of the ACM*, 45(9), 2002, 102–106.
6. M.E. Fayad, H.A. Sanchez, S.G.K. Hegde, A. Basia, and A. Vakil, Software patterns, knowledge maps, and domain analysis, Boca Raton, FL: Auerbach Publications, 2014.

14 Deployment Stable Analysis Pattern

"Sometimes it takes the pain of a deployment to realize how we take the little things for granted and how much we miss them."

Superman

Deployment is an action-filled word. Its means someone being proactive in its original dictionary meaning. In general, deployment is spreading out troops and the military equipments to form an extended frontline. In nutshell, deployment is pooling of available resources in one location or arranging them to achieve a certain objective. The main goal of this chapter is to find a solution to deployment meaning that the pattern created will be more stable, robust, and rugged. In addition, it will also help developers save time, money, and effort that are needed to create a pattern. Repeatability is one of the notable problems in pattern making, and a stable pattern will help avoid this problem altogether. The deployment pattern also helps developers work on the core issue of any problem to find the most appropriate elements of a stable pattern.

The deployment stable analysis pattern includes a main business theme and several BOs, such as resources and types among others. Deployment can be of different types and forms. It can be used to resolve certain issues that arise because of the occurrence of certain events, which can lead to many reasons for different types of resource to be deployed. The reasons for deployment could be ensuing safety to someone or some country, or it could be used to resolve some existing issues. When compared to traditional model, stability model [1–5] provides a most appropriate business theme that is enduring and this main theme could be used to create any numbers or kinds of deployment patterns.

14.1 INTRODUCTION

Defining deployment in terms other than the one given before is tricky and challenging as this word could mean different under diverse contexts. In another scenario, deployment could mean arranging something in a position of readiness or move it strategically or appropriately. Used as a verb, deployment may also mean something that come into a position ready for immediate use. Another definition is putting something into use or action.

In a business context, services like money, brainpower, strategic thinking, brand making, action plans, budgetary allocations, goal setting, and frequent meetings may be used to arrange in a strategy that could be immediately put to use. Similar is the case with many other businesses too. A police force may deploy every possible resource to catch a killer or a criminal. A student may deploy all available resources to write a doctoral thesis or a writer may deploy his writing skills and talent to write the most beautiful poem. Likewise, deployment projects different uses and applications in almost all domains of life. Deployment stable analysis pattern is a versatile stability model [1–5] that will be implemented for all scenarios of deployment.

14.2 PATTERN NAME: DEPLOYMENT STABLE ANALYSIS PATTERN

Deployment is EBT for deployment stable analysis pattern. The definition of deployment mentions it as the movement from one location to another. Deployment has certain rules and regulations that need to be followed. There can be single or multiple resources involved. The resources require structure, and deployment is initiated because of a reason. Deployment is thus an EBT to achieve deployment stable analysis pattern.

14.2.1 Context

As discussed elsewhere in this chapter, deployment is too general to be used as a term that can be applied to different scenarios. However, when developers find out the most important EBTs and BOs for the term, any number of stable patterns could be designed and created. In other words, the pattern for deployment will be applied to many different applications in the context of deployment, while EBT of deployment is used to implement stability model [1–5]. For this chapter, the deployment stable patterns have been considered for three main scenarios: military deployment, resource deployment, skill deployment, or deploying parachute.

14.2.1.1 Scenario 1: Deployment of Military Personnel

Deployment of military personnel (AnyResource) is an example of deployment. A military person is deployed by checking (AnyMechanism) availability, meaning he is sent to a place (AnyMedia) other than their base location. The military (AnyParty) will decide where (AnyStructure) their troops will need to be deployed. This will be based on certain conditions (AnyRule).

14.2.1.2 Scenario 2: Resource Deployment

Deploying available resources (AnyResource) like fuel and gas is necessary, when the need or the demand (AnyReason) arises. When the demand for the resources is received (AnyEvent), the oil companies will work more to extract oil (AnyMechanism) from their reserves. This extracted fuel (AnyEntity) is then deployed to various distributors (Anymedia), so that the fuel can reach consumers.

14.2.1.3 Scenario 3: Deployment of Parachute

Skydiving or jumping from a flying aircraft requires wearing (AnyRule) a parachute, while jumping down from the plane. The skydiver needs (AnyReason) a parachute (AnyResource) to land safely on the ground and the skydiver should carry one before flying. When the descent altitude of the skydiver from the plane is at a certain height from the ground, the skydiver will pull the strings (AnyMechanism) that are attached to the parachute (AnyStructure), so that the parachute can be deployed.

14.2.2 Problem

This chapter plans to solve the problem of lack of applicability of traditional models for deployment to all the scenarios for deployment. A lack of scaling ability has lead to traditional models being redrawn for every scenario of deployment that arises in future. This chapter attempts to find a solution for a model that is reused and scaled by using stability model [2–5].

14.2.2.1 Functional Requirements

1. *Deployment:* Deployment is EBT for deployment stable analysis pattern. The definition of deployment mentions it as the movement from one location to another. Deployment has certain rules and regulations that need to be followed.
2. *AnyParty:* AnyParty is any organization that is directly involved and instrumental in doing deployment process. AnyParty does the deployment for any entity depending on the type and the event, which eventually triggers the deployment.
3. *AnyActor:* AnyActor is a person, who initiates or plans the deployment process. AnyActor can initiate, plan, or proceed with the deployment of AnyEntity by using AnyMechanism.

4. *AnyRule:* AnyRule includes a single rule or a set of rules that AnyParty will follow. AnyRule will be implied to any resource that is part of the deployment process.
5. *AnyMechanism:* AnyMechanism represents the way or method in which the resources will be deployed. Resources can be a person or any material. AnyMechanism is required for AnyEntity and AnyEvent in the scope of deployment.
6. *AnyResource:* AnyResource will be a set of single or multiple resources. AnyResource form the core part of Deployment. Resources are deployed by using any structures. AnyRules that are required to be followed imply to AnyResource.
7. *AnyStructure:* AnyStructure is the structure or a framework, which is a requirement for the AnySource.
8. *AnyReason:* AnyReason will have a reason that is well defined for the actions to be performed. AnyReason defines the need or the request for deployment that in turn determines the type of deployment.
9. *AnyType:* AnyType defines the type of deployment that is being performed. AnyType will then name AnyEntity and AnyEvent that will be in the scope of deployment.
10. *AnyEntity:* AnyEntity includes everything that is part of deployment. The aim of deployment is to deploy the entity on some media. AnyMechanism for AnyEntity will define the method for the entity to be deployed on AnyMedia. AnyEntity has to be defined with much clarity.
11. *AnyEvent:* AnyEvent marks the event or occurrence of AnyEvent, which should be considered as a marker for starting the deployment.
12. *AnyMedia:* AnyMedia is the platform, space, or environment on which the entity is to be deployed. AnyMedia forms a very important part of deployment, as it is the culmination of the deployment process.

14.2.2.2 Nonfunctional Requirements

1. *Timely:* Deployment needs to be performed at the appropriate time to become meaningful. If the deployment is carried out late, it will be meaningful in a given context. For example, if armed forces need deployment to counter insurgency or infiltration, then it should be finished as soon as possible. If there is an unexplained delay in deploying, then the effectiveness of deployment will be reduced. Another example is deploying a parachute, while skydiving from high up in the sky. The parachute should be opened up and deployed at the right time. Otherwise, the entire decent could become too dangerous and life threatening.
2. *Effective:* Deployment should be effective enough to be successful. Some reasons usually drive deployment operation. For that reason to be addressed in a proper manner, the deployment will need to be result-oriented, timely, and organized. For example, deployment of a parachute, while skydiving should see that it is effective and correct. Failure to do so can have catastrophic results.
3. *Achievable:* Deployment should be achievable with a minimum quantum of resources. It is not possible to commit to a deployment that is far away from reaching the state of capability. For example, deploying a handful of armed forces to fight a much larger and well-equipped army will not be achievable. Another example is deploying the parachute to decent just a few hundreds of yards from the ground.

14.2.3 Solution

Pattern Class Diagram, as shown in Figure 14.1.

14.2.3.1 Class Diagram Description

The pattern class diagram shows EBT deployment is realized by using AnyMechanism that takes AnyResources that requires using AnyStructure. Deployment is done due to AnyReason and this

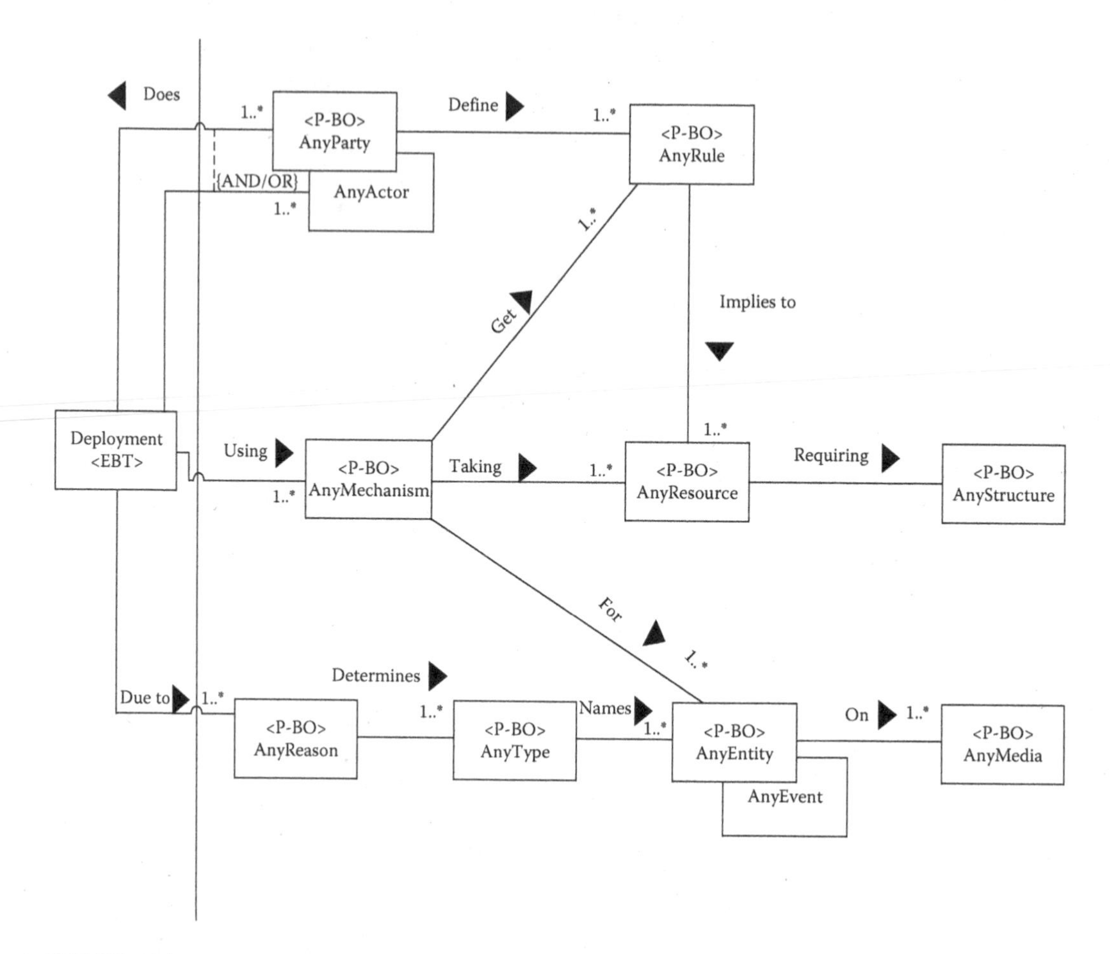

FIGURE 14.1 Deployment stable analysis pattern.

reason will determine AnyType of deployment. The type will further name AnyEntity and AnyEvent that are involved with deployment. The Deployment follows AnyRules and is performed by using AnyMedia.

14.2.4 Five Scenarios

The five scenarios are shown in Table 14.1.

14.3 SUMMARY

Stability model [1–5] is an approach that is quite different from traditional model. It allows patterns that can be reused and is scalable. Traditional models develop models based on a specific situation. By using stability model, the model can be applied across all domains.

14.4 OPEN AND RESEARCH ISSUES

- *e-Deployment:* Utilize stability model [1–4] or knowledge map [5] methodology as a way for developing e-Deployment Engine. Building this engine by using traditional development approaches is not an easy exercise, specifically when several factors can undermine their quality success, such as cost, time, and lack of systematic approaches.

TABLE 14.1
Five Different Applications of Deployment Stable Analysis Pattern

EBT	BOs	Military Deployment	Deploying Skills	Parachute Deployment	Deploying Scientists to Study a Problem	Deployment of Resources
Deployment	AnyParty/ Actor	Military leaders	Communication skills	Person wearing parachute	Government	Organization
	AnyMechanism	Movement of military personnel	Utilization of skills	Pulling the string to open parachute	Designate problem to scientists	Selecting resources
	AnyRule	How military should be deployed	As and when skills are required	Parachutes can be safely opened at certain height	Effective study is needed	Adequate resources
	AnyResource	Military weapons	Using log of skills available	Ropes attached to Parachute	Research	Human resource department or department of resources
	AnyStructure	Hierarchical structures	General voting	Building road through someone's property	Migration policy	Change of religion
	AnyReason	Security	Shortage of skills	Landing safely on earth	Inadequate knowledge of problem	Deploying appropriate resources
	AnyType	Unit wise deployment	Specific skill deployment	Skydiving	Cure for disease	Cleaning crew
	AnyEntity	Military units	skills	Parachute	Problem at hand	Resources
	AnyEvent	Threat to security	Arising of requirement of skills	Jumping out of plane for skydiving	Spreading of disease	Demand
	AnyMedia	Hostile environment	Speech	Air	Publish	Office

- *Deployment-unified software engine (Deployment-USE):* Utilize stability model [1–4] or knowledge map [5] methodology as a way for developing Deployment-USE. The proposed solution attempts to extract the commonality from all the domains and represents it in such a way that it is applicable to a wide range of contexts without trivializing or generalizing the concepts. The engine is a stable structural pattern and it provides a generic engine that can be applied and/or is extensible to any application and can access to deployment of anything and will be applicable to AnyContext by plugging application-specific features. The Deployment-USE can be applied to unlimited context and reusability.
- *Deployment as part of architecture on demand:* Utilize stability model [1–4] or knowledge map [5] methodology as a way for developing architecture on demand using deployment and build unlimited architecture patterns that can be applied to many applications within context, such as military deployment, police deployment, software deployment, system deployment, resource deployment, etc. The proposed solution attempts to extract

the commonality from all the domains and represents it in such a way that it is applicable to a wide range of contexts without trivializing or generalizing the concepts. Each of the stable architecture patterns (architectures on demand) can be used to provide a generic engine that can be applied and/or extensible to any application.

REVIEW QUESTIONS

a. Define deployment and list different meanings of this word.
b. Why is the term deployment special to the context of this chapter?
c. List four scenarios, where this pattern can be applied. Use ones that are highlighted in the chapter.
d. How does deployment pattern help understand the core of the problem?
e. How do you use functional requirements for this chapter?
f. How do you apply nonfunctional requirements for this chapter?
g. Identify the EBT for the word deployment.
h. List some BOs for this term.
i. Discuss some contexts, where you can apply deployment pattern.
j. What are some of the unexpected problems that are likely to come while designing deployment chapter.
k. Create three more patterns for deployment. Use three different scenarios.
l. What are the expected outcomes of creating this pattern?
m. Define some research issues of this chapter.
n. List a brief summary in bullet points.
o. Is this pattern stable over time?
p. Does this pattern the factor of extendibility to it?
q. Is deployment pattern domain and application specific?
r. In the context of the pattern created in this chapter, differentiate between traditional and stable type of patterns.
s. How do different BOs interact with each other and act in combination to create a pattern for deployment.
t. Is this pattern scalable, stable, and reusable? If so, list some of the reasons.

EXERCISES

1. Continuous delivery (CD) is a software engineering approach in which teams produce software in short cycles, ensuring that the software can be reliably released at any time. [1] It aims at building, testing, and releasing software faster and more frequently. The approach helps reduce the cost, time, and risk of delivering changes by allowing for more incremental updates to applications in production. A straightforward and repeatable deployment process is important for CD.
2. System deployment is the deployment of a mechanical device, electrical system, computer program, etc., and its assembly or transformation from a packaged form to an operational working state. Deployment implies moving a product from a temporary or development state to a permanent or desired state.
   a. Select any context (scenario) of each of the above topics.
   b. Describe the scenario of the selected context.
   c. Draw a class diagram based on the deployment pattern to show the application of a selected context in a.
   d. Document a detailed and significant use case as shown in Case Study in Chapters 5 through 12.
   e. Create a sequence diagram of the created use case of c.

PROJECTS

Develop the following systems using the deployment pattern:

1. Software deployment is all of the activities that make a software system available for use. The general deployment process consists of several interrelated activities with possible transitions between them. These activities can occur at the producer side or at the consumer side or both. Because every software system is unique, the precise processes or procedures within each activity can hardly be defined. Therefore, "deployment" should be interpreted as a *general process* that has to be customized according to specific requirements or characteristics. A brief description of each activity will be presented later [6].
2. The military of the United States is deployed in more than 150 countries around the world, with over 150,000 of its active-duty personnel serving outside the United States and its territories and an additional 71,000 deployed in various contingency operations as well as through military attache offices and temporary training assignments in foreign countries [7].
3. A Rapid deployment force is a military formation capable of quick deployment of its forces. Such forces typically consist of elite military units (special ops, paratroopers, marines, etc.) and are usually trained at a higher intensity than the rest of their country's military. They usually receive priority in equipment and training to prepare them for their mission. Quick response force (QRF) should not be confused with rapid deployment forces (US) or rapid response (NATO). QRF units are most often units that react to local or regional issues within their area of jurisdiction, that is, National Guard, militias, Forward Deployed, paramilitary forces, etc.
   a. List all the functional requirements and nonfunctional requirements for each area.
   b. List two challenges for each area.
   c. Name five contexts for each area.
   d. Draw the application of the pattern for each context in c.
   e. Select a significant use case per application and describe each one of them with test cases.
   f. Map each use case in e into a sequence diagram.

REFERENCES

1. M.E. Fayad, and A. Altman, An introduction to software stability, *Communications of the ACM*, 44(9), 2001, 95–98.
2. M.E. Fayad, Accomplishing software stability, *Communications of the ACM*, 45(1), 2002, 95–98.
3. M.E. Fayad, How to deal with software stability, *Communications of ACM*, 45(4), 2002, 109–112.
4. M.E. Fayad, and S. Wu, Merging multiple conventional models into one stable model, *Communications of the ACM*, 45(9), 2002, 102–106.
5. M.E. Fayad, H.A. Sanchez, S.G.K. Hegde, A. Basia, and A. Vakil, *Software Patterns, Knowledge Maps, and Domain Analysis*, Boca Raton, FL: Auerbach Publications, 2014.
6. A. Carzaniga, A. Fuggetta, R.S. Hall, A. Van Der Hoek, D. Heimbigner, and A.L. Wolf *A Characterization Framęwork for Software Deployment Technologies*—Technical Report CU-CS-857-98, Department of Computer Science, University of Colorado, April 1998. http://serl.cs.colorado.edu.
7. L.E. Davis, S.L. Pettyjohn, M.W. Sisson, S.M. Worman, and M.J. McNerney. U.S. Overseas Military Presence: What Are the Strategic Choices? (PDF). Project Air Force. Monographs Series (RAND Corporation), 2012, ISBN 978-0-8330-7340-2. Retrieved 6 November 2012.

15 Change Stable Analysis Pattern

To improve is to change; to be perfect is to change often.

Winston Churchill

The main goal of writing this chapter is to understand the needs and requirements of designing a stable analysis pattern. The stable analysis pattern for change describes how AnyParty or AnyActor, who uses change perform in a system. EBT for this system is change. If any party wants to perform a change, they should have AnyReason to carry it out. A change generally contains events (AnyEvents) and an entity (AnyEntity) [1]. The ultimate aim of using change is to ensure its extendibility over a longer period that eventually leads to impacts (AnyImpact) and outcomes (AnyOutcome) successfully. Changes are performed by using various mechanisms (AnyMechanism) and it takes different forms (AnyForm). Change is always evolving and leads to different aspects of life and situations.

The main goal of this chapter is to design a system by using stability model approach that is applicable to all kind of changes. Other goals of this chapter are to design a stable pattern for "Change" and modeling system over the identified EBT, and to design such a pattern that is stable and can be customized according to application need. In addition, this chapter also demonstrates that we can understand the underlying difference between traditional model for designing systems and stability model, and understand numerous common patterns for any type of change process.

15.1 INTRODUCTION

Change is an enduring concept. It is the ultimate goal of achieving change of any kind. In order to appreciate the degree and nature of change, we must have a level of stability or baseline against which one can measure change. Change is never uniform nor stable and permanent. Not all cultural and social elements change at the same rate. A change usually takes place, when people think of new ways and ideas of doing things. Most of the modern technological innovations like mobile and Internet technology, communication systems, space systems, and others have changed the way men and women live and thrive. All cultural changes are new beliefs, values, and religions and they are adopted in the society with periodic changes.

In this chapter, we plan to demonstrate different aspects of a change stable analysis pattern. The EBT of the pattern is change and the pattern is named as change stable analysis pattern (change). Section 15.1 describes the context of the project and introduces the concepts and ideas that are discussed here. Section 15.2 includes the pattern documentation and explains the EBT, that is, the goal of the ultimate pattern of our system. In Section 15.3, some of the functional and nonfunctional requirements are provided, which are a part of the problem description. Sections 15.4 and 15.5 list different challenges and constraints, of the system this chapter is trying to model. Once the developers understand the problems and challenges, we will try to propose a solution in Section 15.6 by using a class diagram and several CRC cards. Section 15.7 describes applicability and case studies with use cases and sequence diagrams. We will also try to list the applications of the system in different scenarios. Section 15.8 shows the difference between the traditional model and stability model [2–6] and explains how stability model is better than the traditional model.

15.2 NAME: CHANGE STABLE ANALYSIS PATTERN

"Change" is an EBT. According to many online dictionaries, a change is used in many contexts and with different meanings. Used mainly as a verb, it can also be used with or without objects. In essence, a change refers to something that is transforming, modifying, substituting, varying, converting, altering, or reciprocating. A change can be permanent or temporary. A person can change for good or bad. Weather can change its patterns within minutes. An object can change its form from one status to another as in the case of gas, liquid, and solids. A chemical mixture can change itself after adding some catalysts and the change could be permanent. Hence, the word *change* has a multitude of meanings. EBT of change is change. Change is something that happens, which include number of actions that may have occurred and performed by "AnyParty" or "AnyActor." There are a number of change processes happening in this world. It is an enduring and forever living concept. All BOs are prefixed with "Any" to enable generic reusability of the change pattern.

15.2.1 Context

Change stable analysis pattern (change) can be applied to any domain, where several systems or things have to undergo changes by using a certain mechanism. There are different scenarios that use change stable analysis pattern as shown below. Similar analysis methods can be applied to other BOs, thus making stable analysis more reusable and customizable over the other traditional stability models. A change is defined as a series of processes that are carried out to serve a specific purpose. This definition will apply to all the scenarios that are explained below.

15.2.1.1 Change of Workplace Location

Let us assume that Boeing company's headquarters was moved (AnyEvent) to a new location (AnyEntity) such as a new building with more space and light (AnyForm). Since the office space available in the former office building was too small (AnyReason) for people (AnyActor) working there, it was difficult for the employees (AnyParty) to adjust to the new building and environment. The office also hired interior decorators, packers, and freight agencies (AnyParty) to ease the process of preserving the workplace locales and interiors (AnyMechanism). However, productivity decreased (AnyImpact), until workers (AnyParty) adjusted to the new location and its work environs (AnyEntity). Nevertheless, the new location was a better option, rather than sitting in a cluttered workspace (AnyAspect), and putting productivity and discretion of work at risk. Therefore, this change was necessary and helpful in the long run (AnyOutcome).

15.2.1.2 Change of Password

A Gmail user (AnyActor) used it to send and receive (AnyEvent) any email (AnyType). Unfortunately, it was hacked (AnyEvent) and user's account information, personal, and official emails were in serious threat of online fraud (AnyReason). Therefore, the user was forced to change the password (AnyForm) by using the change password application (AnyMechanism). After changing password (AnyOutcome), the user needed to log in to the email account by using the new password. This prevented hackers from using the current login and secured millions of Gmail users account information from getting hacked (AnyImpact). A Gmail account user might may have difficulty while typing password, as he or she is in a habit of writing the old password every time while logging into the Gmail account (AnyAspect).

15.2.1.3 Change Review Meeting

Businesses meetings (AnyEvent) are conducted to resolve pending issues, like for example, a software developer (AnyActor) may incorporate some major changes in the code. Because of the customer's requirement, any change in the code should be updated (Change), and this is the reason for change (AnyReason). This code change (Change) and review discussion contains events (AnyEvent) that need to be discussed minutely in the meeting. This may include different aspects like writing a new set of code (AnyAspect), updating the old code (AnyAspect), or reusing any precreated package. These aspects influence the type (AnyType) of method that is used for code updating, for example, the code change is performed by distributing the existing work (AnyMechanism) among team members (AnyParty). The code change process must ensure that the code execution yields in better efficiency and increased productivity (AnyOutcome), and it also removes existing bugs and errors. This will help the completion of the project quickly and eventually saves time for all parties involved (AnyImpact).

15.2.2 Problem

Change stable analysis pattern (Change) focuses on a specific solution or a model that can be applied across several domains. Therefore, it has to be general enough to handle such applications. The design shall include various patterns. This model illustrates "Enduring Business Themes" to represent an everlasting core concept (Change). Further, it also uses BOs to abstract semi-tangible elements. Finally, it shall use IOs to represent changeable elements of scenario. This way, when we model different scenarios or applications, we do not have to make any major changes to the system except for the IOs.

Therefore, any project or any person can apply this pattern on the problem called "Change" and get a desired solution.

15.2.2.1 Functional Requirement

Change: A functional requirement would suggest "What the system is supposed to do?" All functional requirements are the integral parts of a pattern. Here, change means the act or instance of making or becoming different. BOs define the functional requirements of the system in stability model [2–6]. Functional requirements also describe the set of inputs, outputs, and the behavior of the system.

Given below are some functional requirements for change

- AnyParty: AnyParty is any organization or a corporate firm that undergoes change or an entity, who initiates change in any of its processes.
- AnyActor: AnyActor is any person, who initiates change or who wants change.
- AnyAspect: AnyAspect in change is something that influences AnyMechanism in the system. There could be aspects that could influence the selection of mechanism to introduce a change. For example, a certain aspect of project management within an IT firm may influence the board members to choose a specific type of change mechanism that is most appropriate for the situation.
- AnyMechanism: AnyMechanism that helps a change takes place. Any type of mechanism could be used to initiate changes within an entity or a system. It could be a simple board meeting in the case of a corporate firm. It could be a lengthy counseling session to introduce positive changes in a person. Global warming could force someone or an institution to set in mechanisms that result in reduced green house gases and dangerous emission.
- AnyForm: It is something on which AnyEntity or AnyEvent can act. Changes could be of any form. Based on the type of form, an entity or an event can act on to initiate changes.
- AnyOutcome: AnyOutcome is a set of changes, desired in most cases, within a certain system or person. An outcome always leads to a result that could affect a certain entity to accept a change.

- AnyImpact: AnyImpact is the impact or effect of changes that takes AnyForm. Impact could be the result of a change and it could be positive or negative.
- AnyReason: AnyReason is a well-defined reason due to which change takes place. There should be a specific reason for introducing change.
- AnyType: AnyType is a type of reason that leads to change. The reason should define the type of reasons that are attributable for any changes.
- AnyEntity: AnyEntity are the things that are part of being changed. It could be anything—a human, an organization or a firm, an office, an institution, or it could even be global weather system.
- AnyEvent: AnyEvent is any event, which is part of evolving, modification, and change. Any event or situation may force someone to introduce changes within a system.

15.2.2.2 Nonfunctional Requirement

Change: Nonfunctional requirements for a system refer to something that is related to the domain and they are enduring. The requirements have to act as per the quality factors of that system. Given below are some of the nonfunctional requirements for *change.*

1. *Acceptable:* Any change in the system or domain, or any living or nonliving thing should be acceptable to all parties concerned. The main goal of any change is to be accepted by the system on which change is performed. The change is never complete, if AnyActor or AnyParty do not accept the changes.
2. *Evolving:* A change should be continuously evolving; it has to develop gradually and lead to a new form. The ultimate goal of change is to evolve from anywhere it occurs and that gives a new form and characteristics.
3. *Timely:* Changes are bound to happen and they take their own time and duration to complete and be accepted. The criteria for change should be time and any system meeting the requirements must be considered changed and evolved.
4. *Renewal:* Any change that occurs in a person or a system should be renewable. Without this attribute, one cannot accept change to take place permanently. In other words, someone should be able to make repeated attempts and continued efforts to introduce a positive change.
5. *Novelty:* A change should be new, modern, novel, and fresh. Novelty denotes something that is refreshing and cajoling. A novel change in a system takes rigorous efforts and dedicated work. For example, shifting office from one location to the other may introduce novelty to the system in the form of new interiors and changed atmosphere.
6. *Transformation:* A change should lead to a permanent transformation. Transformation is a process of evolving and it is slow and gradual. For example, transforming a convict to accept clean and normal societal behavior takes a long time and continued effort. However, when a positive change takes place, the transformation process would be complete leading to the person's changes in social behavior.
7. *Revolution:* All changes that occurred in the past were either positive or negative. However, a great many changes have been revolutionary and trend setting. In other words, change if any takes place, they should be trend setting and revolutionary. For example, changing the highly polluted global climate could be revolutionary and trend setting because of a series of positive effects that it brings to this world. A simple change in an IT process could be revolutionary and ever remembered. For example, changes in hardware and software technology of the past have made the virtual word an exciting place to work.
8. *Innovation:* All changes are innovative and interesting. Without someone being innovative, changes may not succeed nor may they result in a positive change. Innovative changes

in online communicative technologies have resulted in online social media platforms such as FaceBook and Twitter.

9. *Diversity:* All changes result in a marked diversity. In fact, positive changes bring diverse changes to an entity or an institution and the world; essentially speaking, diverse changes themselves are a big change to the system. A positive climate change would lead to a diverse world where everyone, including plants and animals, live in perfect harmony and peace.
10. *Transition:* Sometimes, some changes are transitive in nature before ending in a big change. In other words, they can bring a noticeable transition to the system. A change is both a linear and vertical process, where certain events and minor changes are in a transition phase and an intermediary mode.

15.2.3 Solution

15.2.3.1 Class Diagram

Class Diagram as shown in Figure 15.1.

15.2.3.2 Class Diagram Description

1. One or more AnyParty or AnyActor undergo a change.
2. Execution of change is introduced by using AnyMechanism.
3. There has to be AnyReason for a change to happen.
4. AnyReason defines the type of change to be done.
5. AnyType contains one or more AnyEvent or AnyEntity.
6. AnyEvent or AnyEntity acts on AnyForm to make changes.
7. AnyActor and/or AnyParty look for AnyAspect in the system to change.
8. AnyAspect always influences the change mechanism.
9. AnyForm of change produces AnyOutcome, which can be a positive or negative change.
10. The outcome of change may have AnyImpact.

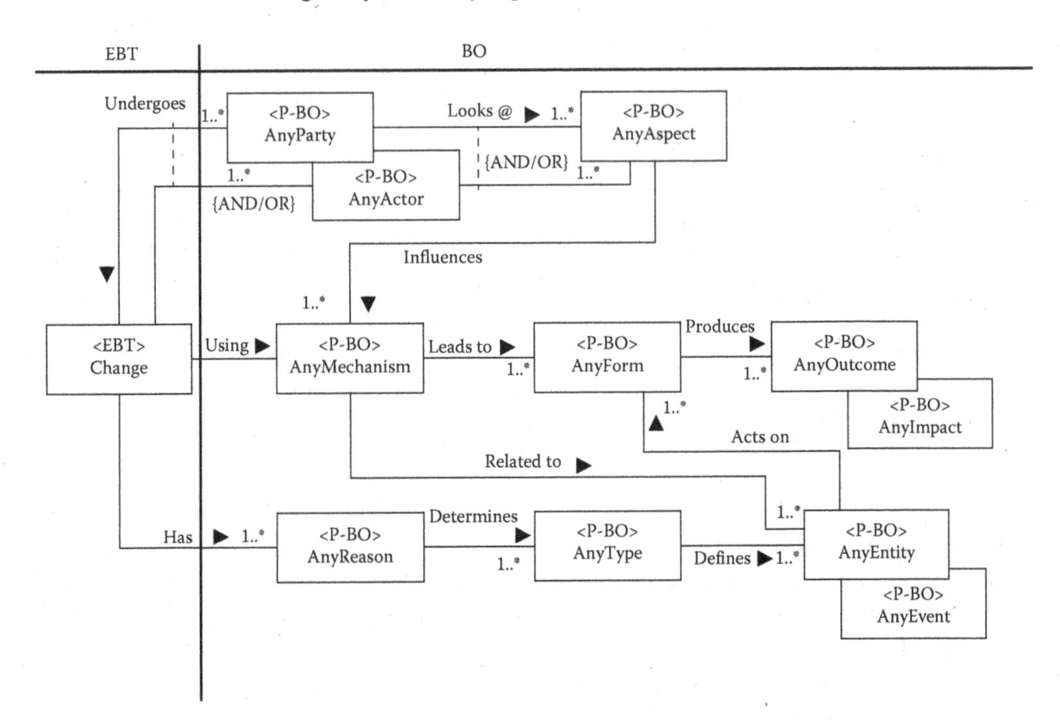

FIGURE 15.1 Change stable analysis pattern.

TABLE 15.1
Five Different Applications of Change Stable Analysis Pattern

EBT	BO	App-1	App-2	App-3	App-4	App-5
Change	AnyAspect	Finance	Business	Career	Study	Behavior
	AnyParty	University	IT Company	202-Class	Project team	Friend
	AnyMechanism	Introducing new rules	Discussion	Canvas	Project discussion	Friend conversation
	AnyEvent	Graduation	Meeting	Quiz evaluation	Project submission	Exams
	AnyForm	Application	Code review	Marks list	Requirement list	Exam review sheet
	AnyReason	Revise tuition fee	Change in application	Change in grades	Updated template	Stress
	AnyOutcome	Less enrollment	Updated code	Grade shifted to A from B	Updated final document	Differences between friends during exams

15.2.3.3 Applicability

In Table 15.1, there are five applications where a "Change Stable Analysis Pattern" is applicable.

15.3 SUMMARY

Stability model [2–5] enables systems to achieve the desired outcomes with greater efficiency. Stable architectures aim to unify common aspects of different systems. Identification of EBTs ensures that architectures are not built for a particular application, but are built in a universal way to facilitate reuse and extend. Stability model [2–5] focuses on the core and essential requirements and it overcomes the limitations of traditional models, which are built around complete requirements for a particular system.

Traditional models are quicker to develop but they are not reusable. For better and efficient software development, we must ensure that we invest time, money, and effort to develop stable models that are adaptable and reusable across many applications and domains. In comparison, stable analysis patterns are definitely the better choice for implementation in an industry that is demanding extensive reusability and customization.

15.4 OPEN RESEARCH ISSUES

1. *e-Change:* Utilize stability model [3–6] or knowledge map [7] methodology as a way for developing e-Change Engine. Building this engine by using traditional development approaches is not an easy exercise, specifically when several factors can undermine their quality success, such as cost, time, and lack of systematic approaches.

REVIEW QUESTIONS

1. Define *change* and list different meanings of this word.
2. Why is the term change special to the context of this chapter and how does it influence the scope of the chapter?
3. List four scenarios where change pattern can be applied.
4. How does change pattern help understand the core of the problem?
5. What are the functional requirements for this chapter?

6. What are nonfunctional requirements for this chapter? List some more requirements that are not covered in the chapter.
7. Identify the EBT for the word *Change* and give reasons for this choice.
8. List some BOs for this term and explain their role.
9. Discuss various contexts where you can apply this pattern.
10. What are some of the anticipated problems that are likely to emerge while designing change pattern.
11. Create three more patterns for change. Use three different scenarios.
12. What are the expected outcomes of creating this pattern?
13. Define some major and minor research issues of this chapter.
14. Can you provide a summary in bullet points?
15. Is this pattern stable and reusable over time? Why and give reasons.
16. Does this pattern give the factor of extendibility to it?
17. Is change pattern domain-, context-, and application-specific? Provide valid reasons, if the answer is yes.
18. Can you differentiate between traditional and stable type of patterns. Highlight some points and provide arguments for each answer.
19. How do BOs interact with each other and act in combination to create a pattern for deployment. Provide a list of BOSs for patterns that you did while answering Question 11.
20. Is this pattern scalable? If so, list some of the reasons.

EXERCISES

1. Change Speed
   a. Explain change speed and select any context (scenario).
   b. Draw a class diagram based on the change pattern to show the application of selected context in a.
   c. Document a detailed and significant use case as shown in Case Study 1.
   d. Create a sequence diagram of the created use case of c.
2. Change Control
   a. Explain change control and select any context (scenario).
   b. Draw a class diagram based on the change pattern to show the application of a selected context in a.
   c. Document a detailed and significant use case as shown in Case Study 1.
   d. Create a sequence diagram of the created use case of c.
3. Change a Subject
   a. Explain change as subject and select any context (scenario).
   b. Draw a class diagram based on the change pattern to show the application of a selected context in a.
   c. Document a detailed and significant use case as shown in Case Study 1.
   d. Create a sequence diagram of the created use case of c.

PROJECTS

Develop the following systems using the change pattern:

1. *Accounting change*: A change in accounting principles, accounting estimates, or the reporting entity. A change in an accounting principle is a change in a method used, such as using a different depreciation method or switching from LIFO to FIFO. An example of an accounting estimate change could be the recalculation of machine's estimated life due to wear and tear. The reporting entity could change due to a merger or a break up of a company.

2. *Change agent*: A role to help members of an organization adapt to organizational change or to create organizational change.
3. *Attitude change*: Where attitudes are associated beliefs and behaviors toward some object [7]. They are not stable, and because of the communication and behavior of other people, are subject to change by social influences, as well as by the individual's motivation to maintain cognitive consistency when cognitive dissonance occurs—when two attitudes or attitude and behavior conflict. Attitudes and attitude objects are functions of affective and cognitive components. It has been suggested that the inter-structural composition of an associative network can be altered by the activation of a single node. Thus, by activating an affective or emotional node, attitude change may be possible, though affective and cognitive components tend to be intertwined [8]. There are three bases for attitude change: compliance, identification, and internalization. These three processes represent the different levels of attitude change [9].
   a. List all the functional requirements and nonfunctional requirements for each area.
   b. List two challenges for each area.
   c. Name five contexts for each area.
   d. Draw the application of the pattern for each context in c.
   e. Select a significant use case per application and describe each one of them with text cases.
   f. Map each use case in e into a sequence diagram.

REFERENCES

1. R. Churchill, *Winston S. Churchill: Young Statesman*, Concord, CA: C & T Publications, 1967, 287–289.
2. M.E. Fayad, and A. Altman, An introduction to software stability, *Communications of the ACM*, 44(9), 2001, 95–98.
3. M.E. Fayad, Accomplishing software stability, *Communications of the ACM*, 45(1), 2002, 95–98.
4. M. E. Fayad, How to deal with software stability, *Communications of ACM*, 45(4), 2002, 109–112.
5. M. E. Fayad, and S. Wu, Merging multiple conventional models into one stable model, *Communications of the ACM*, 45(9), 2002, 102–106.
6. M.E. Fayad, H.A. Sanchez, S.G.K. Hegde, A. Basia, and A. Vakil. *Software Patterns, Knowledge Maps, and Domain Analysis*, Boca Raton, FL: Auerbach Publications, December 2014.
7. W. McGuire, Attitudes and attitude change, in G. Lindzey, and E. Aronson, eds. *Handbook of Social Psychology: Special Fields and Applications*, vol. 2, New York, NJ: Random House, 1985, 233–346.
8. A. Eagly, and Chaiken, S. Attitude strength, attitude structure and resistance to change, in R. Petty and J. Kosnik, eds. *Attitude Strength*, Mahwah, NJ: Erlbaum, 1995, 413–432.
9. H.C. Kolman, Compliance, identification, and internalization: Three processes of attitude change, *Journal of Conflict Resolution*, 2(1), 1938, 51–60.

16 Propaganda Stable Analysis Pattern

> Propaganda, to be effective, must be believed. To be believed, it must be credible. To be credible, it must be true.
>
> **Hubert H. Humphrey [1]**

Propaganda is possibly one of the most commonly used (147th in ranking) English word today. Propaganda looks very sinister and deceiving! The age of propaganda stretches back to World War II when Allied forces, Germans and Italians, carried out a sustained campaign based on propaganda that yielded rich dividends for them. Propaganda is also used today and it will be used tomorrow too by institutions, business firms, and governmental agencies. In essence, propaganda is a kind of communication usually based on highly skewed, biased, and misleading campaign of advertisement and promotion. It is focused on influencing or manipulating the mass attitude of a population toward some causes and reasons that are deceiving in nature. Propaganda could be purely political in nature and could also be strategic.

Propaganda is a highly focused and specific word. Largely, it has one well-defined meaning. Hence, it makes sense to use this word as a basis for creating a stable software pattern for different contexts and scenarios. However, we may need to ascertain that the word, *Propaganda*, is applicable to all contexts that exist within the ambit of this concept. One of the most significant advantages of a pattern built based on the term *Propaganda* is its ruggedness and shelf life. It also provides a lasting and stable solution to promote different applications that are useful for different contexts. One need not spend huge amounts of money, time, and effort as this pattern is highly flexible and easily doable. In addition, this pattern is repeatable and workable in a quick time frame. Lastly, this pattern is also stable and capable of providing a lasting solution to the core problem related to different contexts of propaganda.

16.1 INTRODUCTION

Propaganda is a systematic mode of disseminating information that is impartial, biased, and one-sided. It is used to influence a larger audience to promote a hidden agenda more often by presenting facts, data, and information to produce a hypnotic effect. In other words, propaganda is irrational in its approach and often deceiving in nature. Propaganda is not progressive nor is it well meaning. In essence, the main idea behind using propaganda is to tarnish one's image to self-promote as in the case of the World War I.

There are too many examples of propaganda in human history. It may also occur in a personal capacity when a person involves himself/herself in an operation of propaganda. A politician may try to build a mental image for himself and present it to the outside world. He or she may like to present a great image by disseminating a wrong picture to the public. One of the most common propaganda operations is by firms and businesses that try to spread a message like, "everyone is doing and why can't you" or "I did that successfully and so can you." Politicians may try to build false images of themselves doing common work that all common people usually do in life. Allied countries during World War I spread a fear in the minds of people by showing that they were in a terribly dangerous situation that they were the real solution for all problems. Someone may try to spread wrong information about others by resorting to propaganda whisper talk. Occurrences of propaganda are also very common in offices, institutions, and firms.

Hence, propaganda is a truly focused term that can be elaborated to study and device a stable software pattern—propaganda. This stable pattern also helps us in deciphering the core problems and its pitfalls while using propaganda pattern. As explained earlier, the ultimate goal of propaganda is propaganda itself as it is too specific in its understanding and nature. Hence, the core concept of *Propaganda* remains the same for all contexts where the word is used. This chapter provides a brief glimpse of a stable pattern for *Propaganda* that works as a common template for all domains and contexts where one uses it.

16.2 PATTERN NAME: PROPAGANDA STABLE ANALYSIS PATTERN

In this chapter, we intend to establish a working model for stable analysis pattern for propaganda. Propaganda is taken as the central enduring business theme (EBT). Propaganda is usually carried out for AnyReason, which defines the types of events or entities the propaganda would work on. The central theme of propaganda is duration for which the propaganda goes on forward and AnyDuration defines it, and it can be analyzed/measured using *AnyMeasure*. AnyParty or AnyActor initiates propaganda who later lists a set of AnyCriteria. The latter would influence AnyMechanism used for disseminating propaganda. AnyMechanism works on AnyEntity that resides on AnyMedia, usually right through the propaganda process, which eventually generates some sort of AnyImpact, which could be permanent or short timed.

16.2.1 Context

Propaganda stable analysis pattern can be applied across diverse domains, where the process of propaganda is used. This process should have some measuring metrics and must go on forward for duration of time to create AnyImpact. Given below are some scenarios that would help us understand the Propaganda analysis pattern.

16.2.1.1 Political Propaganda

North Korean military junta (AnyParty) engages in a statewide propaganda to keep the population in control using fear and fright (AnyMechanism). It believes that to keep the population loyal to it (AnyReason), the entire population must be fed a sustained propaganda based on a cult of personality (AnyType). This propaganda carried out eventually reaches every citizen of the state of North Korea (AnyEntity) living in the state (AnyMedia). The ruling party has its own political ideology called Juche (AnyCriteria), which influence how and in what effect the mechanism of dissemination of propaganda takes place. The party measures the effectiveness of propaganda based on the turnout (AnyMeasure) during all state-sponsored events, for example, the Pyongyang Marathon. This process has been carried on since 1953 (AnyDuration) and has made the country a pariah in international community (AnyImpact).

16.2.1.2 Business Manipulation

Amway (AnyParty) is a pyramid scheme that disseminates propaganda to attract more people to work for them (AnyImpact). They believe that they are a multilevel-marketing scheme, which gives opportunities to people to start their own business or make a little cash on the side (AnyReason). Using this reason they select their propaganda material such as pamphlets, recording, anecdotes (AnyType) to manipulate young entrepreneurs and housewives (AnyEntity) living around the area (AnyMedia). They take the prospective employee to business seminars and related events (AnyMechanism) for couple of weeks (AnyDuration) and they measure the interest of the person by calling them (AnyMeasure) over and over again. If the person is interested he is supplied a kit, which contains dead-weight items for which he/she becomes indebted to the company (AnyImpact).

16.2.2 Problem

The main problem that one might face is making the stable analysis pattern for propaganda reusable and applicable across all domains and for all contexts. Current modeling based on traditional techniques only takes care of a couple of contexts/scenarios. We intend to make this model usable for any number of applications.

16.2.2.1 Functional Requirements

Propaganda: Propaganda is taken as the central EBT. Propaganda is usually carried out for AnyReason, which defines the types of events or entities the propaganda would work on.

AnyParty: Any party refers to an individual, a political party, or even any user that initiates the process of propaganda. This party defines rules and/or criteria, which influences the kind of mechanism that will be used for the process of propaganda. AnyParty has AnyCriteria to set a well-sustained propaganda process and AnyParty uses some quantifying measure (AnyMeasure) to determine the effectiveness and reachability of propaganda.

AnyMechanism: AnyMechanism is any technique or process that will be used for manipulating an event or an entity. It is decided by AnyParty based on the criteria and/or roles that the party has in mind. This mechanism must be measurable and be a process going over a duration of time. Propaganda sets in motion AnyMechanism for a predefined duration to have maximum impact (AnyImpact). Anyparty initiates propaganda for some definite reasons (AnyReason).

AnyDuration: Along with AnyMeasure, AnyDuration is the central theme of propaganda, since propaganda has continued over certain duration to make any measure of the impact. The methods used (AnyMechanism) by AnyParty for any amount of time (AnyDuration) will have a definite impact (AnyImpact).

AnyMeasure: Propaganda needs to have a certain kind of measure that can be compared to or analyzed in its own space, to determine the effectiveness and reachability of the propaganda being disseminated. This has to be in consonance with time duration to have any type of influence (AnyImpact).

AnyImpact: AnyMechanism, over time duration will result in some kind of influence or impact. They can be positive or negative depending on the context. For a propaganda to succeed, the method or mechanism (AnyMechanism) used should be effective and the person or entities who want to succeed (AnyParty) will choose the most effective one.

AnyContext: Propaganda usually takes place for a specific reason. The reason determines the type of entities the mechanism would work on. Propaganda process (Propaganda) always depends on some specific reasons (AnyReason).

AnyType: AnyType refers to variations in types of mechanisms that can be employed in the process of propaganda. The reason for which propaganda takes place, names the types of entities or events on which manipulation would be used. It will single out target entities on which propaganda needs a focused effort.

AnyEntity/AnyEvent: AnyEntity or AnyEvent refers to the collective domain on which propaganda will be carried out. AnyEntity contextualizes the types of mechanism and the process itself. One entity or event may reside over multiple media.

AnyMedia: Media refers to anything that has the capacity to carry out propaganda. Entities and events may reside on one media or spread across various media. Most common media platforms for carrying out propaganda like TV, radio, social media, internet, declassified files, secret archives, whisper campaigns, and oration in a public place in front of a large audience have the ability to create instant impact and influence (AnyImpact).

16.2.2.2 Nonfunctional Requirements

Measurable: Propaganda should be measurable. In other words, it should have specific quantifying metrics. The measurability could be in the form of the number of people who are converted as in the case of World War II, or the number of consumers who switch from using one type of product to another.

Effective: Propaganda is about creating a definite impact, however or whatever minute its eventual effect may be. For this reason, propaganda should be effective enough. An effective propaganda may be converting floating voters into assured ones for a political party, or converting the minds of entire population to believe in a specific political ideology. A propaganda process is said to be effective when it achieves its intended goal.

Achievable: AnyKind of manipulation should be achievable, since altering the course of events is critical for measuring the impact of manipulation. Disseminating too many lies that are obvious for a population may not achieve anything for the person who initiates propaganda campaign. A propaganda campaign should have a definite set of goals that are achievable in future.

Repeatable: Once a person or an entity finds success after conducting propaganda, the entire process should be capable of repetition on another set of target. Propaganda campaign carried out after repeated attempts may bring hugely positive results for the party that initiates it, as shown in Figure 16.1.

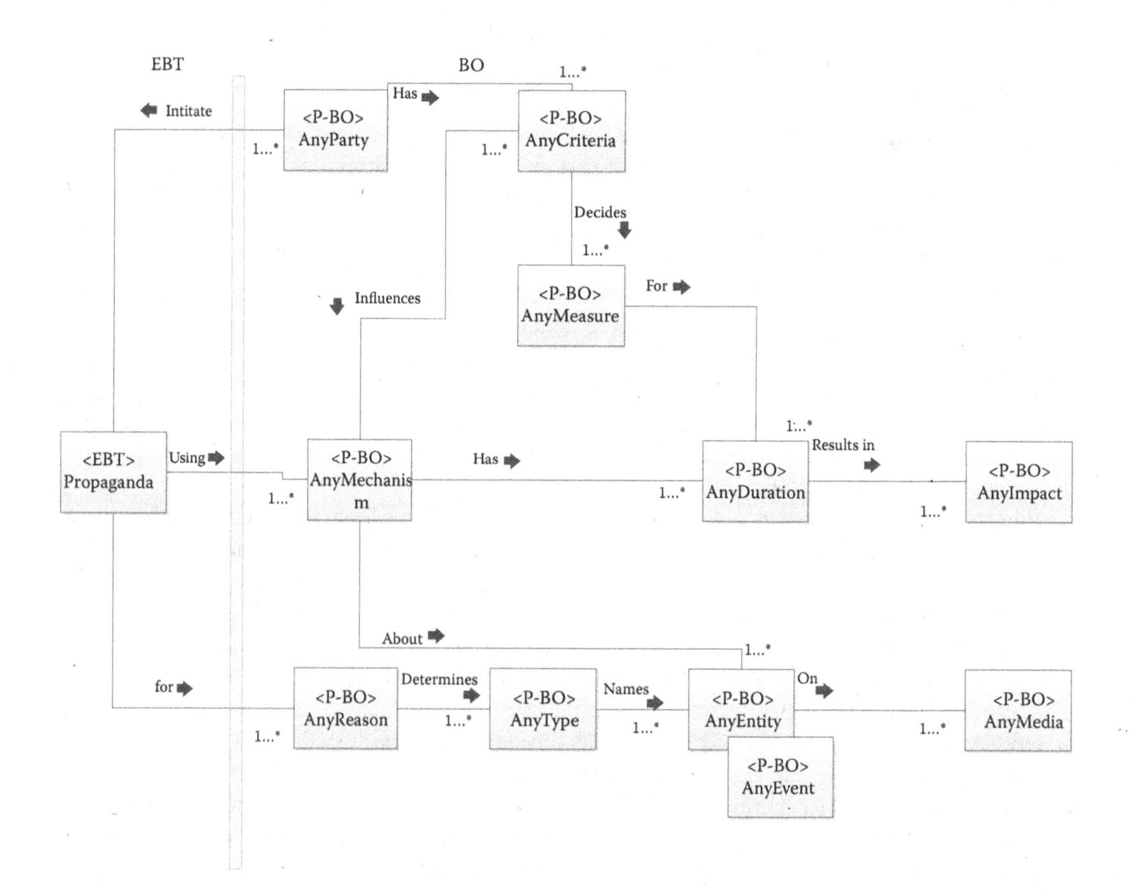

FIGURE 16.1 Propaganda stable analysis pattern.

16.2.3 Solution

16.2.3.1 Class Diagram Description

1. Propaganda should not take place without a valid reason.
2. One or more AnyParty initiates the process of propaganda.
3. AnyParty defines AnyCriteria and/or AnyRules that would be used.
4. AnyCriteria determines AnyType of AnyEntity.
5. Propaganda is facilitated using AnyMechanism.
6. AnyMechanisms are influenced by AnyCriteria and/or AnyRule defined by AnyParty.
7. AnyMeasure has AnyDuration.
8. AnyMechanism results in AnyImpact that can be measured.
9. AnyMechanism is used on or is about AnyEntity/AnyEvent.
10. AnyEntity/AnyEvent reside on AnyMedia.
11. AnyCriteria decided AnyMeasure being used.
12. AnyMechanism goes on for AnyDuration, which is tangible with AnyMeasure.

16.3 SUMMARY

In conclusion, this chapter elaborately analyzed stability model by using propaganda as the EBT. It also compared stability model with a traditional model. Stability models are far more flexible and stable that anyone can reuse it continuously without having to reinvent the wheel repeatedly, thereby saving precious time, money, and effort. A stable pattern built on propaganda is also easy to understand when compared to a traditional pattern because it focuses on the principles of deploying BOs and EBT. By employing them in the stability models, one can build huge numbers of applications by using the same framework. Long-term benefits of using a stability model also provide several benefits like code reusability, scalability, flexibility, and robustness.

16.4 OPEN RESEARCH ISSUES

1. *e-Propaganda:* Utilize stability model [2–6] or knowledge map [7] methodology as a way for developing e-Propaganda Engine. Building this engine by using traditional development approaches is not an easy exercise, specifically when several factors can undermine their quality success, such as cost, time, and lack of systematic approaches.
2. *Propaganda as part of architecture on demand:* Utilize stability model or knowledge map methodology as a way for developing architecture on demand using propaganda and build unlimited architecture patterns that can be applied to many applications within context, such as media propaganda, specific country propaganda, art propaganda, music propaganda, etc. The proposed solution attempts to extract the commonality from all the domains and represent it in such a way that it is applicable to a wide range of contexts without trivializing or generalizing the concepts. Each of the stable architecture patterns (architectures on demand) can be used to provide a generic engine that can be applied and/or extensible to any application.

REVIEW QUESTIONS

1. Why is propaganda a famous English word? List your answers.
2. Why is the term propaganda special to the context of this chapter?
3. List some of the most notable instances of propaganda in the history of humanity.
4. List five scenarios, where you can apply propaganda pattern. Do not use the ones that are highlighted in the chapter.

5. How does propaganda pattern help understand the core of the problem?
6. Is the word propaganda a specific term or is it too general to be used in an array of different contexts.
7. What are the functional requirements for this chapter? Are they the BOs for the pattern?
8. How do you apply nonfunctional requirements for this chapter? List more and explain their relevance.
9. Identify the EBT for the word propaganda.
10. List some BOs for this term and define each one of them in your own words.
11. Discuss some contexts, where you can apply this pattern.
12. What are some of the unexpected problems that are likely to come while designing the pattern on propaganda?
13. Create five more patterns for propaganda. Use three different scenarios.
14. What are the expected outcomes of creating this pattern? Compare them with a traditional pattern design.
15. Define some research issues of this chapter and give reasons for how one can solve them.
16. List a brief summary in bullet points.
17. Is this pattern stable and rugged over time?
18. Is propaganda pattern domain- and application-specific?
19. How do different BOs interact with each other and act in combination to create a pattern for deployment. Write an exhaustive scenario where different BOs play their roles.

EXERCISES

1. *Space propaganda* is about achievements in space science and technology that are used as propaganda. Space propaganda was used during the cold war, and is also used today.
   a. Explain space propaganda and select any context (scenario).
   b. Draw a class diagram based on the propaganda pattern to show the application of selected context in a.
   c. Document a detailed and significant use case as shown in Case Study 1.
   d. Create a sequence diagram of the created use case of c.
2. *Hate Propaganda or Hate speech*, outside the law, is speech that attacks a person or group on the basis of attributes such as gender, ethnic origin, religion, race, disability, or sexual orientation [7].
   a. Explain hate propaganda and select any context (scenario).
   b. Draw a class diagram based on the propaganda pattern to show the application of a selected context in a.
   c. Document a detailed and significant use case as shown in Case Study 1.
   d. Create a sequence diagram of the created use case of c.

PROJECTS

Develop the following systems using the propaganda pattern:

1. *Atrocity propaganda* is a term that refers to the spreading of deliberate fabrications or exaggerations about the crimes committed by an enemy, constituting a form of psychological warfare.
2. A *propaganda film* is a film that involves some form of propaganda. Propaganda films may be packaged in numerous ways, but are most often documentary-style productions or fictional screenplays, that are produced to convince the viewer of a specific political point or influence the opinions or behavior of the viewer, often by providing subjective content that may be deliberately misleading [8,9].

3. *Corporate propaganda* refers to propagandist claims made by a corporation (or corporations), for the purpose of manipulating market opinion with regard to that corporation, and its activities. Just as the use of these products and services can provide pluses that outweigh the minuses to society and individuals, their advocacy may function more positively than negatively. Common euphemisms for corporate propaganda are advertising and public relations.
4. *Self-propaganda* is a form of propaganda and indoctrination performed by an individual or a group on oneself.
   a. List all the functional requirements and nonfunctional requirements for each area.
   b. List two challenges for each area.
   c. Name five contexts for each area.
   d. Draw the application of the pattern for each context in c.
   e. Select a significant use case per application and describe each one of them with text cases.
   f. Map each use case in e into a sequence diagram.

REFERENCES

1. C.L. Garrettson, *Hubert H. Humphrey: The Politics of Joy*, New Brunswick, NJ: Transaction Publishers, 1993.
2. M.E. Fayad, and A. Altman, An introduction to software stability, *Communications of the ACM*, 44(9), 2001, 95–98.
3. M.E. Fayad, Accomplishing software stability, *Communications of the ACM*, 45(1), 2002, 95–98.
4. M.E. Fayad, How to deal with software stability, *Communications of ACM*, 45(4), 2002, 109–112.
5. M.E. Fayad, and S. Wu. Merging multiple conventional models into one stable model, *Communications of the ACM*, 45(9), 2002, 102–106.
6. M.E. Fayad, H.A. Sanchez, S.G.K. Hegde, A. Basia, and A. Vakil. *Software Patterns, Knowledge Maps, and Domain Analysis*, Boca Raton, FL: Auerbach Publications, 2014.
7. J.T. Nockleby, Hate speech, in *Encyclopedia of the American Constitution*, 2nd ed., L.W. Levy and K.L. Karst, ed., Detroit: Macmillan Reference US, vol. 3, pp. 1277–1279. Cited in "Library 2.0 and the Problem of Hate Speech," by Margaret Brown-Sica and J. Beall, *Electronic Journal of Academic and Special Librarianship*, 9(2), Summer 2008.
8. T. Bennett, The celluloid war: State and studio in Anglo-American propaganda film-making, 1939–1941, *The International History Review 24.1*, 64(34), 2002.
9. J. Combs, *Film Propaganda and American Politics*, New York, NY: Garland Publishing, 1994. p. 32.

17 Fairness Stable Analysis Pattern

Fairness is what justice really is.

Potter Stewart [1]

A most often used English word, fairness, denotes a wide spectrum of use in different domains and circumstances. Used at least in ten different contexts, fairness could be pinned down to four important types of applications in English language: (a) having a disposition that is free of favoritism or bias, (b) equitable to all parties concerned, (c) being in tandem with relative merit or significance, and (d) consistent with rules and logic or ethics. The inference that one can make from these different interpretations of the word suggests that the word fairness denotes someone who is unbiased, consistent, reasonable, fair-minded, impartial, and nondiscriminatory. Nonetheless, fairness also represents an array of other meanings spread mostly in the domain of skin color, beauty, law, personal integrity, and others. In other words, fairness is a diversely used word that stretches across many domains, contexts, and perspectives.

It is quite challenging to use this word in the world of software pattern making. Still, developers could use it as a base to create a stable software pattern called fairness. The larger purpose will be served when this pattern is applicable to all situations where the word fairness is used. Invariably, this pattern will ensure longer shelf life, extendibility, and stability apart from its universal applicability.

17.1 INTRODUCTION

Fairness, a widely used English word, connotes different meanings and understandings under diverse circumstances. A person is said to be fair in his/her attitude toward others, when he/she is unbiased and impartial. When one is consistent with set rules and logic, he/she is fair while in legal jargon, someone will play fair in a proper and legal manner. A girl looks very fair when she has light-colored skin that is free of blemishes and spots. When someone's hair is light in color, she is said to be fair-haired. However, the ultimate meaning of the word is all about someone being impartial, non-favoring, equitable, and unbiased.

In connection with this chapter, fairness could be extended to create a well-designed and stable software pattern called fairness. In the process, pattern developers will succeed in solving existing core problems with the patterns created by using a traditional method. The core concept of fairness is similar for all domains where it is used and this unique attribute makes it an interesting pattern that is free of all existing pitfalls and lacunae. This chapter elaborates on the possibilities of creating a stable software pattern for fairness that also works as a common template for use in all domains and situations where the word fairness is extensively used and worked.

17.2 PATTERN NAME: FAIRNESS STABLE ANALYSIS PATTERN

The pattern name that is taken for study is fairness stable analysis pattern. It is a stable analysis pattern because fairness is EBT. Fairness is treating everyone equally without discrimination and giving equal opportunity and rights. AnyParty or AnyActor, especially humans, can show fairness of be fair to others through AnyVerdict that can be provided by one or more of AnyVerdict. Fairness can arise only through some contexts or situations (AnyContext). AnyActor can show a range of fairness attributes depending on the situation (AnyType) and it names one or more numbers of AnyEntity or AnyEvent. AnyActor or AnyParty will display some sort of evidence of showing fairness to others (AnyEvidence). A range of tools, methods, or mechanisms is used by AnyActor to produce some instances of outcome and results (AnyOutcome). In turn, these mechanisms (AnyMechanism) are influenced in the process by some evidence that shows fairness being displayed (AnyEvidence).

17.2.1 Context

Fairness is the quality that can be used to judge a scenario by providing a decision that is discrimination-free and that abides by all rules and standards. Fairness is also a standard for treating everyone equally without any discrimination. Fairness stable analysis pattern can be used across diverse situations, where fairness is displayed by an entity, which is usually a human. This process should have some quantifying metrics and should go on to create fairness pattern. The impact or result of fairness on other entities could be love, affection, consideration, sympathy, empathy, impartiality, bias-free, equality, and justice. Listed below are some of the scenarios that would help us understand the fairness pattern analysis.

17.2.1.1 Sportsmanship

Sportsmanship is an aspiration that a sport or activity will be enjoyed for its own sake, with proper consideration for fairness, by following all the rules of the game with ethics, respect, and a sense of fellowship with one's competitors. Here is an example of fairness in a match of football that is played by many countries in the world. Two teams are currently playing a match. The match is played smoothly without any problem, until a bad referee call (AnyContext) over a goal (AnyEvent), results in fierce argument between two captains (AnyParty/AnyActor). To check whether the call was made correctly or not, the organizers of the game decides to replay the entire sequence of the referee call by playing a video playback (AnyMechanism). The playback suggests that a bad call was made in favor of the team B (AnyEvidence). In addition, an additional proof that the call was made in haste is also found out (AnyEntity). In the end, the evidence of truth prevails over the captain of team A, and he decides that the team B is the actual winner of the team and agrees to forego and lose the game (AnyOutcome). A goal in favor of the team B is announced (AnyVerdict) and eventually the captain of the team A displays fairness, magnanimity, and humility of accepting defeat and the outcome in the spirit of the game.

17.2.1.2 Legal Judgment

In a court of law, both the judge (AnyActor) and juries (AnyParty) are hearing plea of not guilty by a murder accused under trial. The current argument on the feasibility of awarding death to the under trial (AnyContext) over a purported accusation of heinous murder (AnyEvent) has too many loopholes and weaknesses that could be exploited by the state (AnyParty). The previous judgment of death pronounced by a lower court (AnyOutcome) as one-sided and was not fair, which was a fact that was put forward by many legal luminaries. After a round of fierce arguments and counterarguments by the defense and state, the higher court decides to hear the case again on a priority basis and pronounces a new set of rules to be followed by all parties in the case concerned (AnyMechanism). The defense argues the case and provides a set of solid evidence in front of the chair that may acquit the accused (AnyEvidence). Furthermore, the defense also supplies enough proof to show that the earlier judgment was made in haste (AnyEntity). As a result, the juries accept all evidences forwarded by the defense and eventually decide that the earlier judgment was not fair (AnyOutcome) and acquits the accused (AnyVerdict).

17.2.2 Problem

Fairness can have a variety of applications, ranging from multiple domains to some specific contexts. Therefore, a working solution to fairness stable analysis pattern must be general enough to handle these general applications. The design will include EBT as fairness. It is essential that the solution of fairness stable design pattern be flexible to represent different scenarios of fairness. There is a need to design and develop a pattern for "Fairness" that captures the core concept of fairness design pattern and which adheres to the context in which can be used for general applications. To manage a diverse set of scenarios, the fairness stable design pattern will use EBTs to represent the core concepts, and employ BOs to abstract the externally stable, but internally adaptable elements. Finally, it will use IOs to represent changeable elements of the scenario. The IOs can be replaced and changed according to the applicability of a given scenario.

17.2.2.1 Functional Requirements

Functional requirements define "What the system is supposed to do?" These are the integral parts of the pattern. BOs set out the functional requirements of the system in a stability model [2–5]. Functional requirements also describe the set of inputs, outputs, and the behavior of the system.

1. *Fairness*: Fairness is the quality that can be used to judge a scenario, by providing a decision that is discrimination-free abiding by all rules and standards. Fairness is also a standard for treating everyone equally without any discrimination.
2. *AnyVerdict*: A decision on an issue in a civil or criminal case or an inquest of all concerned. AnyOutcome is a product of AnyVerdict, while AnyVerdict uses the services of different tools and techniques (AnyMechanism) that help to announce any verdict.
3. *AnyParty/AnyActor*: The representation of a group of any living (human being or creature) or nonliving things (hardware and software). AnyActor is the main user of the system. Actor has four types: hardware, software, people, and creature. They represent an organization or group. AnyParty/AnyActor presents evidence (AnyEvidence) by using a suitable mechanism (AnyMechanism) to process a favorable result (AnyOutcome).
4. *AnyMechanism*: An established process by which something takes place or is brought about. AnyMechanism uses a set of tools and procedures to pronounce a verdict (AnyVerdict), which eventually classified as a result (AnyOutcome).
5. *AnyType*: A category of anything having common characteristics. In a pattern document diagram, a context (AnyContext) determines the type of common attributes (AnyType).
6. *AnyEntity*: The direction being imposed on a thing with distinct and independent existence; is an entity.
7. *AnyEvent*: The scenario or situation, where something occurs that leads to a demand for fair outcome. In a football match, a bad call may be termed as an event. Similarly, an unfair trial may result in a false announcement of judgment against a nonguilty.
8. *AnyContext*: The context brings in the related constraints and influences the course of action.
9. *AnyOutcome*: The outcome represents the way a thing turns out or a consequence.
10. *AnyEvidence*: The available body of fact or information that indicates whether a belief or proposition is true or valid.

17.2.2.2 Nonfunctional Requirements

Nonfunctional requirements define "How a system is supposed to be?" They describe the criteria to gauge the operation of the system externally rather than focusing on its internal structure or behavior. They are the quality factors of the system.

- *Absolute:* Every process should be executed fairly by treating all with impartiality and equality by conforming to established rules and standards [2] wherein decisions should be based on objective criteria, rather than on bias, prejudice, or preferring benefit to one person over another for improper reasons [3].

- *Valid*: Fairness should be executed in compliance with rules and law, thus making it legally bound and accepted. Nothing should be done against laws and regulations. Fairness should be reasonable and well founded.
- *Reliable*: Fairness should be consistently good in quality or performance, and it should be justifiable, understandable, and acceptable. It should also be able to be trusted too. It should be able to perform its functions abiding by rules and for a specified period.
- *Effective*: Fairness should produce the desired output or intended result thereby increasing the chances of success or effectiveness. It should be constructive tending to serve a useful purpose.

17.2.3 Solution

17.2.3.1 Class Diagram

Class diagram as shown in Figure 17.1.

17.2.3.2 Class Diagram Description

1. One or more AnyParty or AnyActor provides fairness through AnyVerdict.
2. Fairness can be furnished through one or more AnyVerdict.
3. Fairness must be within one or more AnyContext.
4. AnyContext determines one or more AnyType.
5. AnyType names one or more AnyEntity and AnyEvent.
6. AnyParty uses one or more AnyEvidence.
7. AnyVerdict uses one or more AnyMechanism.
8. AnyMechanism produces one or more AnyOutcome.
9. AnyVerdict is about of one or more AnyEntity.
10. AnyMechanism is influenced by one or more AnyEvidence.

17.3 SUMMARY

Stability model offers better stability and scalability, since it provides a holistic way of modeling, because it can be applied to every application irrespective of the domain or context. While modeling a system in traditional approach, the system becomes more specific for a particular application for

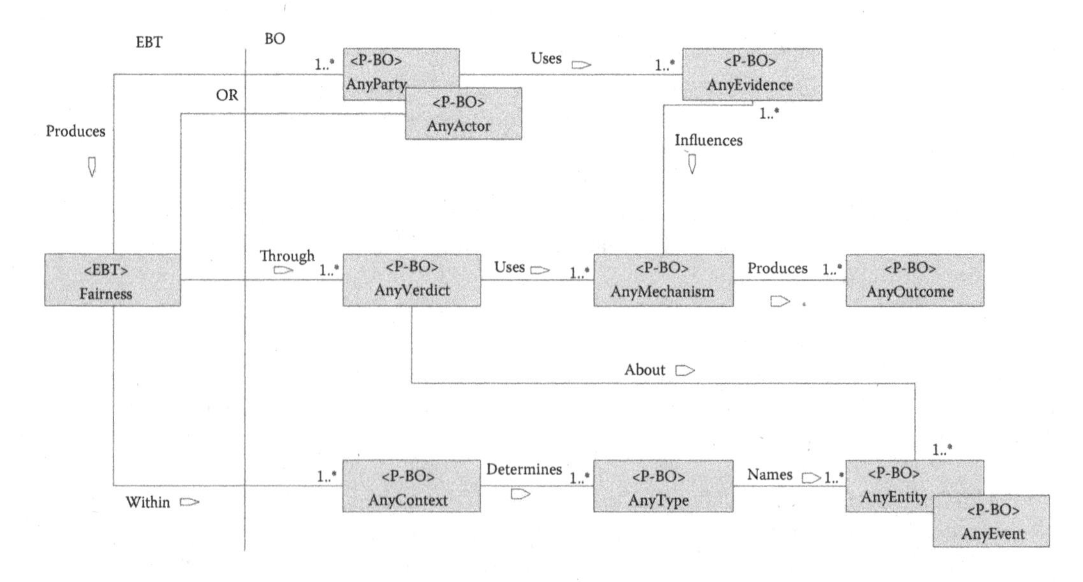

FIGURE 17.1 Fairness stable analysis pattern.

which it is being modeled. Nevertheless, stability model focuses on the ultimate goal of the application and it models the system accordingly. This makes stability model more flexible and maintainable. Stability model also points out different flaws in traditional modeling and hence it can build better software models with its stable patterns.

17.4 OPEN RESEARCH ISSUES

1. *e-Fairness:* Utilize stability model [2–5] or knowledge map [6] methodology as a way for developing e-Fairness engine. Building this engine by using traditional development approaches is not an easy exercise, specifically when several factors can undermine their quality success, such as cost, time, and lack of systematic approaches.

REVIEW QUESTIONS

1. What is fairness? Can you list different meanings of this word?
2. Is the word fairness the most appropriate candidate for creating a stable pattern? If so, give reasons.
3. What is the main business theme (EBT) for fairness?
4. Can you highlight different applications of the word fairness?
5. Could fairness pattern be applied to all domains?
6. Identify and list different BOs for this pattern.
7. Do they work as a team to make the pattern stable and robust?
8. List four, where you can design a pattern for fairness.
9. Is it possible to write a class diagram to represent each one of these patterns?
10. Fairness is a specific word. Do you agree? If yes, give reasons.
11. Can you list some of the identified nonfunctional requirements for this pattern?
12. Can you highlight three important benefits of designing this pattern?
13. List five main disadvantages of this pattern, if any.
14. Compare this stable pattern with the ones that are designed by using a traditional method.
15. What are some of the industrial applications of this pattern? Describe them briefly.
16. What are some of the critical issues that you can raise out of this chapter?
17. Is fairness pattern stable and application specific?
18. Is it possible to extend it to different applications and domains?
19. Is it reusable and extendible?
20. BOs make this pattern stable and extendible. Do you agree?

EXERCISE

1. Distributive fairness.
   a. Explain distributive and select any context (scenario).
   b. Draw a class diagram based on the fairness pattern to show the application of selected context in a.
   c. Document a detailed and significant use case as shown in Case Study 1.
   d. Create a sequence diagram of the created use case of c.

PROJECTS

Develop the following systems using the fairness pattern:

1. Tax fairness: A tax platform based on an ideal that aims to create a system of taxation that is fair, clear, and equivalent for all taxpayers. Overall, tax fairness looks to limit the amount of tax legislation and rules that benefit one segment of the tax-paying population over another.

2. Political fairness or political egalitarianism: Is where members of a society are of equal standing in terms of political power or influence. It is a founding principle of various forms of democracy. It was an idea, which was supported by Thomas Jefferson. It is a concept similar to moral reciprocity and legal equality.
   a. List all the functional requirements and nonfunctional requirements for each area.
   b. List two challenges for each area.
   c. Name five contexts for each area.
   d. Draw the application of the pattern for each context in c.
   e. Select a significant use case per application and describe each one of them with text cases.
   f. Map each use case in e into a sequence diagram.

REFERENCES

1. L. Friedman, *The Justices of the United States Supreme Court: Their Lives and Major Opinions*, vol. V, New York, NY: Chelsea House Publishers, 1978, 291–292.
2. M.E. Fayad, and A. Altman, An introduction to software stability, *Communications of the ACM*, 44(9), 2001, 95–98.
3. M.E. Fayad, Accomplishing software stability, *Communications of the ACM*, 45(1), 2002, 95–98.
4. M.E. Fayad, How to deal with software stability, *Communications of ACM*, 45(4), 2002, 109–112.
5. M.E. Fayad, and S. Wu, Merging multiple conventional models into one stable model. *Communications of the ACM* 45(9), 2002, 102–106.
6. M.E. Fayad, H.A. Sanchez, S.G.K. Hegde, A. Basia, and A. Vakil, *Software Patterns, Knowledge Maps, and Domain Analysis*, Boca Raton, FL: Auerbach Publications, 2014.

18 Anxiety Stable Analysis Pattern

Anxiety, it just stops your life.

Amanda Seyfried

Anxiety is a well-known English word that is used almost every day by millions of people. Originating from both French and Latin, it means *anxiété* or *anxietas*, respectively. Used as a form of noun, the term anxiety denotes something that is psychological and mental in nature. In its common usage, anxiety means a feeling of worry, tension, fickle mindedness, nervousness, and unease about something with no definite outcome. Anxiety has many meanings and definitions most of which work differently according to contexts in which they are used. Anxiety is used entirely differently in the context of psychiatry while in a personal sense its meaning denotes a short burst of fear that might disappear very quickly.

Whatever the case, anxiety is a typical English word that is used commonly depending on any given situation and contexts. However, anxiety still relates to one word that is common to all forms of usage—Anxiety. This aspect makes it a challenge to use the word in the domain of software pattern making. However, it could be used as a solid platform to create a stable software pattern for anxiety. Before creating a pattern for the word, software pattern developers should ascertain and guarantee that the word is applicable to all contexts that occur within the domain where anxiety is used. Stable patterns designed and composed based on a stable technique ensures longer shelf life and extendibility, apart from its robustness. This pattern is also extremely useful in the sense that it can be applied to promote different applications that work under any type of situation. Better flexibility and doable nature makes it much more desiring to improve it further to reflect the demand for a stable pattern.

18.1 INTRODUCTION

The word anxiety is a multifaceted word that is used in different contexts and domains such as personal relationship, psychology, psychiatry, sports, medicine, and other fields where one wants to indicate panic and fear. Some words that rhyme with anxiety are contrariety, dubiety, impiety, impropriety, inebriety, notoriety, piety, satiety, sobriety, ubiety, and variety. In the domain of psychiatry, anxiety means a nervous disorder that is marked by excessive uneasiness and apprehension specifically with a compulsive form of behavior and panic attacks. In other words, anxiety means a mental affliction that needs immediate medical treatment. In a personal sense, anxiety is a strong desire and concern to do something or for something to occur in future. An employee's undue anxiety to please the superior may backfire at times or it might even please eventually resulting in a salary hike. Just before an athlete prepares to run for a sprint race, he/she may develop attacks of panic and anxiety. Similarly, a student who is facing an imminent exam may experience cold feet and attacks of anxiety.

Hence, anxiety could be elaborated and well defined to create a stable software pattern—Anxiety. This stable pattern will also solve existing core problems and its pitfalls, while using it. As highlighted earlier, the ultimate goal of anxiety is anxiety itself, as it is too specific in its understanding. Hence, the core concept of *anxiety* remains similar for all contexts where the word is used extensively. This chapter provides details on a stable pattern for *anxiety* that works as a common template for all domains and contexts where the word anxiety is used.

18.2 NAME: ANXIETY STABLE ANALYSIS PATTERN

In this chapter, we aim to create a working model for stable analysis pattern for anxiety. Anxiety is taken as the central enduring business theme (EBT).

Both humans (AnyParty) and animals (AnyParty) display bouts of anxiety. The party affected by anxiety shows evidence (AnyEvidence) of being anxious. Evidence influences the mechanism of anxiety. Anxiety lasts for a specific time (AnyDuration) and has consequences (AnyConsequence). Anxiety is caused for a specific reason (AnyReason) and can be of many types (AnyType). Mechanism is about an event (AnyEvent) or (AnyEntity). An event or entity is recorded in any media (AnyMedia).

18.2.1 Context

Anxiety stable analysis pattern can be used across diverse domains, where anxiety is encountered by an entity, which could be an animal or human. This process should have some measuring metrics and must go on forward for a duration of time to create AnyImpact. The impact of anxiety could be mental affliction, depression, cold sweat, fear, and panic. Given below are some scenarios that would help us understand the AnyPropaganda pattern analysis.

18.2.1.1 Scenario 1: Work Place Anxiety

An employee (AnyParty) is affected by anxiety. The employee shows evidence of anxiety like nausea (AnyEvidence). Stress (AnyReason) is the major reason for anxiety. Anxiety arises by using a mechanism (AnyMechanism) that triggers it. Anxiety lasts for certain duration (AnyDuration). An anxious person does things like over-reacting (AnyConsequence). The reason for anxiety determines its type (AnyType). The mechanism of anxiety is about the occurrence of an event like meeting deadlines (AnyEvent). The scenario of anxiety happens within the environment of office (AnyMedia).

18.2.1.2 Scenario 2: Anxious Wait for Test Results

A student (AnyParty) is affected by anxiety about impending announcement of test results (AnyEvent). A student is anxious as he/she is very eager (AnyReason) to know the score. The student shows anxiety by being nervous (AnyEvidence). The student expresses anxiety by using a mechanism like fidgeting or pacing forward and backward (AnyMechanism). Anxiety lasts for a certain time till the results are announced (AnyDuration) and results in not paying attention to any other task (AnyConsequence). The reason for anxiety determines the type (AnyType). The mechanism of anxiety is about knowing the results of the test with the classroom (AnyMedia).

18.2.2 Problem

The main problem is making the stable analysis pattern for anxiety reusable and applicable across all domains and for all contexts. Current modeling based on traditional techniques only takes care of one of a couple of contexts/scenarios. In addition, it also consumes more money and additional effort to create a pattern that could result in future problems. We are intending to make this model usable for any number of applications. Additionally, the stable analysis diagram should be such that it applies to all contexts and situations of anxiety without distorting the fundamental structure of the anxiety concept.

18.2.2.1 Functional Requirements

Anxiety: Anxiety is taken as the central Enduring Business Theme (EBT).

AnyActor/AnyEntity: AnyActor refers to any person or animal that is capable of being anxious. The actor explicitly shows evidence of being anxious. AnyActor is anxious in various situations and under different contexts. AnyActor/AnyEntity shows evidences of anxiety in some form or other (AnyEvidence).

AnyEvidence: AnyEvidence is an indication that AnyActor is anxious. The anxious actor shows signs that influence the mechanism of anxiety. Evidence for anxiety is manifested in the form of some specific symptoms and signs.

AnyMechanism: AnyMechanism is a method in which the actor expresses anxiety. The mechanism is about AnyEvent that the Actor is anxious about. The actor could be under a panic attack, or in the state of mental symptoms.

AnyDuration: Duration refers to a time. Anxiety in AnyParty lasts for a certain duration of time (AnyDuration). After this duration, the party can be anxious again. For some people, duration might stretch for a lengthy time (in the case of people who are inflicted by mental conditions) or it could be short lived as in the case of a student who is taking exams. Longer the time duration, deeper will be the consequence (AnyConsequence).

AnyConsequence: In the case of anxiety, as an effect of anxiety, there is always a consequence (AnyConsequence). The anxiety in AnyParty is bound to produce some consequence. Consequences could be in the form of depression and schizophrenia in mental patients, while students may fail in their examinations due to excessive anxiety syndrome.

AnyReason: Anxiety occurs for a reason. AnyActor can only be anxious for a specific reason (AnyReason). A reason explains what made AnyActor show anxiety. AnyActor cannot be anxious without a valid reason. Reasons for anxiety could range from fear of failure as in the case of students and athletes. It could also be in the form of panic attack experienced by an employee when he is under constant stress and pressure of work.

AnyType/AnyEntity: Anxiety can be of various types. AnyReason determines the type (AnyType) of anxiety. The reason for anxiety names the type of the anxiety. AnyType names AnyEntity/AnyEvent and is influenced by AnyMedia that is the environment where the anxiety is felt and experienced. An entity is the domain in which anxiety occurs. The entity uses AnyMechanism.

AnyEvent: AnyEvent refers to the events that make the actor show anxiety by using AnyMechanism. AnyEvent occurs in an environment that we refer to as AnyMedia.

AnyMedia: AnyMedia refers to an environment, in which the actor is anxious and shows evidence of being anxious.

18.2.2.2 Nonfunctional Requirements

Motivation: Anxiety should act as a motivating factor to the actor. Because of being anxious, the actor should develop the desire to do something positive. In fact, anxiety comes from deep within the heart and motivating factors could range from simple fear to extreme stress and pressure. Without internal motivation, anxiety may never appear and act as an external stimulus to create some consequences.

Urgency: Anxiety should make the actor do tasks quickly. Anxiety should lead the actor to think or perform tasks quickly. The factor of urgency is inbuilt with the motivating factors. For example, a person who is suffering from panic syndrome and extreme anxiety could take decisions faster than one can imagine. In patients with mental syndrome, this could lead to suicidal syndrome. In the case of students and athletes who are appearing for some events, such as examinations and sprint, the urgency to finish off the events quickly and emerge as victorious may develop.

Alertness: Anxiety should create alertness in the actor. Anxiety about the event or entity should ensure that the actor is constantly aware and alert of any changes. Mental alertness is common among all actors who experience anxiety. Mentally challenged people are extremely alert, while students show undue haste in writing their final examinations. In other words, anxious people are mentally alert too.

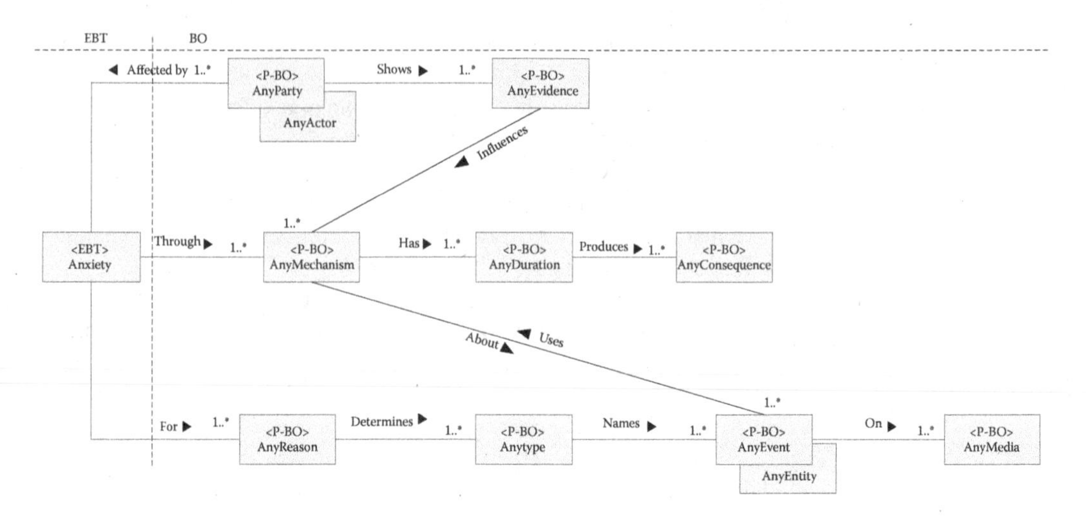

FIGURE 18.1 Anxiety stable analysis pattern.

Apprehension: People, who experience anxiety, are apprehensive and worried about what will happen in the immediate future. Apprehension is the expectation of adversity or misfortune, misery or it is the suspicion or fear of future trouble or evil. Apprehension is more related to the factor of "understanding" and it is more specific and just a form of anxiety. In other words, a fine thin line differentiates anxiety and apprehension.

18.2.3 Solution (Figure 18.1)

18.2.3.1 Class Diagram Description

1. One or more AnyParty or AnyActor are affected by anxiety
2. AnyParty or AnyActor show one or more AnyEvidence of anxiety
3. Anxiety is expressed through one or more AnyMechanism
4. AnyEvidence influences one or more AnyMechanism
5. AnyMechanism has one or more duration
6. AnyDuration of AnyMechanism produces one or more AnyConsequence
7. Anxiety occurs for one or more AnyReason
8. AnyReason determines one or more AnyType of anxiety
9. AnyType name one or more AnyEvent or AnyEntity
10. AnyEvent or AnyEntity uses or more AnyMechanism
11. AnyEvent or AnyEntity occurs in one or more AnyMedia

18.3 SUMMARY

Anxiety stable pattern as elaborated in this chapter, manages several intricate issues in one attempt. In many applications of anxiety, the main problem that pattern makers face is its generality and universality. In other words, anxiety denotes different meanings under different scenarios that make any pattern making very challenging. However, deciphering the core meaning of anxiety makes it easy for developers to find out the main business theme—Anxiety. This chapter has provided a brief summary of how anxiety could be used to create a pattern that is stable, robust, extendible, repeatable, and applicable while its wide applicability makes it a desirous solution to pattern makers.

18.4 OPEN RESEARCH ISSUES

1. *e-Anxiety:* Utilize the stability model [1,2,3,4] or knowledge map [5] methodology as a way for developing e-Anxiety Engine. Building this engine by using traditional development approaches is not an easy exercise, specifically when several factors can undermine their quality success, such as cost, time, and lack of systematic approaches.
2. *Anxiety as part of Architecture on Demand:* Utilize the stability model [1,2,3,4] or knowledge map [5] methodology as a way for developing architecture on demand using anxiety and build unlimited architecture patterns that can be applied to many applications within context, such as: Society Anxiety, Performance Anxiety, Trail Anxiety, Psychological Anxiety, etc. The proposed solution attempts to extract the commonality from all the domains and represent it in such a way that it is applicable to a wide range of contexts without trivializing or generalizing the concepts. Each of the stable architecture patterns (Architectures on Demand) can be used to provide a generic engine that can be applied and/or extensible to any application.

REVIEW QUESTIONS

1. Define the real meaning of anxiety. What are the different meanings of this word?
2. Is the word anxiety a candidate for creating a stable pattern? If so, give reasons.
3. What is the main business theme (EBT) for anxiety?
4. Can you list different applications of the word anxiety?
5. Do you agree to the suggestion that anxiety pattern is applicable to all domains where it plays a major role?
6. Identify different BOs for this pattern. Provide a brief explanation for each one of them.
7. Do these BOs work in tandem to make the pattern stable?
8. List five or six areas, where you can design a pattern for anxiety.
9. Can you write a class diagram to represent each one of these pattern.
10. Do you feel that functional requirements identified for this pattern are important to separate them from the EBT?
11. What are some of the identified nonfunctional requirements for this pattern?
12. Can you list six important benefits of this pattern?
13. Can you list five pitfalls of this pattern, if any?
14. Can you compare this stable pattern with the one that is designed by using a traditional method?
15. What are some of the industrial applications of this pattern? Consider the practical aspects of usage.
16. What are some of the issues that you can make out from this chapter?
17. Is anxiety pattern stable and application specific?
18. Can you extend it to different applications?
19. Is this pattern reusable?

Anxiety can give rise to different consequences. Is it true? If so, write down entire sequence of how the latter arise. Hint: Refer to the BOs of this pattern.

EXERCISES

1. Anxiety Control
   a. Explain anxiety control and select any context (scenario).
   b. Draw a class diagram based on the anxiety pattern to show the application of selected context in a.

   c. Document a detailed and significant use case.
   d. Create a sequence diagram of the created use case of c.
2. Glossophobia or speech anxiety is the fear of public speaking or of speaking in general [6].
   a. Explain Analysis in Business and select any context (scenario) of analysis in business.
   b. Draw a class diagram based on the Analysis pattern to show the application of a selected context in a.
   c. Document a detailed and significant use case.
   d. Create a sequence diagram of the created use case of c.

PROJECTS

Develop the following systems using the anxiety pattern:

1. *Panic attacks* are periods of intense fear or apprehension of sudden onset accompanied by at least four or more bodily or cognitive symptoms (such as heart palpitations, dizziness, shortness of breath, or feelings of unreality) of variable duration from minutes to hours [7].
2. *Dental fear* (also called *dental phobia, odontophobia, dentophobia, dentist phobia*, and *dental anxiety*) is the fear of dentistry and of receiving dental care. However, it has been suggested that use of the term *dental phobia* should not be used for people who do not feel that their fears are excessive or unreasonable, and instead resemble individuals with post-traumatic stress disorder, caused by previous traumatic dental experiences [8].
3. *Hypochondriasis*, also known as *hypochondria, health anxiety* or *illness anxiety disorder*, refers to worry about having a serious illness. This debilitating condition is the result of an inaccurate perception of the condition of body or mind despite the absence of an actual medical condition [9].
   a. List all the functional requirements and non-functional requirements for each area.
   b. List two challenges for each area.
   c. Name five context for each area.
   d. Draw the application of the pattern for each context in c.
   e. Select a significant use case per application and describe each one of them with text cases.
   f. Map each use case in e into a sequence diagram.

REFERENCES

1. M.E. Fayad, and A. Altman, An introduction to software stability. *Communications of the ACM*, 44(9), 2001, 95–98.
2. M.E. Fayad, Accomplishing software stability. *Communications of the ACM*, 45(1), 2002, 95–98.
3. M.E. Fayad, How to deal with software stability. *Communications of ACM*, 45(4), 2002, 109–112.
4. M.E. Fayad, and S. Wu, Merging multiple conventional models into one stable model. *Communications of the ACM*, 45(9), 2002.
5. M.E. Fayad, H.A. Sanchez, S.G..K. Hegde, A. Basia, and A. Vakil, *Software Patterns, Knowledge Maps, and Domain Analysis*. Boca Raton, FL: Auerbach Publications, 2014.
6. C. Hamilton, *Communicating for Results, a Guide for Business and the Professions* (eighth edition). Belmont, CA: Thomson Wadsworth, 2008/2005.
7. MedlinePlus Encyclopedia Panic disorder.
8. H.S. Bracha, E.M. Vega, C.B. Vega, Posttraumatic dental-care anxiety (PTDA): Is dental phobia a misnomer? (PDF). *Hawaii Dent J*, 37(5), 2006, 17–9. PMID 17152624.
9. M.D. Avia, and M.A. Ruiz, Recommendations for the treatment of hypochondriac patients. *Journal of Contemporary Psychotherapy*, 35(3), 2005, 301–313.

19 Future Work and Conclusions

The future belongs to those who prepare for it today.

Malcolm X, 1962 [1]

This book has presented a new and pragmatic approach for understanding the problem domain and in utilizing SAPs for any field of knowledge and modeling the right and stable software systems, components, and frameworks.

The book provides 3 different and unique templates that is used for documenting all aspects of stable analysis patterns: Detailed; Mid-Size, and Short documentation template.

Along with the core value of reusing the presented patterns, this book also helps readers attain the basic knowledge that is needed to analyze and extract analysis patterns for their own domains of interests. Moreover, readers will also learn and master ways to document their own patterns in an effective, easy, and comprehensible manner.

In addition, the book has also answered the following questions:

1. How can we achieve or reach the objective of software stability [2–5] over a period and later build SAPs that can be effectively reused?
2. How can the SAP be captured and used to model the core knowledge of the problem.
3. How can we achieve the necessary level of abstraction that makes the resulting analysis patterns effectively reusable, yet easy to comprehend and understand?
4. What are the necessary details that analysis patterns must provide in order to ensure a smooth transition from the analysis phase to the design phase?
5. What is the most practical way to describe and narrate analysis patterns, to make them easy to understand and reuse?

Throughout this book, we have furnished a number of answers to them, and practical approaches to follow clear-cut processes that arise from these answers. The Software Stability Concepts acted as the major backbone to all these questions [2–7, and 8].

By applying concepts of stability model to the assumptions of analysis patterns, we suggest the concept of SAPs. The main idea behind using SAPs is to analyze the overall problem under question, in terms of its EBTs and the BOs, mainly with the goal of increased stability and broader reuse. By examining the ensuing problem in terms of their EBTs and the BOs, the resulting pattern will form the core knowledge of the problem. The ultimate goal of this new concept is achieving ample stability. Accordingly, these stable patterns could be easily comprehended and reused to model the same underlying problem, under any given situation.

Throughout this book, we were constantly highlighting the fact that the essence of SAPs is two-fold: A clear paradigm and a precise visual representation. For the paradigm approach, we have presented a set of guidelines, heuristics, and quality factors that will ease the process of creating SAPs, along with a well-defined and reusable documentation. On the other hand, for visual representation, we have offered visual gadgets or symbols that convey what the SAPs are and how to apply them.

19.1 FUTURE WORK

Presented throughout the course of this book are innumerable, open research issues that might throw some light on possible research work for the future. In addition to these open research issues, utilizing SAPs as a tool for understanding, constructing stable and concurrent software development from true and tested requirements, design, and architecture, the following list, will only represent an initial draft of work that will be carried out to extend the scope and enhance the usage of SAPs:

1. Unified Assets Programming Engine (UAPE) is a stable programming environment, where Stable Software Patterns will be used to *provide a true and definite understanding of the problem space and focus users' requirements analysis accurately* [9,10], *and build software applications very quickly.* The book shows that this new formation approach of discovering/creating SAPs [9] concurs with Alexander's current understanding of architectural patterns. This agreement is not accidental, but truly fundamental. The SAPs are a kind of knowledge patterns that underline human problem-solving methods and they may entice the pattern community to seek patterns by building a foundation of architectures on demand from a broader system perspective. Each chapter of the book concludes with open research issues, review questions, exercises, and projects.
2. SAPs Catalog: Generate a catalog of stable analysis patterns fully documented and implemented to be used as part of new stable programming environments, and show how to generate sophisticated applications very easily and quickly, and perform dynamic analysis of each of the generated applications on top of the SAPs. This will ultimately lead to comparative studies and real-time data about the dynamic analysis, and allow the developers and users of the applications to give a concrete results based on real running systems or applications.
3. *Standardize and formalize SAPs' building process.* Another possible project would be the standardization and formalization of the process of building SAPs by using formal languages, such as Object Z, and Z++ (Object-Z; Z++ Language Syntax Chart). The utilization of formal languages will empower SAPs creation processes with validity, integrity, efficiency and authority, so that software developers with different levels of technical expertise may use them wisely and properly.

19.2 SUMMARY

Software analysis patterns play a major and decisive role in reducing the overall cost and in condensing the time duration of software project lifecycles. However, building reusable and SAPs is still a major challenge. SAPs are the new and fresh tools for building stable and reusable analysis patterns based on the concept of software stability.

Software stability concepts have demonstrated great promise and immense hope in the area of software reuse and lifecycle improvement. In practice, stability models apply the concepts of "Enduring Business Themes" or "Goals" (EBTs) and "Business Objects" or "Capabilities to achieve the Goals" (BOs). These revolutionary concepts have been shown to produce and yield models that are both stable over time and stable across various paradigm shifts, within a given domain or application context. By applying the enduring concepts of stability model to the notion of analysis patterns, this book proposes the concept of SAPs. Here, an attempt is made to analyze the problem under consideration, in terms of its EBTs and the BOs, with the ultimate goal of reaching increased stability and broader reuse. By analyzing the problem in terms of its EBTs and BOs, the resulting pattern models constructs the core knowledge of the problem. The ultimate goal, therefore, is achieving *stability.* As a result, these stable patterns could be easily understood and reused to model the same problem in any context [2–6].

The work reported in this book brings truly significant contributions to the computing field for several important reasons. It is the first and the only *complete reference manual* on the topic of SAPs. It is also the first book on handling the true understanding of the problem space, and it will teach a reader methods and processes to analyze the user's requirements accurately, and ways to build a myriad of cost-effective and highly maintainable systems by using SAPs.

Major advantages of SAPs:

1. Highlight major aspects, techniques, and processes that are related to software problem understanding.
2. Illustrate many delicate problems with existing analysis patterns.
3. Provide workable solutions to the most controversial and debatable question that software analysis patterns face today.
4. Provide a diversity of domain-less SAPs that a developer can easily comprehend and reuse to model similar problems in any given context.
5. Show how to link the analysis pattern to the design phase. It will also highlight the main design issues that are necessary for a smooth transition between the analysis and design.
6. Provide a new template for improving the communication of analysis patterns among all developers. This template aims to capture the static and dynamic behavior of the pattern, while maintaining the simplicity of reading and understanding the pattern. Subsequently, the reusability of the patterns will be enhanced further.

REVIEW QUESTIONS

1. What are the advantages of using SAPs?
2. Present ways of enhancing SAPs usage.

EXERCISES

1. Research ways of enhancing SAPs usage (other than pointers given in this chapter).
2. Research existing applications and discuss problems with them. Explain whether one can fix these problems, by using SAPs.

REFERENCES

1. G.L. LaFollette, CPA.CITP. 'The Future Belongs To Those Who Prepare For It Today' ~ Malcolm X, 1962, December 2008.
2. M.E. Fayad, and Altman, A. Introduction to software stability. *Communications of the ACM*, 44(9), 2001, 95–98.
3. M.E. Fayad, Accomplishing software stability, *Communications of the ACM*, 45(1), 2002a, 111–115.
4. M.E. Fayad, How to deal with software stability, *Communications of ACM*, 45(4), 2002b, 109–112.
5. M.E. Fayad, and S. Wu, Merging multiple conventional models into one stable model. *Communications of the ACM*, 45(9), 2002, 95–98.
6. H. Hamza, and M.E. Fayad. A pattern language for building stable analysis patterns, *9th Conference on Pattern Language of Programs (PLoP2002)*, IL, 2002.
7. D.C. Schmidt, M.E. Fayad, and R. Johnson. Software patterns, *Communications of the ACM*, 39(10), 1996, 37–39.
8. M.E. Fayad, H.A. Sanchez, S.G.K. Hegde, A. Basia, and A. Vakil. Software Patterns, Knowledge Maps, and Domain Analysis. Boca Raton, FL: Auerbach Publications, 2014.
9. H. Hamza, and M.E. Fayad. A pattern language for building stable analysis patterns. In *Proceedings of 9th Conference on Pattern Languages of Programs 2002 (PLoP02)*, Monticello, IL, September 2002.
10. H. Hamza, and M.E. Fayad. Applying analysis patterns through analogy: Problems and solutions. *Journal of Object Technology*, 3(3), 2004.

Index

B

C

D

G

H

I

J

K

L

M

N

O

P

Q

R